ENERGY AND US

Sources, Uses, Technologies, Economics, Policies and the Environment

Glenn A. Gibson
Electrical and Computer Engineering Department
The University of Texas at El Paso

ISBN number 978-1-4609-3037-3

LCCN number 2011904864

To my grandchildren

Ashley, Kayla, Katelynn, Paul, Garett, Matthew, Jake and Ryan

112ᵀᴴ CONGRESS 2011-2012

Senate Committee on Energy and Natural Resources

Democrats

Jeff Bingaman (NM), Chairman
Ron Wyden (OR)
Tim Johnson (SD)
Mary L. Landrieu (LA)
Maria Cantwell (WA)
Bernard Sanders I (VT)
Debbie Stabenow (MI)
Mark Udall (CO)
Jeanne Shaheen (NH)
Al Franken (WV)
Christopher A. Coons

Republicans

Lisa Murkowski (AK)
Richard Burr (NC)
John Barrasso (WY)
James E. Risch (ID)
Mike Lee (UT)
Rand Paul (KY)
Daniel Coats (IN)
Rob Portman (OH)
John Hoevan (ND)

112TH CONGRESS 2011-2012

House Committee on Energy and Commerce

Republicans

Fred Upton (MI), Chairman
Joe Barton (TX)
Cliff Stearns (FL)
Ed Whitfield (KY)
Jon Shimkus (IL)
John R. Pitts (PA)
Mary Bono Mack (CA)
Greg Walden (OR)
Lee Terry (NE)
Mike Rogers (MI)
Sue Myrick (NC)
John Sullivan (OK)
Tim Murphy (PA)
Michael Burgess (TX)
Marsha Blackburn (TN)
Brian P. Bilbray (CA)
Charles F. Bass (NH)
Phil Gingrey (GA)
Steve Scalise (LA)
Bob Latta (OH)
Cathy McMorris Rodgers (WA)
Greg Harper (MS)
Leonard Lance (NJ)
Bill Cassidy (LA)
Brett Guthrie (KY)
Pete Olson (TX)
David McKinley (WV)
Cory Gardner (CO)
Mike Pompeo (KS)
Adam Kinzinger (IL)
Morgan Griffiith (VA)

Democrats

Henry A. Waxman (CA)
James D. Dingell (MI)
Edward J. Markey (MA)
Edolphus Towns (NY)
Frank Pallone Jr. (NJ)
Bobby L. Rush (IL)
Anna G. Eshoo (CA)
Eliot L. Engel (NY)
Gene Green (TX)
Diana DeGette (CO)
Lois Capps (CA)
Michael F. Doyle (PA)
Jane Harman (CA)
Jan Schakowsky (IL)
Charles A. Gonzalez (TX)
Jay Inslee (WA)
Tammy Baldwin (WI)
Mike Ross (AR)
Anthony D. Weiner (NY)
Jim Matheson (UT)
G. K. Butterfield (NC)
John Barrow (GA)
Doris O. Matsui (CA)

CONTENTS

PREFACE

It is the purpose of this book to provide the reader with information pertaining to energy, its uses and sources and its effects on our environment, economy, politics, and lives in general. It examines the technologies that promise to make better, more efficient use of energy, reduce our dependence on fossil fuels and avoid their undesirable side effects. The thrust of the book is to present the facts as they are known at the time of its writing. However, some mathematical projections have been included based on these facts and possible future scenarios. It is hoped that the book will not only increase the reader's understanding of the nature of energy, but also its limitations, the finiteness of our most commonly used fuels, its contribution to global warming, its downsides in terms of pollution and its relationship to international politics. The reader should gain an appreciation of energy's importance to business and our living standards.

Most of the data presented has been obtained from government information agencies such as the Energy Information Administration (EIA), which is an agency within the United States Department of Energy (DOE), and the Bureau of Transportation Statistics (BTS), which is within the United States Department of Transportation (DOT). The EIA and BTS collect these data from national and international government reports, corporations and research institutions. They revise the data as new, more reliable data becomes available. Most of their data may be found on the websites *www.eia.doe.gov* and *www.bts.gov*. The information on the types of energy sources and the current technologies has been obtained from a variety of industrial and scientific journals, reports and websites.

Although little knowledge of science is required to understand the bulk of the material, a very limited number of equations and chemical formulas do appear and the material may be better appreciated if the reader has a rudimentary understanding of physics and chemistry. For this reason, Appendix A is included to provide the basic science of energy for those who need a refresher. Also, one of the greatest difficulties in reading material on energy is assimilating and relating the numerous units that are used to quantify the various energy related amounts. Another is remembering all of the abbreviations and acronyms. For convenience, Appendix B summarizes the units of measure and the conversion factors between them and Appendix C provides a list of the acronyms that appear throughout the book.

This book has been made possible by the information supplied by a number of United States governmental agencies. Its writing has relied heavily on the data made available through their websites and publications. Also, credit goes to the staff at CreateSpace. In particular, the author would like thank John Rieck for helping with the introduction and submission of my manuscript. Finally I would like to thank my wife for her patience and proofreading.

1

ENERGY

To most of us, energy is what makes our cars go, lights our lights and heats our homes. To an engineer, it powers our electronics and provides the pushes and pulls that drive our machinery. But to a physicist, it is one of the two fundamental constituents of the universe, energy and matter. Energy cannot be precisely defined without delving into the complex equations that govern our universe, so no definition of it is given here, although some discussion of its definition is given in Appendix A. As indicated in Appendix A, energy is basically a measure of the motion of matter or something capable of imparting such motion. For now, we shall rely on our intuitive notion of energy based on its everyday uses. It is evident from these uses that energy exists in several different forms. Its basic forms are heat, mechanical, electromagnetism (including light), chemical and nuclear, and for them to be useful it is almost always necessary to convert one form of energy into another. In fact, it is often necessary to perform a series of conversions before the energy is finally used. When coal is used to generate electricity to power a light bulb, the chemical energy in the coal is converted to heat in the form of steam, which is then used to drive a mechanical turbine, which, in turn, drives an electrical generator, whose electricity is finally converted to the electromagnetic energy of light by the light bulb.

Regardless of how energy is viewed, our ability to extract and use it is what makes our modern way of life possible. By seeking out and utilizing other sources of energy, we have made it possible to do more in less time and with less effort. No longer do we have to spend our waking hours hunting for food or even hitching up the buggy and spending half a day riding it to town. As we progressed from using basic energy sources such as fire to cook food and warm our bodies, wind to fill our sails, animals to do the heavy work and provide transportation and water and wind to grind grain, to more complex sources such as chemical energy to power the industrial revolution and nuclear energy to generate electricity, our standard of living has improved correspondingly. Modern mankind has learned to leverage his or her body's energy by a factor of 20 by using other energy sources. Whether viewed over time or space, the relationship between energy and the way we live is clear. Today, one need only compare the industrialized nations to those of the third world to see the correspondence between energy usage and quality of life.

The purpose of this chapter is to give an overall introduction to energy and some of the aspects and effects that are common to all types of energy. The chapters that follow are to provide the detailed information regarding the various characteristics and ramifications related to energy usage. This chapter briefly introduces the:

- Units for measuring energy and the rate at which energy is produced or consumed.
- Effects of energy usage on our standard of living.
- Relationship between energy and mass and the different forms of energy.
- Nonrenewable and renewable energy sources.
- Energy sources and uses.
- Consumption of energy and population.
- Stages through which energy passes between its origin and consumption.
- Energy storage and the importance of volumetric and weight energy densities.
- Definition of energy efficiency.
- The origins of energy sources and international politics.

Included at the end of the chapter is a discussion of the sources of the data quoted throughout the book.

Amounts of energy are not always noted using the same units of measure. Scientists most often use joules and dieticians most often use calories, but engineers and industry normally use British thermal units (Btus) or, when discussing electrical energy, kilowatt-hours (kWhs). One Btu is 1055.1 joules and is the energy contained in approximately one thousandth of one ounce of gasoline. A million Btu (MBtu) is equivalent to the heat energy produced by burning 97 pounds of medium grade coal or 8 gallons of gasoline. Each American man, woman and child, on average consumes 325 MBtu each year. This is the energy equivalent of 17 tons of coal, enough to fill a bin that is 7 feet wide, 12 feet long and 5 feet high. The average person in Japan uses about half this amount and the average African uses less than one-twentieth this amount. One kWh is equivalent to 3,412 Btu and is the energy consumed by ten 100 watt light bulbs in one hour. In 2007, the average per day electricity consumption of an American household was 34.3 kWh, or the equivalent energy of 14.3 such light bulbs being on for 24 hours. When comparing energy sources the unit of energy used is quite often *barrels of oil equivalent (boe)*, one boe being the energy obtained by burning a typical barrel of petroleum. Although the energy in petroleum varies according to its quality, the Bureau of Transportation Statistics (BTS) uses the conversion 138,000 Btu per gallon (Btu/gal), or about 5.8

million Btu per barrel (MBtu/bbl). This is the amount assumed in this book. (See Appendix B for a summary of units and conversions between units.)

Power is the rate at which energy is created or consumed and is typically measured in British thermal units per hour (Btu/h), kilowatts or watts. If a device continuously consumed energy at the rate of one Btu/h for one hour, it would have consumed one Btu of energy. A pickup truck that consumes two gallons of diesel in one hour would consume about 276,000 Btu/h of power. If a generator continually outputs one kilowatt for two hours, it would have output two kilowatt-hours of energy. If the rate of energy consumption of a cell phone is five joules per second, then the power consumption of the phone is five watts. A 100 watt light bulb consumes 100 joules per second. If the energy rate, i.e. power, is not constant, then it is often the average power that is indicated even though the word power is normally used.

The discrepancy between the per capita energy consumption of the richer versus the poorer nations is apparent from Fig. 1-1, which compares some of the world's most affluent societies with some of its poorest. But it is not just the amount of energy a nation consumes, a nation's prosperity also depends on how efficiently the energy is used. *Energy intensity* is the amount of energy consumed by a nation per dollar of its *gross domestic product* (*GDP*), where its GDP is the value of all goods and services the nation produces (or, more accurately, the total value of its

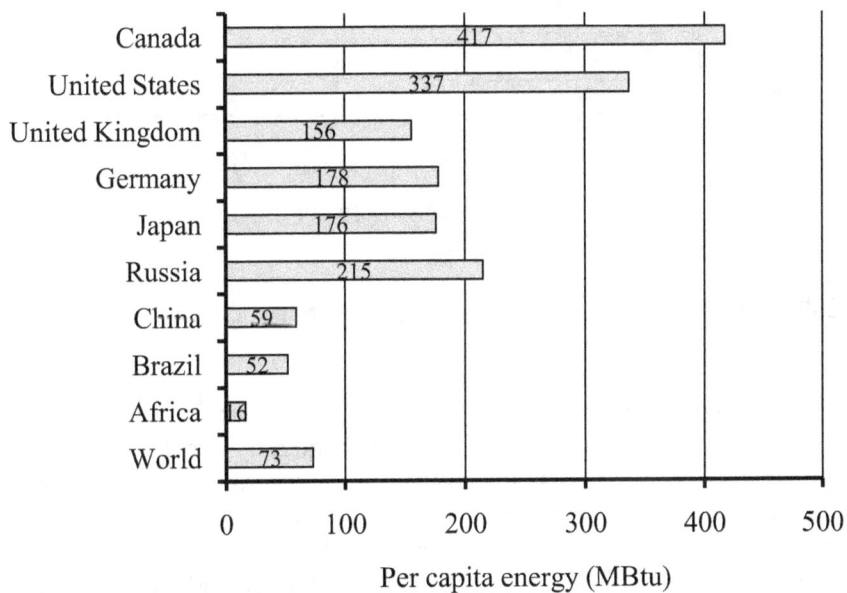

Fig. 1-1 2007 energy consumption per capita per year [EIA].

consumer, investment and government spending plus the difference in the values of its exports and imports). Figure 1-2 gives the energy intensity of several of the world's countries in Btu per 2005 United States dollar of GDP (Btu/dollar). When a dollar amount is given it is usually accompanied by the year for which the amount applies. Because of inflation, the purchasing power of the dollar normally declines from year to year (although it could increase). For a later year one needs to adjust the amount according to the inflation that has taken place in the intervening years. For Fig. 1-2, the amounts were given in 2005 dollars even though the Energy Information Administration (EIA) data was for the year 2007.

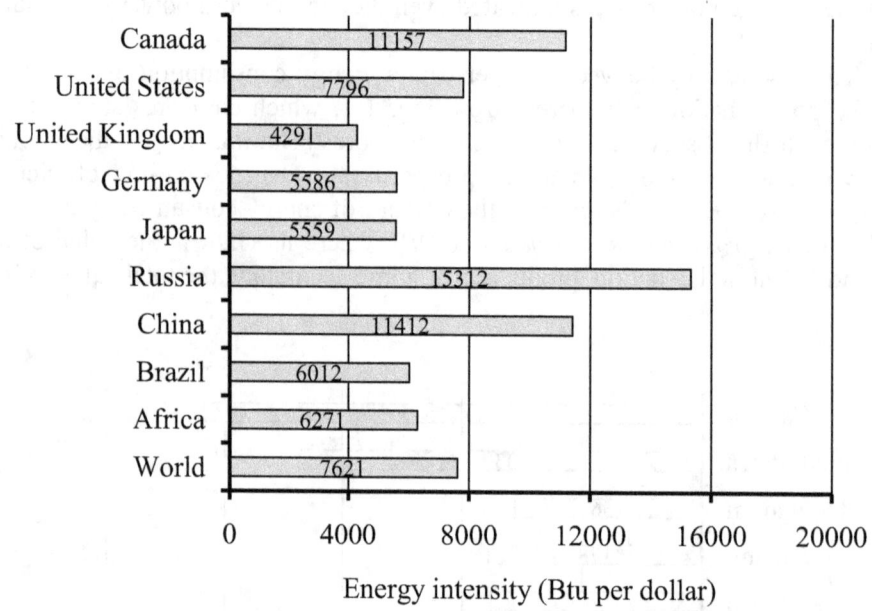

Energy intensity (Btu per dollar)

Fig. 1-2 2007 energy intensity in Btu per 2005 United States dollar [EIA].

Not only do the richer nations utilize more energy, but they have learned to do so more effectively as well. In 1980, the energy intensity for the United States was 56 percent more than it was in 2002. Largely due to America's increased productivity with respect to energy, the percent of the average American's disposable (i.e., non-essential) income spent on energy dropped from eight percent in 1980 to four percent in 2002. By 2006 this percentage had increased to six, primarily due to a tripling of the price of petroleum. Economics and energy are inseparably intertwined. Unfortunately, our ever-increasing use of energy has been accompanied by some undesirable side effects that have partially negated its benefits and may have a significant impact on our future economy, environment and living standards.

While the GDP of a nation does indicate a nation's well-being to some extent, the United Nations uses a measure known as the *Human Development Index* (*HDI*), which includes a nation's average life expectancy, literacy rate and educational level as well as its GDP. In 2004 the United Nations Development Program (UNDP) gave a report on the HDI for most of the world's countries. Figure 1-3 gives the HDI versus the per capita energy consumption for a few example countries. As with Fig. 1-2, this figure makes apparent the relationship between energy usage and the quality of life of a nation. Of the nations shown in Fig. 1-3, Norway had the highest HDI at 0.965 and highest energy use per capita at 424.5 MBtu/yr, and Niger had the lowest HDI at 0.311 and the lowest per capita energy usage at 1.4 MBtu/yr. The world average was an HDI of 0.741 and a per capita energy consumption of 70.1 MBtu/yr. The United States (the third dot from the right) had the eighth highest HDI at 0.949 with a per capita energy consumption of 342.7 MBtu/yr.

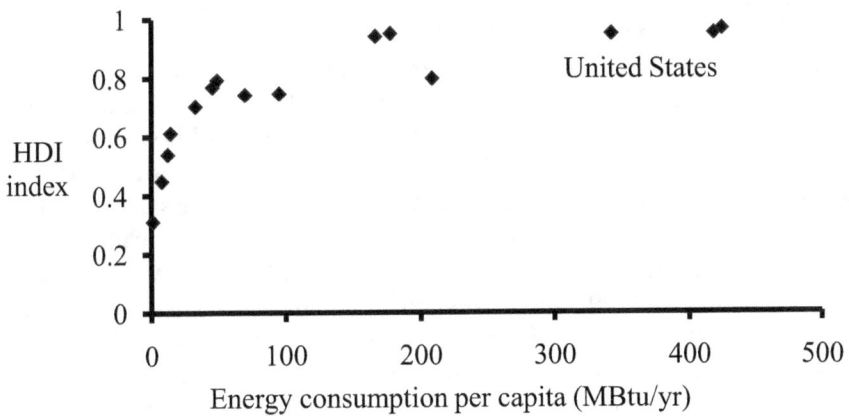

Fig. 1-3 Human development index vs per capita energy consumption
          [UNDP,EIA].

There are extenuating circumstances that cause some nations to use more energy per capita than others even though their energy intensity is the same. Space heating is a major contributor to energy usage in colder climates and population density and distribution affects average travel distances. Canada is cold and sparsely populated, while India is warm and densely populated. Unfortunately, another factor related to consumption is human nature which is such that, the more one has, the more he or she is apt to waste.

Although most of what has been said and written implies energy is in short supply, the problem is not that the Earth is in danger of running out of energy. Energy is everywhere. According to Einstein's famous equation, energy is equal to mass times the speed of light squared ($E=mc^2$). This means that everything that has mass can be viewed as stored energy. Because the speed of light is approximately 300,000,000 meters per second and its square is approximately 90 quadrillion (90,000,000,000,000,000), even a small amount of mass contains a huge amount of energy. It does not matter what form the mass is in, be it uranium, cheese or garbage. Although it is not possible, if all the mass in ten barrels of oil could be turned into energy, it would more than supply the United States entire energy needs for one year.

The problem is that of extracting energy from mass in order to put it to useful work. All energy must originate from the process of converting mass to energy, whether it is a nuclear reaction or a chemical reaction such as burning coal. Surprisingly, the number of extraction processes that are currently known and in use is rather small. They are primarily the burning of organic materials, mainly fossil fuels (coal, natural gas and petroleum), nuclear fission of uranium and nuclear fusion of hydrogen. Together these three processes are responsible for producing almost all of our energy needs. It is not possible to turn all of the mass of a material into energy and none of these processes is very efficient. For fusion of hydrogen, only about 0.7% of the mass is converted to energy, and for fission of uranium less than 0.1% is converted. A nuclear reaction is typically ten million times more efficient than a chemical reaction. When burning a fossil fuel, very little of the mass is converted to energy. If it were possible to burn petroleum with the same efficiency as the fusion of hydrogen, then the petroleum energy needs of the United States could be satisfied with less than 500 barrels of oil per year as opposed to the 6.8 billion barrels consumed in 2009, which satisfied less than 40 percent of our needs.

Even though fusion is by far the most efficient conversion of mass to energy, currently the only known way of producing a useful amount of energy from it on Earth is in the form of the violent explosion of a hydrogen bomb. Hydrogen fusion is what powers the sun and we must depend on the small fraction of the sun's energy that strikes the Earth to provide our fusion energy. The fusion of other elements takes place in older stars, but such fusion requires tremendous pressures and temperatures and is beyond human capabilities.

According to the big bang theory, mass was formed by the reactions that occurred following the enormous explosion that created our universe. As the universe cooled, this mass collected into stars, planets and other astronomical entities.   While much of the energy produced by the big bang went into the formation of mass, some of it was stored in the residual heat and motion of these entities, including Earth. Also, Earth's interior is heated by the fission of the radioactive elements potassium, thorium and uranium deposited in its core at the time it was formed. Today, a very small percentage of the energy we use comes from the heat (i.e., geothermal) energy stored under the Earth's crust or the mechanical rotation of Earth (which causes Earth's tidal activity), but this percentage may increase in the future.

There are two major categories of sources from which energy is obtained, those that are nonrenewable and those that are renewable. The *nonrenewable sources* are those that cannot be replaced in a reasonable period of time and the *renewable sources* are those that can be replaced within a century or so or are so vast that they will never be depleted by human activity, e.g., the rotation of Earth. The nonrenewable sources are the fossil fuels and uranium. It has taken millions of years to form the fossil fuels found in Earth's crust and our uranium is that which was deposited during Earth's formation.

The renewable sources are those that are directly or indirectly obtained from the energy radiating from the sun, the motion of the Earth or the heat radiating from within the Earth. Solar energy, which is obtained from the sun's electromagnetic radiation, may be used directly to generate electricity or heat air, water or other substances, or indirectly by means of wind for generating electricity or water in rivers for producing mechanical power or generating electricity. The sun's rays also create biomass, such as wood, corn or animal waste, which can be burned or processed to produce liquid synthetic fuels. Although only a small part of the sun's energy reaches the Earth, we still receive a huge amount of energy from the sun each day. Geothermal energy can be used directly to heat buildings or to generate electricity, but its use is currently economically practical only where it is readily available near the Earth's surface. It is considered renewable only if its heat can be locally replenished within a reasonable time. Electricity can also be generated by tidal motion. However, as with geothermal activity, there are a limited number of locations where this form of generation is currently economically feasible. In general, turning a renewable source into useful work tends to be a relatively inefficient and costly process, but the economic viability of renewable sources will improve as fossil fuels become less available and more expensive. Renewable sources tend to have fewer undesirable side effects.

Figure 1-4 gives the important sources of energy, their relationships to the energy types and the more prominent agents of conversion from one type to another. There are five basic types of energy, nuclear energy, electromagnetism, chemical energy, mechanical energy and heat. Note that a battery converts both electromagnetic energy to chemical energy (while charging) and chemical energy to electromagnetic energy (while discharging), and electromagnetic radiation produces heat and heat produces electromagnetic radiation. Also, there are various means of changing the manifestation of energy within a given type (e.g., a solar cell changes radiated energy to an electric current). From this figure, one can see the various paths from energy sources to end uses. For example, one may trace the conversion of the sun's electromagnetic radiation into the chemical energy of wood by photosynthesis and then into heat by burning. The conversions in the figure are discussed in detail in the chapters that follow.

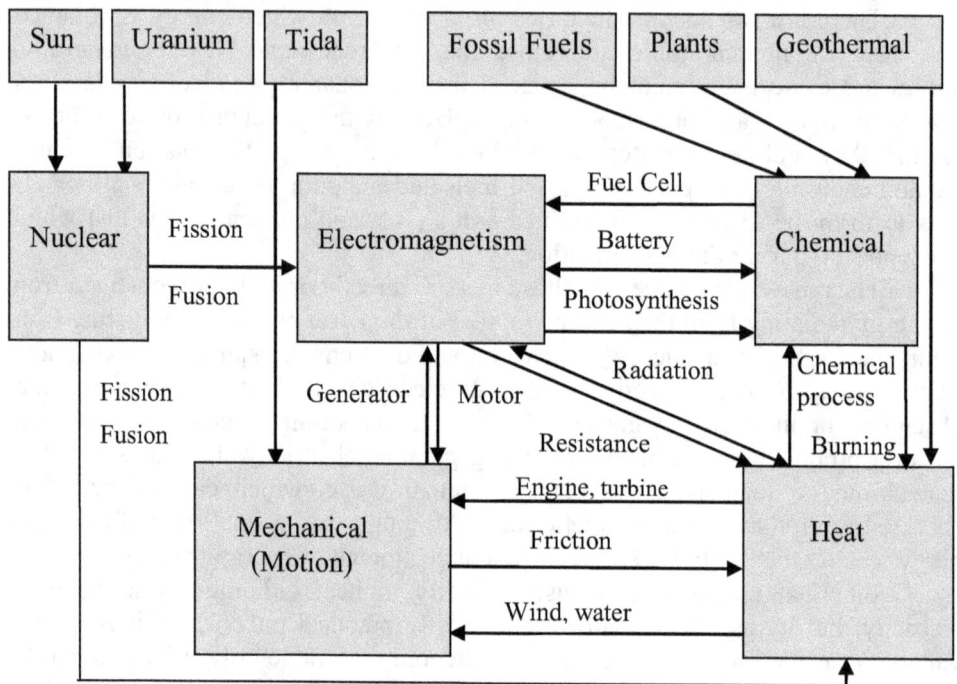

Fig.1-4 Energy sources and types and conversions between types.

There is an important law of physics known as the *Conservation of Energy*, which essentially states energy cannot disappear. It can change form or be transmitted into space, but it cannot cease to exist. Because there have been high-energy experiments that have very briefly converted energy into mass, the law is not strictly true unless mass is considered a form of stored energy, but for all practical purposes the law is considered to be true. This law means that in the conversions in Fig. 1-4 there is no loss of energy. However, although not indicated in the Fig. 1-4, all of the conversion processes involve some conversion to heat which cannot be usefully retrieved. This fact is called the *Second Law of Thermodynamics* and means that no energy conversion process is 100 percent efficient. (There is more discussion of these two laws in Appendix A.)

Given in Fig. 1-5 are the main categories from which energy originates and how they relate to our everyday uses of energy. Although most of these uses may be satisfied by the sources without resorting to electricity, it is often more expedient to first use them to generate electricity and then use the electricity to provide the needed energy.

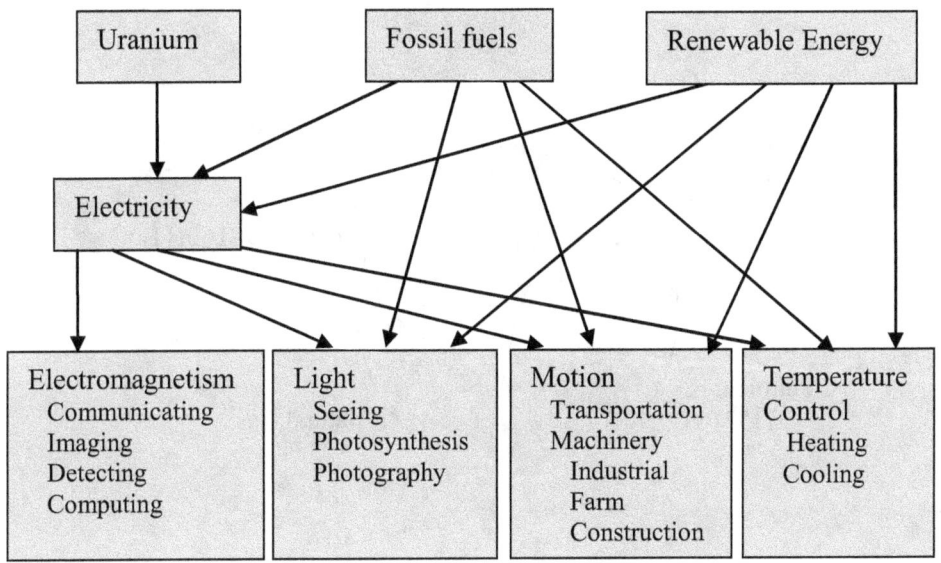

Fig. 1-5 Relationships among energy sources and their common uses.

The amounts and percentages of the sources used in the United States in 2008 are shown in Fig. 1-6. The amounts are given in parentheses in trillions of British thermal units (TBtus). A trillion Btu is equivalent to the energy content of 7.25 million gallons, or 172,533 barrels, of oil. Our considerable reliance on the burning of fossil fuels is clear. Petroleum provides 37 percent of our energy, natural gas 24 percent and coal 23 percent, making the total contribution of fossil fuels 84 percent. Uranium, the other nonrenewable source, contributes another 9 percent, leaving only 7 percent of our 2008 energy requirements being satisfied by renewable means.

The principal uses of energy are lighting, heating and cooling, transporting goods and people, manufacturing, construction, farming and materials refining, including the refining of fuels. The EIA places the American end users into four major sectors, residential, commercial, industrial and transportation. This agency also supplies data on the generation of electricity, but as seen in Fig.1-5, generating electricity is just one of the steps in supplying energy and generating plants are not considered end users. About 40 percent of the energy consumed by the four sectors is first converted into electricity. Figure 1-7 shows the amounts and percentages of energy consumed by the four sectors within the United States in 2008 in TBtu.

Together Figs. 1-6 and 1-7 show where our energy comes from and where it is used. The small difference between the totals given in Figs. 1-6 and 1-7 is due to net electricity imports into the United States from Canada.

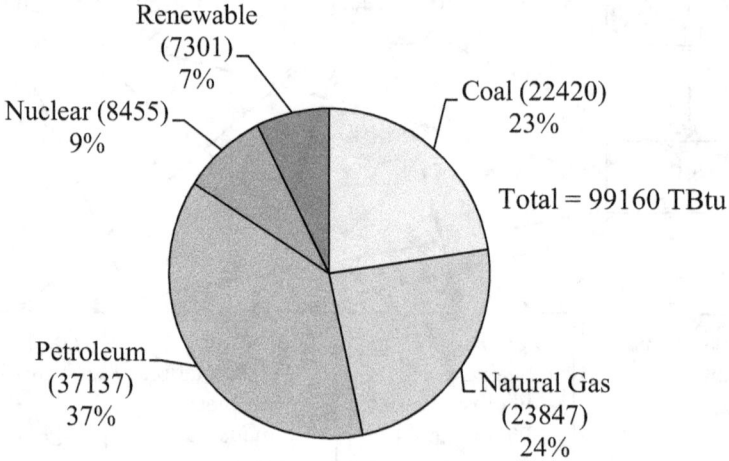

Fig.1-6 Energy sources in the United States in 2008 in TBtu [EIA].

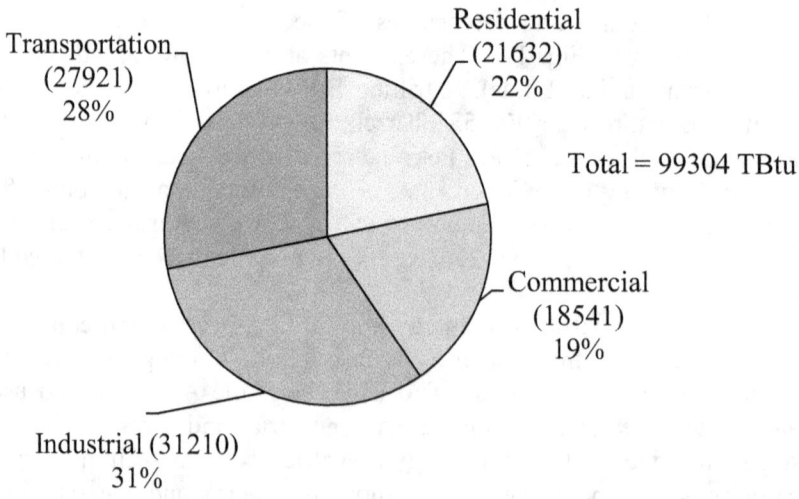

Fig. 1-7 Sector energy consumption in the United States in 2008 in TBtu [EIA].

Figure 1-8 illustrates the energy consumption in both the world and United States in quadrillions of Btu, or peta-Btu (PBtu), during the period 1980 through 2008. Superimposed on the actual data curves are the exponential approximation of the world data and the linear approximation of the United States data. An *exponential growth* is one for which the growth increases a quantity by an equal percentage each year such as the growth of a bank account that draws compound interest. A *linear growth* is one that increases a quantity by an equal amount each year and can be represented by a straight line. The exponential approximation of world data shows an annually compounded 1.95 percent increase. This growth implies that in the succeeding years the worldwide energy consumption would increase by 5.48 PBtu, 5.58 PBtu, 5.69 PBtu and so on, and for the year 2040 the energy consumption would be 895 PBtu. The United States has increased consumption at a linear rate of 1.094 PBtu per year. If this were to continue, our consumption in 2040 would be 139 PBtu. Much of these increases have been and will continue to be due to population increases, but for the world as a whole much of the increase will be caused by the industrialization taking place in the developing countries, especially China, India and Brazil. Between 1980 and 2008 these countries increased their consumption by factors of 4.5, 4.7 and 2.5, respectively, far exceeding the world average factor of 1.7.

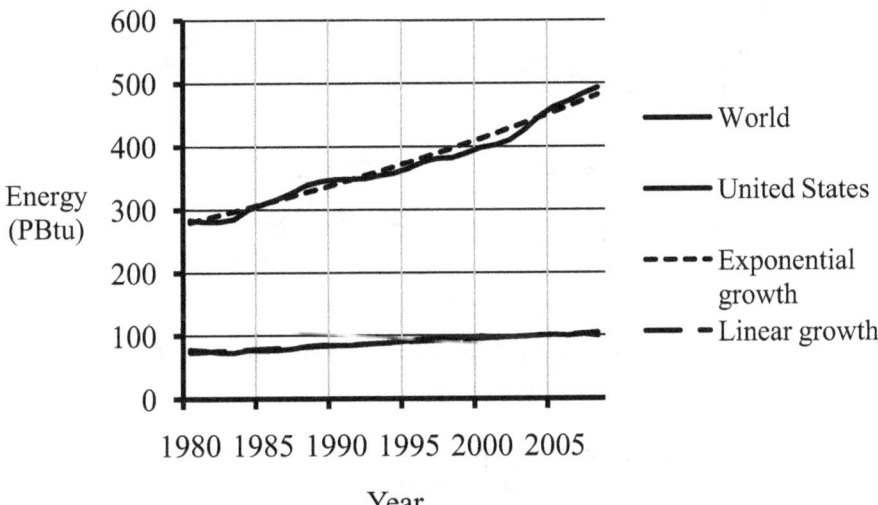

Fig. 1-8 Energy consumption of the world and United States in PBtu [EIA].

The dominant pressures on the world's energy supply are the increase in population and the desire to improve living standards through the use of energy, particularly by those in the poorer nations who want to attain the same station as those in the United States and Europe. The population and anticipated population in the United States and world for the 1900 through 2050 period are given in Fig. 1-9. The world projection through 2050 was not made by exponential or linear extrapolation, but was made using a complex procedure involving many factors including birth and death rates. It is predicted that within the next 200 years the world's population will stabilize between ten and twelve billion people. Making a projection for the United States is even more complex because of immigration. How reliable these projections are is open to debate, but they are likely to be reasonably accurate for the next 40 years.

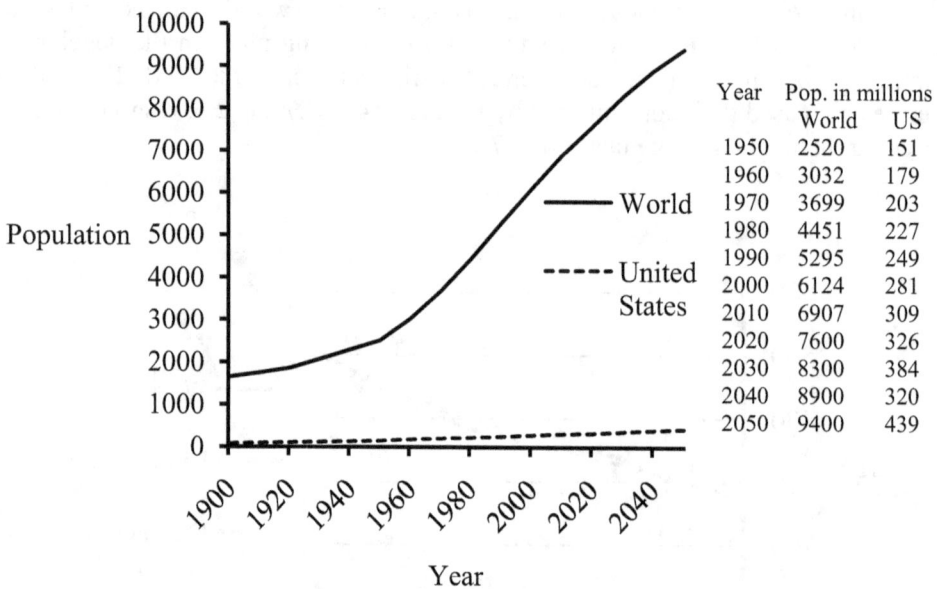

Fig. 1-9 Population of the world and United States [USCB, UN].

Figure 1-10 shows the energy per capita for the United States and world for the years 1980 through 2010. Although the energy per capita in the United States has been more or less flat, the energy per capita has increased from 63 MBtu to 73 MBtu for the world as a whole because of both the rise in population and the rapid expansion of the economies of some of the larger third world countries. Countries such as the United States are approaching the saturation point for which an increase in the per capita energy usage does not significantly improve living standards, but

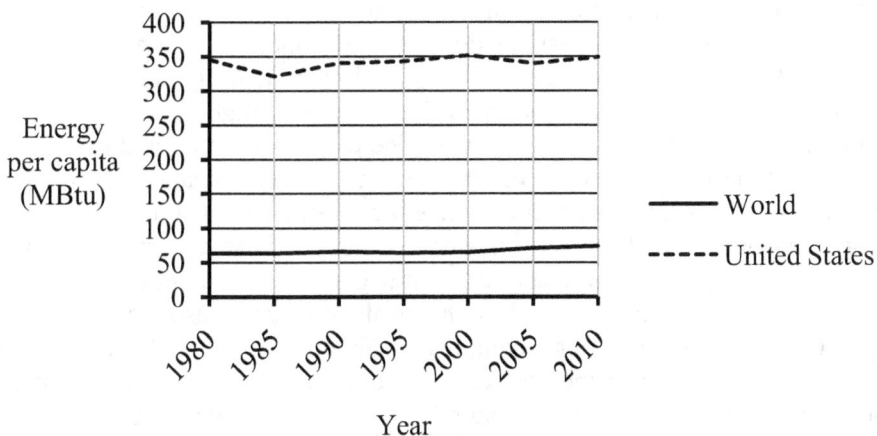

Fig. 1-10 Energy use per capita for the world and United States [EIA, USCB, UN].

nations such as China and India have not reached that point and, understandably, want to continue improving their standards through energy usage. Note that energy per capita is computed by dividing energy by population and is increased by either an increase in energy usage or a decrease in population.

Energy may pass through several stages before it is put to useful work. Its original stage is normally referred to as its *production stage*, even though it may be simply extracted from the ground (e.g., coal is mined). The stage at which it is finally used for its intended purpose is called its *consumption stage* or *usage stage*. There are usually several *storage, transportation, refining* and *conversion stages* between the production and consumption stages. Each stage results in energy losses (i.e., that portion of the input energy that is wasted) and, perhaps, polluting byproducts. However, a refining stage may also result in useful byproducts. For example, the petroleum that is used to power gasoline automobiles typically passes through the following stages:

Production→Storage→Transportation→Storage→Refining→Storage
→Transportation→Storage→Consumption

The resulting undesired byproducts include sludge ponds, gas vapors, oil spills, and carbon dioxide, which pollute the ground, air and water. Besides gasoline, the refining stage results in several useful byproducts such as propane, diesel fuel, heating oil, jet fuel, tar and asphalt.

Each stage in the process of putting an energy source to useful work presents its problems. Ocean transportation of natural gas is dangerous and difficult because it is explosive and must be liquefied to reduce its volume. Storage of electricity is difficult, costly and inefficient. This limits the usefulness of electricity generation from unsteady sources such as the wind or sunlight. Not only does the cost and efficiency of a stage need to be considered, the pollution and dangers that may be created must be taken into account. Any stage or combination of stages may render the process of using a particular source for a particular use infeasible. Storage of energy, no matter what form it is in, is needed not just in stationary locations, but also while the energy is being transported or drawn off for consumption by a moving vehicle. Most often, energy is stored in chemical form (e.g., gasoline or a charged battery) or mechanical form (e.g., compressed air or a spinning flywheel). When energy is being stored on a moving vehicle, its energy density in terms of both volume and weight are important. Natural gas must be reduced in volume (i.e., compressed) in order for it to be reasonably transported, and the energy density by weight must be high when energy is used to power an airplane.

The *volume energy density* is the energy per unit volume of a substance and the *weight energy density* is the energy per unit weight of a substance. The volume and weight energy densities of several common fuels are given in Table 1-1. These densities are typical only and may vary widely depending on the quality of the fuel.

Table 1-1: Energy densities by volume and weight.

Fuel	Volume (MBtu/m^3)	Weight (MBtu/short ton)	Weight (MBtu/t)
Coal	20	21	23
Gasoline	34	40	44
Diesel	37	38	42
Biodiesel	33	33	36
Natural gas	0.036	44	48
Liquid	22	44	48
Propane	23	39	43
Ethanol	21	24	26
Methanol	15	17	19
Hydrogen gas	0.006	112	123
Liquid	8	112	123
Wood	5.3	16	18
Uranium	1.5×10^9	70×10^6	77×10^6
Car battery	0.16	0.1	0.11

This is particularly true of the densities for wood and coal. Also, the energy density of a fuel varies depending on the way it is measured. If all of the heat in the products of the combustion is included in the computation, then a higher value results than is found if only some of this heat is taken into account. If all heat is included, then the computed amount is the *high heat value* (*HHV*) energy density. If only the heat of the products after they cool to 150 °C (302 °F) is considered, then the determined amount is the *low heat value* (*LHV*). Normally, the amount quoted is some practical value between the LHV and HHV. In this book, a value that is representative of the values found in the various references is assumed.

    Although the table indicates the energy densities of the fuels, the volume and weight of the equipment required to contain and use the fuel must also be taken into account. Natural gas turns into a liquid at -162°C (-260°F), propane becomes a liquid at −42°C (−44°F) and hydrogen becomes a liquid at -253°C (-423°F), But under pressure they can be maintained as a liquid at higher temperatures. Therefore, liquid propane can be kept in a tightly sealed steel tank at normal air temperatures, but it is not practical to store liquid natural gas or hydrogen without refrigeration. Because it takes a lot of heavy equipment to cool and contain liquid natural gas or hydrogen, it is not practical to use these liquids for energy storage in most vehicles. However, compressed natural gas is used to fuel some vehicles, including automobiles, and compressed hydrogen may be used in future automobiles that use fuel cells. Because the equipment needed to both safely store and use uranium is heavy, bulky and very expensive, the only vehicles currently fueled by uranium are military ships and submarines.

    The term "efficiency" as it applies to an energy system appears throughout the literature on energy. The nature of the system may be chemical, mechanical, electrical or nuclear, or some combination thereof. *Energy efficiency*, or simply *efficiency*, may have different meanings depending on the context and, unfortunately, the word is not always clearly defined. It is often left to the reader to ascertain the meaning from the discussion. Usually it means the useful energy output by the system divided by the useful energy input to the system. A high efficiency implies the energy losses relative to the useful output energy are small. But, how are the input and output energies to be computed? If coal is burned to drive a steam turbine that is attached to an electric generator, how should the input energy be computed? Obviously, the mass of the coal times the speed of light squared would not be of practical value because only the energy released by a chemical reaction (i.e., combustion) is utilized. More likely, the input energy is taken to be the theoretical maximum heat energy that can be obtained by burning coal, even though the coal may not be completely burned. On the other hand, one may prefer to use the average amount of heat that is produced when coal is burned.

    Also in question is the output energy. Although the energy out of the generator could be easily measured under a variety of conditions, what set of conditions should be assumed? In addition, after the steam exits the turbine, it may be used to heat a building and not be wasted. In such case, the heat energy that is output by the radiators in the building should also be considered useful output energy. By including this heat in the efficiency computation, the efficiency would be much higher. But this heat should not be counted unless it is used. Finally, if part of the time no one is in the building, should this heat be considered lost? Usually, this is not the case.

In a large system, there may be several sources of input and several outputs and there will be energy losses at each stage in the system. Consider the production and transportation of natural gas across the ocean. The gas must be liquefied to reduce its volume and then transported, perhaps thousands of miles, by ship. Both the compression and transportation stages would consume appreciable amounts of energy that should not be ignored when making an efficiency calculation.

Although computing a precise efficiency may be desirable, it is seldom possible, or even practical. The efficiency found may be accurate for precisely stated conditions only. It normally makes more sense to assume average conditions and use efficiency as an approximate measure of the system. When you buy a car, you probably take into account the miles per gallon. Even though it may not reflect your driving conditions, miles per gallon is still informative when comparing cars. Knowing the efficiencies of alternative processes is most useful when making comparisons. In this case, the exact efficiencies are not as important as ordering them from best to worst. When making quantitative comparisons, it is necessary to use the same methodology when computing the efficiencies. The efficiencies quoted in energy-related literature, including this book, should be taken as estimates. The author has made a concerted effort to give efficiencies as accurately as possible and will accompany them with clarifying comments and assumptions whenever it seems helpful.

As with all decisions, tradeoffs are involved in the decisions regarding the production and use of energy. Availability of the sources and international politics must be considered as well as economics, efficiency and pollution. Availability of a source depends on its location and whether or not it is renewable. A nonrenewable source is finite and eventually may be used up. Clearly, how long a source lasts depends on how fast it is consumed, which, in turn, depends on how much it is needed and the feasibility of its replacements. Uranium may last for several centuries and coal may last a few centuries, but both petroleum and natural gas could become almost depleted by the year 2100. Some will remain because their prices will go up as they become harder and more expensive to produce, and their uses will decrease as their prices increase. If uranium and coal replace petroleum and natural gas as energy sources, the depletion of their deposits will be accelerated.

The fact that nonrenewable sources are more readily available in certain countries than in others is the reason energy plays such an important role in international politics. The heavy users of energy are North America, Europe and the Orient, which are the places where petroleum and natural gas are being rapidly depleted. Most of the remaining petroleum and natural gas reservoirs are in countries with unstable governments or governments that are unfriendly to the industrialized nations or both, and conflict between these two sets of countries could, and sometimes does, impact the world's economy dramatically.

Location is important with respect to renewable sources as well. Most of Norway's electricity is produced by hydroelectric power plants because of its abundance of streams and rivers in its deep valleys. In the United States, the Great Plains and Great Lakes are best suited for wind-powered electricity generation and the Southwest is best suited for generating electricity by harvesting direct solar radiation. Also, the location of a source relative to where it is needed causes transportation and its associated costs and problems to become major considerations. As noted above, the difficulties in transporting natural gas by ship makes shipping natural gas across the ocean expensive and dangerous. The transport and disposition of nuclear waste is a major consideration in the use of nuclear power.

Conservation, changing energy sources, and technological innovation are the primary means by which we can insure our energy future and reduce unwanted environmental effects. *Conservation* is simply a matter of using less energy by reducing the amount of energy we waste. It may be accomplished by constructing better buildings, turning off lights and equipment that are not in use, creating better mass transit systems, carpooling and so on. Technological innovations could reduce the amount of energy used and pollution by making machines more efficient, finding new energy sources, making renewable sources economically feasible and improving pollution control equipment. Ultimately we shall be forced to make better, more efficient use of our natural resources and rely more on renewable sources. It is still possible that we may discover a method for using hydrogen fusion. Finding a hydrogen fusion process that can be used to generate useful energy on Earth is considered the ultimate solution to providing energy. The Earth contains enough hydrogen of the proper form (i.e., deuterium) to supply our energy requirements indefinitely. Although there has been a vast amount of research directed toward this end, a controllable fusion process has been elusive. At present, and in the foreseeable future, hydrogen fusion should not be relied on to provide our energy needs.

Location, finiteness, politics, economics, pollution, technical capabilities and environmental factors make the future energy puzzle an extremely difficult one to solve. There are too many unknowns for us to make our decisions regarding energy with a high degree of certainty. The best that we can do is base our decisions on what we understand about our past, present and most probable future situations. It is not the primary purpose of this book to present possible solutions to this puzzle; rather it is intended to present the facts to the extent they are known and point out the principal realistic possibilities and obstacles that lie ahead. The succeeding chapters more thoroughly discuss the topics and problems briefly introduced in the preceding paragraphs.

Caution should be exercised when making predictions based on past and present economic conditions. There are simply too many unknowns for such predictions to be accurate and they provide very crude estimates at best. It is impossible to predict the effects of a yet unknown advance in technology. Certainly a means of providing useful nuclear fusion would have a profound effect on everything we do, but to a lesser extent so would cheaper and more efficient solar cells or batteries with much higher energy densities. Energy is so engrained in the way we live, that the ramifications of a change in the price of one source of energy to a particular aspect of our lives is frequently impossible to predict. How does the price of petroleum or natural gas affect the education of our children? Not only do the prices of these fuels change the cost of heating, cooling and lighting, they also change the cost of busing and school lunch programs, and the schools' tax revenues.

Despite the numerous inaccuracies associated with energy-related predictions, we should not blindly proceed without regard for the future. It is useful to study future scenarios based on past and present data and various assumptions. Even though the amount of petroleum left to be discovered is not known, upper and lower limits have been estimated using computer models and the resulting range of possibilities provides valuable insights for planning our future. If the assumptions are realistic and well-defined, then studying their implications is important. As time passes, the assumptions should be updated and the predictions changed accordingly. If a new technology becomes available, the cost of an energy source changes radically or the continued use of a source is found to be particularly detrimental, then a whole new set of assumptions may be needed.

Most of the data presented regarding energy sources and uses have been taken from or derived from data published by the Energy Information Administration (EIA), an agency within the United States Department of Energy (DOE), and the Bureau of Transportation Statistics (BTS), an agency within the United States Department of Transportation (DOT). In most cases, these agencies have compiled their data from other sources, such as professional, government and industrial reports and journals. It is these data that are used to indicate where we are and the trends that indicate the direction we are going. Most of the data used is the data for the years up to and including 2007, 2008 or 2009, because these years are the last years for which revised data was published at the time of writing. It should be understood that the collection of huge amounts of data does not produce exact results regardless of the amount of effort put into it. In fact, the EIA and BTS data are not always consistent. Keep in mind that these agencies collect their data from numerous sources and these sources do not always agree. In particular, the international data were taken from the EIA website which included a disclaimer stating that they had gotten the international data from outside sources and do not certify their accuracy. The EIA obtained much of its international oil and gas data from the *Oil and Gas Journal* and *World Oil*. A summary of the agencies from which data has been taken and their acronyms and web addresses is given in Table 1-2. In this chapter and the remainder of the book, when a figure is derived from data published by one or more of these agencies, the agencies' acronyms are included in brackets in the figure's caption. These notations do not necessarily mean that the data originated at the agencies. Normally, an agency will accompany its data with references to it sources. Other important sources, which were primarily used to research the various available technologies, include *Scientific American*, *IEEE Spectrum* (published by the Institute of Electrical and Electronics Engineers, Inc.), MIT's *Technology Review* (published by the Massachusetts Institute of Technology) and *Wikipedia (www.wikipedia.org)*. Information obtained from Wikipedia and other nongovernmental websites was primarily used in describing various systems and was crosschecked with other sources.

Table 1-2: Agencies and organizations used as sources.

Agency	Abbr.	Web address
Energy Information Administration	EIA	*www.eia.dot.gov*
Bureau of Transportation Statistics	BTS	*www.bts.gov*
United States Census Bureau	USCB	*www.census.gov*
United States Geological Survey	USGS	*www.usgs.org*
International Monetary Fund	IMF	*www.imf.org*
Bureau of Labor Statistics	BLS	*www.bls.gov*
United Nations	UN	*www.un.org*
United Nations Development Program	UNDP	*www.undp.org*
International Atomic Energy Agency	IAEA	*www.iaea.org*
World Nuclear Association	WNA	*www.world.nuclear.org*
Nuclear Regulatory Commission	NRC	*www.nrc.gov*
Nat. Oceanic and Atmospheric Adm.	NOAA	www.noaa.gov
Environmental Protection Agency	EPA	www.epa.gov
Intergov. Panel on Climate Change	IPCC	www.ipcc.ch

Most of the figures and tables in this book are derived from data for large areas or categories. For example, in Fig.1-1 the average energy amounts per person for various nations are displayed. It is frequently true that averages over large populations can be collected or computed more accurately than the same amounts for small populations or individuals in a population. Although data for small populations are sometimes useful, this book concentrates on national and regional issues that require national or regional data only. In particular, many of the graphs were derived from data taken from the EIA website. For worldwide data the EIA has collected data not only for nations, but has also summarized this data by dividing the world into the regions defined in Table 1-3. In this book, the "Central and South

America" region is referred to as "South America" and "Asia" includes Australia and Oceania, but some of both Europe and Asia is included in "Eurasia."

Table 1-3 EIA region definitions.

Region	Definition
North America	United States, Canada, Mexico, Greenland, Bermuda, St. Pierre
Cent. and So. America	Central America, South America, Antarctica, West Indies
Europe	Europe excluding Eurasia, Turkey, Iceland, Mediterranean
Eurasia	Former USSR including Lithuania, Latvia, Estonia
Middle East	Arabian Peninsula, Iran, Iraq, Syria, Jordan, Israel
Africa	Africa, Seychelles, Madagascar
Asia	Asia not included above, Australia, Oceania

The remaining chapters proceed by studying the direct sources of energy, electricity, energy consumption, the unwanted side effects of energy usage and the legislation and agreements related to energy. Chapters 2 through 5 discuss the nonrenewable sources petroleum, natural gas, coal and uranium, respectively. Chapter 6 examines the renewable sources biomass, geothermal, solar thermal, solar photovoltaic, water and wind, and Chap.7 examines electricity. Chapters 8 and 9 discuss the usage by the EIA sectors and the pollution due to this usage, and Chap. 10 reviews the last one hundred years of energy-related policy with special attention being given to legislation enacted in the United States.

This book does not assume more than a basic knowledge of science. However, an appendix has been included for those who want to know something of the history of energy and gain a deeper understanding of the more technical subject matter. If one chooses to read Appendix A, it is suggested that he or she do so before proceeding further. In addition, there is an appendix that summarizes the units of measure, their conversions, their abbreviations and their prefixes. If the reader is not familiar with the various abbreviations and prefixes, it would be worthwhile to study Appendix B at this point. Finally, for convenience, Appendix C contains a list of acronyms and their meanings.

2

PETROLEUM

We begin with the energy source of most concern, particularly American concern. We begin with the energy source that has serious ramifications with regard to pollution, geopolitics, economics and its own finiteness. We begin with petroleum, often referred to as crude oil or simply oil, or in the heyday of the Texas oil boom, black gold. *Petroleum* is an oily liquid that is extracted from the ground or seabed. Its color varies from yellowish-gray to black and its viscosity varies from that of very thin syrup to that of an almost tar-like substance known as heavy crude. Sometimes it gushes from the earth and other times it must be pumped, and ultimately it may need to be coaxed from the earth by applying chemicals, pressurized gas or steam.

Petroleum is believed to have formed from plant and animal organisms that have been buried in the Earth's crust over the last 600 million years. These organisms were first converted to matter called kerogen, but as the burial depth increased, so did the temperature and pressure. When the temperature and pressue reached a critical values, the kerogen was converted to natural gas and petroleum. In locations with the right geology, the natural gas and petroleum rose up toward the surface, and even breached the surface in a few places.

The petroleum that has come to the surface on its own has been used for heat, light, construction and medicine since ancient times. But in the last two centuries, science and technology has expanded its uses to the point where it is now considered indispensable. The first known oil well, the present means of getting almost all petroleum, was drilled by the Chinese in the fourth century. The American petroleum industry began in earnest in 1859, when Edwin Drake drilled a shallow oil well in Pennsylvania. During the last half of the nineteenth century the petroleum industry was sustained primarily by the demand for kerosene, which was used in lamps. Petroleum production increased from 2000 barrels in 1859 to 57 million barrels in 1899. The demand for petroleum mushroomed when the mass production of automobiles began in the early twentieth century, and in 2006 Americans consumed about 7.7 billion barrels. (This 2006 amount dropped to about 6.8 billion barrels by 2009 because of a severe recession.)

Petroleum is, as are all fossil fuels, primarily a mixture of hydrocarbons, but does contain some sulfur and other compounds. A *hydrocarbon* is a compound made up of molecules that contain only hydrogen (H) and carbon (C) atoms. The hydrocarbon molecules that have the simplest structures have chainlike constructions, chemical formulas of the form C_nH_{2n+2} and Lewis formulas of the form

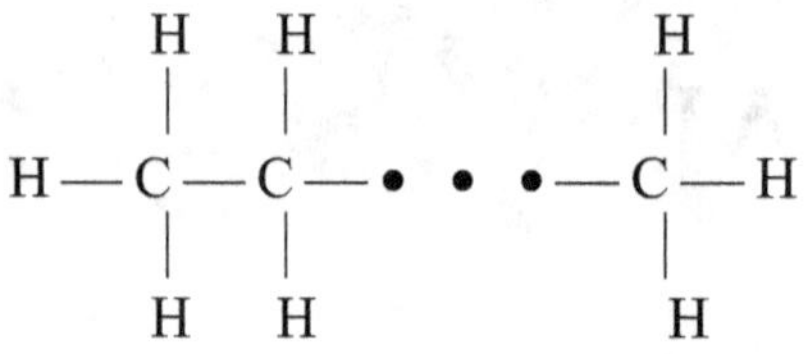

$$
\begin{array}{ccccccc}
 & H & H & & & H & \\
 & | & | & & & | & \\
H- & C- & C- & \bullet \; \bullet \; \bullet & - & C- & H \\
 & | & | & & & | & \\
 & H & H & & & H &
\end{array}
$$

However, a hydrocarbon molecule may be quite complex and may include a complicated array of bonds that form rings of atoms. Hydrocarbon compounds are categorized by the number of carbon atoms they contain. The main categories of substances that can be obtained from petroleum are summarized in Table 2-1.

Table 2-1: Hydrocarbons in petroleum.

Name of product	No. of carbon atoms	Boiling point °C
Methane	1	-162
Ethane	2	-89
Propane	3	-42
Butane	4	-0.5
Gasoline	5-10	36-175
Kerosene, jet fuel	10-18	175-275
Diesel	12-20	190-330
Fuel oil	14-22	230-360
Lubricating oil	20-30	Above 350
Heavier oils, paraffin	22-50	Above 350

As anyone who follows the trading of energy futures knows, not all petroleum is equal. Petroleum is graded according to its density and sulfur content. It is classified as light, medium and heavy and is traded under such names as West Texas Intermediate, Oklahoma Sweet and Nigerian Bonny Light. Of interest is its energy content per barrel and ability to be refined into diesel, gasoline and the lighter oils. Light sweet crude fetches the highest price and heavy crude the lowest price. What is most often quoted in the news is the New York Mercantile Exchange (NYMEX) futures prices.

The process of producing, transporting and refining petroleum is depicted in Fig. 2-1. Most presently produced petroleum comes from oil wells that are thousands of feet deep and many are drilled into the seabed. Because these wells are so deep and more and more tend to be located in inhospitable places, they have become very expensive to drill and maintain. No longer do oil companies blindly drill wells in likely locations to determine whether or not oil is present, a practice known as wildcatting. Today, before drilling begins, thorough research employing modern, high-tech equipment that includes high-speed computers is used to determine where a test well should be drilled in order to maximize its success. As time goes by, finding and developing new fields has become increasingly expensive. Recently, billions of dollars have been spent in an attempt to establish a deep-water field in the Gulf of Mexico. Exxon has spent over 100 million dollars to drill a single, 4-mile deep well in the Caspian Sea. In the future, more and more will be spent to discover less and less.

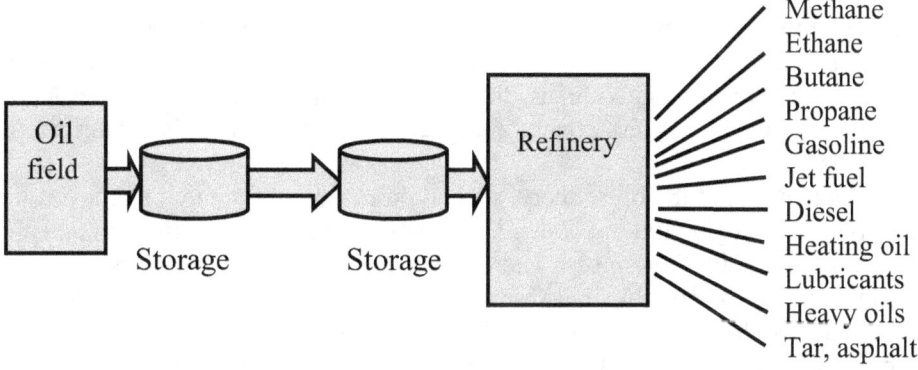

Fig. 2-1 Production to refining stages.

At first, the pressure at the bottom of the well forces the oil up to the top of the well and no pumping is needed. On average, this *primary* recovery produces about 20 percent of the oil in a reservoir. A *secondary* method that involves pumping, perhaps aided by forcing a gas (usually air, natural gas or carbon dioxide) or water into the reservoir, normally produces another 15 to 25 percent, and a *tertiary* method may be used to extract an additional 5 to 15 percent. Tertiary methods reduce the oil's viscosity. Most often this is done by increasing the oil's temperature with steam or by simply burning some of the oil, known as *in-situ burning*, to heat up the other oil. Less often, a detergent is employed to reduce the oil's viscosity. Both secondary and tertiary methods increase the cost of recovery and tertiary methods may be prohibitively expensive depending on the current price for oil. As discussed in Chap. 7, steam injection is usually done when a steam-driven electric power plant is located over the oil field and the steam it produces in generating electricity can be forced into the reservoir. This means that the energy used to heat the water is used for two purposes instead of one.

After the oil has been brought to the surface, it must be collected and stored before being transported to an oil refinery. Figure 2-1 shows only two transportation stages and two storage stages, but there may be several alternating transportation and storage stages before the oil reaches a refinery. Transportation may be accomplished by a combination of pipelines, ships, trucks and trains, each offering the possibility of pollution-causing accidents. Pipeline and ship transportation are the least expensive, but tend to result in the most serious accidents because of the quantity of oil that may by spilled by a single accident. However, one of the principal advantages of petroleum is that, because of its high energy density and non-volatility, it is relatively easy and safe to store and transport. Storage facilities may consist of a single tank or group of tanks called an *oil depot* or *tank farm*. Unless storage facilities are properly inspected and maintained, they eventually develop leaks that pollute the ground water. Because pipelines and storage facilities often pass through or are in isolated areas, they are difficult to guard and are prone to vandalism and sabotage. These attacks and guarding against them also contribute to the increasing price of oil.

Upon arrival at the oil refinery, several processes are required to convert the oil into the products given in Fig. 2-1. First, the oil is fed into a desalination unit and then sent to a *distillation tower* where the oil is boiled off into its principal components. This is possible because, as shown in Table 2-1, these different components have different boiling points. The sludge that does not boil off is used to produce asphalt or, since it still has energy content, as a low-grade fuel. After distillation, there is further cleansing to remove sulfur.

Distillation alone does not yield either the quantities or qualities of the desired products. Further processing is needed to increase the purity and combustion properties of these products. This is particularly true for gasoline and diesel, which are burned in the internal combustion engines that power our cars and trucks. The chainlike molecules having the formulas C_nH_{2n+2} are not, by themselves, good enough. Gasoline and diesel are given octane and cetane ratings, respectively, to indicate their performance levels. Some of the molecules in gasoline and diesel must be reformed in order to increase their performance ratings. Also, it is possible to use *cracking units* to break the more complex molecules into molecules having fewer carbon atoms, and *isomerization units* to combine molecules with only a few carbon atoms into heavier molecules. Even asphalt may be further processed into gasoline, diesel or coke by sending it to a *coking unit*. Lastly, there are compression and liquefaction units to reduce the volume of the gaseous compounds, methane, ethane, propane and butane. There may be units for adding oxygen to the methane, ethane and propane to obtain methanol (methyl alcohol), ethanol (ethyl alcohol), and propanol. Methanol is sometimes used to power hydrogen fuel cell vehicles and ethanol is used as a gasoline additive to raise its octane rating.

Because molecules can be broken down or combined, a refinery has a certain amount of latitude in adjusting the relative amounts of its products. Normally, the highest demands are for gasoline, diesel, jet fuel and heating oil and refineries are designed and adjusted accordingly. In winter, the demand for gasoline and diesel goes down and refineries are adjusted to produce more heating oil, while in summer more gasoline and diesel are needed and refineries are readjusted to produce more gasoline and diesel. European refineries are designed to yield a higher proportion of diesel than American refineries because more of the cars and light trucks in Europe run on diesel. The 2008 breakdown (a typical breakdown) of the output produced by American refineries is given in Fig. 2-2. The numbers in parentheses indicate the amounts of the products obtained from 42 gallons (one barrel) of oil. Forty-two gallons converts into approximately 44.6 gallons (1.062 barrels) of products due to the fact that a chemical reaction may result in products that are less dense than its reactants, even when both are in liquid form (e.g., gasoline is less dense than petroleum). This increase is referred to as *refinery gain*.

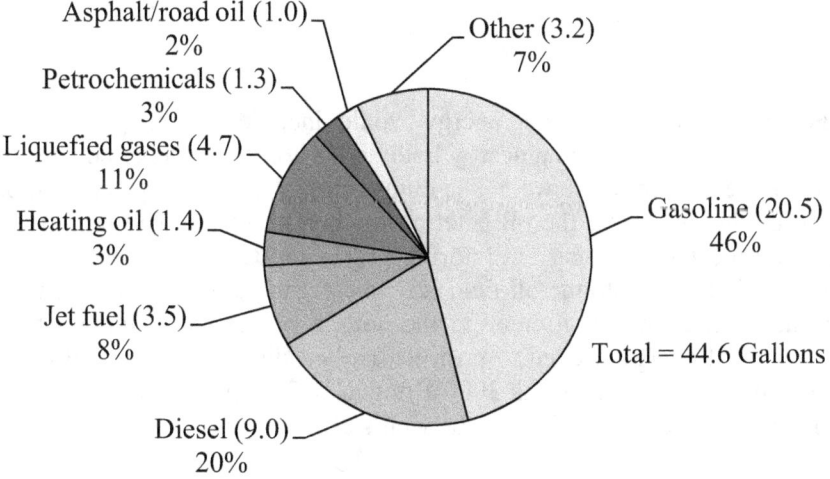

Fig. 2-2 2008 products from 42 gallons of petroleum.

The percentage of gasoline tends to vary with time between 42 and 47 percent, but the amount of diesel and gasoline together are almost always near 67 percent. The total amount of products in 2008 was 19.5 million barrels per day.

Theoretically, the *energy efficiency* of a system is the useful energy gotten out of the system divided by the useful energy put into the system. Useful energy normally means the maximum energy one can reasonably expect to get from an input. For burning petroleum or a petroleum product, the useful energy varies according to its quality. In presenting its data, the United States Bureau of Transportation Statistics (BTS) uses British thermal units per gallon (Btu/gal) values given in Table 2-2.

Table 2-2: Useful energy densities for petroleum
and its products [BTS].

Product	Energy content in Btu/gal
Petroleum	138,000
Jet fuel	135,000
Aviation gasoline	120,200
Automobile gasoline	125,000
Diesel	138,700
Heating oil	138,700
Residual fuel	149,700

For an oil well, the input energy would include the energy required to extract the oil and transport it to a nearby holding tank and the energy content of the oil that enters the bottom of the well. The output energy is the energy content of the oil fed into the holding tank, the oil entering the bottom of the well minus any oil that is lost during the pumping and storing processes. Consider a well in which carbon dioxide is forced into the oil field and an electric pump is used to lift the oil to the surface. For each gallon entering the bottom of the well, suppose that the average amount of energy needed to push the carbon dioxide into the oil reservoir and to operate the electric pump is 3000 Btu/gal. If 0.01 percent of the oil is lost before it reaches the holding tank, then the well's efficiency is

$$0.9999 \times 138,000/(138,000+3000) = 0.979,$$

or 97.9 percent. The energy lost through leakage by a single production, storage or transportation stage is likely negligible, but the average cumulative leakage may or may not be negligible from an efficiency standpoint.

The overall efficiency of several stages in a sequence is the product of the efficiencies of the individual stages. For the sequence of stages shown in Fig. 2-1, assume that the efficiency for getting the oil to the surface is 97.9 percent, transporting the oil to the refinery is 98.1 percent, storing it at the refinery is 100 percent and refining the oil is 95 percent, then the overall efficiency is

$$0.979 \times 0.981 \times 1 \times 0.95 = 0.912,$$

or 91.2 percent.

Although it is easy to declare that efficiency is the output energy divided by the input energy, it is somewhat less easy to determine the efficiency of a practical system. In the above example, the electrical energy supplied to the pump was considered to be one of the inputs, but some energy experts insist that this input should be traced back to the source of the energy used to generate the electricity, or at least to the energy supplied to the steam turbine that drives the generator. The truth is that as one traces backwards into an energy supply system, a complex, interlocking array of supply points is likely to be encountered. It is almost impossible to collect all the data needed to determine an accurate value for an average amount of input energy. However, by limiting the size of the system, breaking it into manageable parts and finding an approximate average efficiency for each part, a meaningful average or range for the overall efficiency can usually be found.

Figure 2-3 depicts a gasoline supply process from a refinery to an automobile's fuel tank. As before, an approximate efficiency could be computed for each stage, and the product of these efficiencies would be the refinery-to-automobile-tank efficiency. By multiplying the well-to-refinery output efficiency by this value, an overall efficiency for getting gasoline to an automobile, called *well-to-tank (WTT) efficiency*, is determined. As discussed in Chap. 8, the WTT efficiency is commonly used to compare various types of energy sources used by vehicles and the

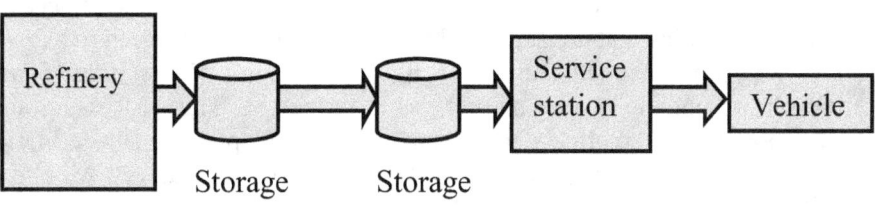

Fig. 2-3 Refining to user stages.

most frequently quoted average WTT efficiency for petroleum is 88 percent. This efficiency will decrease as wells become deeper and more inaccessible and the quality of oil decreases.

The 2008 oil refining capacities of the Earth's major regions are given in Fig. 2-4. Again, the figures in parentheses are the amounts in millions of barrels per day Mbbl/day. Australia, New Zealand and the Pacific islands are included in Asia/Oceania. The combined capacities of the world's oil refineries in that year was 85.4 Mbbl/day, of which America's refining capacity was 17.6 Mbbl/day, or 20.6 percent of the world's capacity. Russia had the second largest capacity at 5.43 Mbbl/day, which was less than one third that of the United States. But our oil refineries still do not output enough to meet our enormous appetite for petroleum products, which often exceeds 20 Mbbl/day.

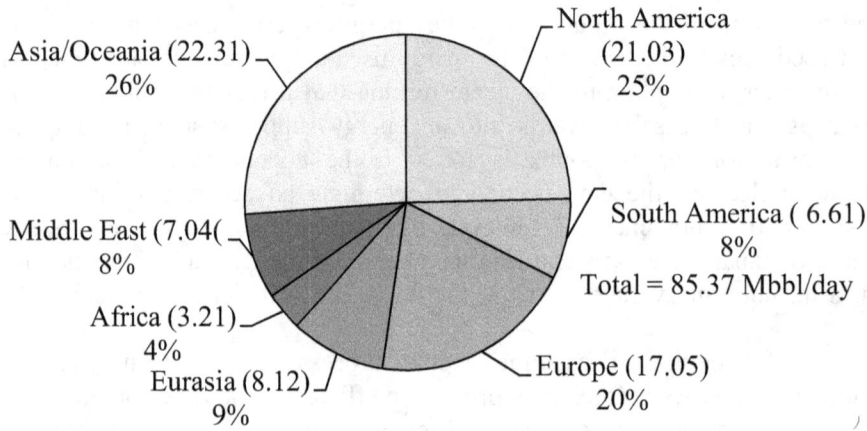

Fig. 2-4 2008 petroleum refining capacities (Mbbl/day) [EIA].

Because of the way both petroleum and natural gas are formed, both petroleum and natural gas reservoirs contain a mixture of liquid and gaseous hydrocarbons. The liquids that are separated from natural gas are referred to as *lease condensates*, which are extracted at the wellhead, and *natural gas plant liquids (NGPL)*, which are obtained when natural gas is processed. This processing adds about 2.16 Mbbl/day to America's petroleum products supplied by its refineries. Therefore, our total petroleum product supply is slightly less than 20 Mbbl/day, which is still short of our usual demand. North America's 2008 petroleum refining capacity was 21.03 Mbbl/day, and most of the United States shortfall was made up for by the refineries in Canada and Mexico. Importing refined gasoline by ship is hazardous, so the refining should be done as closely as possible to where it ultimately is used.

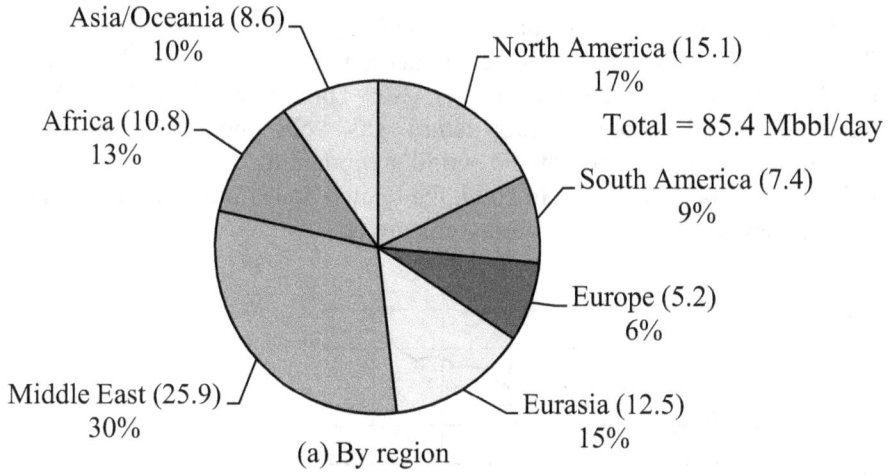

(a) By region

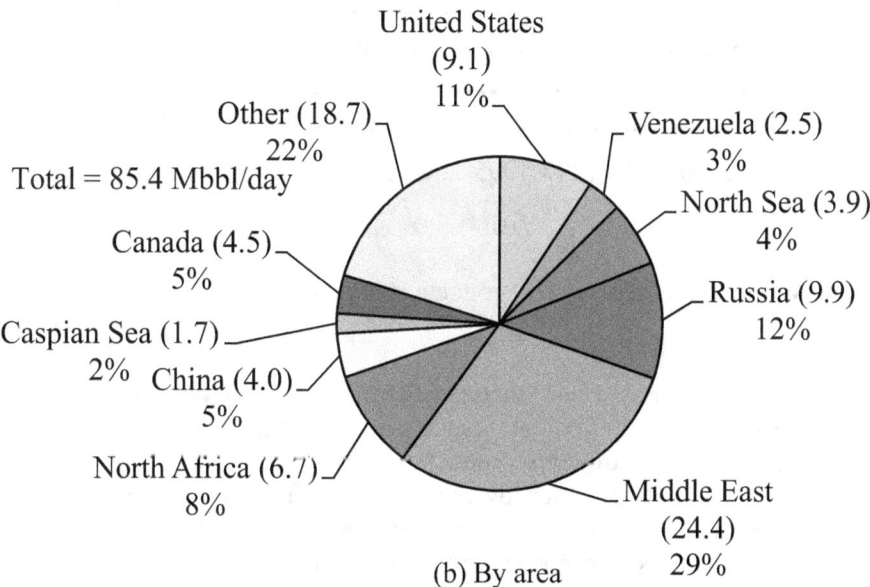

(b) By area

Fig. 2-5 2008 petroleum production by major region and area [EIA].

The average 2008 petroleum production of both the world's major regions and areas are summarized in Fig. 2-5. The data used in constructing these charts includes the lease condensates and NGPLs, but do not include the refinery gains. Of the 85.4 Mbbl/day, about 45 percent came from what one might call the oil crescent consisting of northern Africa, the Middle East, the Caspian Sea, Southeast Asia and Indonesia. One tip of the crescent lies in Nigeria, the other tip lies in Indonesia and the middle extends from Saudi Arabia north to the Caspian Sea.

The history of petroleum product production from 1980 to 2009 for the world and United States is shown in Fig. 2-6. This figure includes refinery processing gains, liquids obtained from natural gas wells and petroleum liquids obtained by other means. Although the world's production increased from 63.0 Mbbl/day in 1980 to 84.2 Mbbl/day in 2009, the United States production decreased from 10.8 Mbbl/day to 9.1 Mbbl/day during this period.

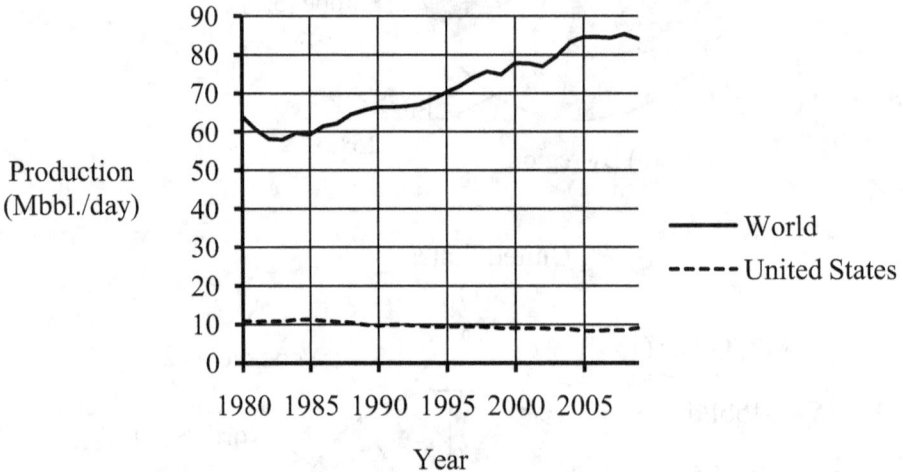

Fig. 2-6 World and United States production of petroleum [EIA].

In a paper presented at an American Petroleum Industry meeting in 1956, M. King Hubbert, the chief geology consultant with the Shell Oil Company, accurately predicted that petroleum production from the ground (this excludes processing gain and liquids obtained by other means) in the United States would reach its peak in the early 1970s. A curve similar to his, often referenced as Hubbert's curve, is shown superimposed with the actual United States production in Fig. 2-7. The United States production, in fact, peaked at 9.64 Mbbl/day in 1970. Hubbert argued that, because exhaustible materials such as petroleum are finite, their rate of extraction must, in time, approach zero, and this implies their production must reach a peak and then drop off at some point. Based on past data and his hypothesis that the decrease would be approximately symmetric with the increase, he arrived at a curve similar to the one shown. Others argued that, in the foreseeable future, an increasing price of oil would spur production, thereby causing supply to keep up with demand. His response was, "There is a different and more fundamental cost of exploration and production. So long as oil is used as a source of energy, when the energy cost of recovering a barrel of oil becomes greater than the energy content of the oil, production of a barrel of oil will cease no matter what the monetary price may be." In other words, if oil is extracted solely for its energy and it takes more energy to extract it than the energy it contains, then there is no reason to extract it. Of course, oil is used for purposes other than energy (e.g., making plastics and fertilizer) and a relatively small amount may continue to be produced for these purposes even though there is a net energy loss.

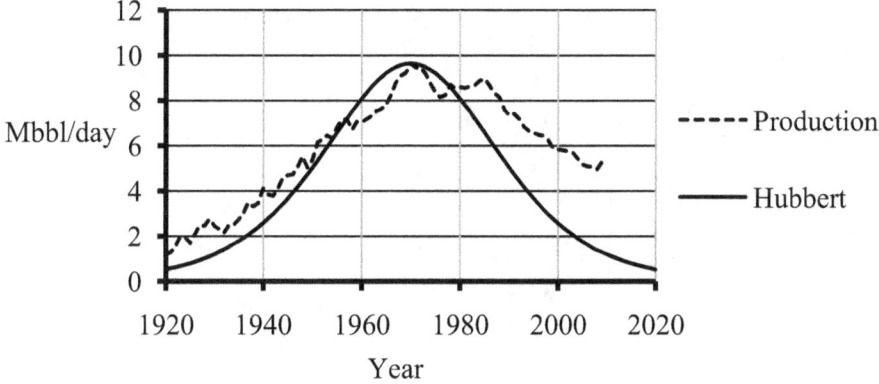

Fig. 2-7 United States production and Hubbert curve [EIA].

From Fig. 2-7 it is seen that the actual production curve is not symmetric. The falloff is not as rapid as the rise. This is probably due to the fact that the United States has been able to increase its imports and conserve its reserves. But the world as a whole does not have this option, so once it reaches its peak, worldwide production is likely to decrease more rapidly. Numerous studies have used Hubbert's ideas, although many have assumed different curves, both symmetric and asymmetric. One such study is considered toward the end of this chapter.

The known reserves that are considered recoverable using current technology are called *proven reserves*. These reserves are broken down according to the major oil producing regions in Fig. 2-8. At the end of 2009, the world total was estimated to be 1,342 billion barrels, i.e. gigabarrels (Gbbls). It is very difficult to accurately estimate reserves, even in the United States where data are readily available. In other countries we must rely on the data supplied by them or the oil companies operating in them, and there are sometimes incentives for them to overestimate their reserves. The members of the Organization of Petroleum Exporting Countries (OPEC), which includes the Middle East, northern Africa, Indonesia and Venezuela, base their agreed upon export quotas on the amount of reserves in each country. In 1987, several of these countries increased their reserve estimates dramatically so that they could export more oil. The historical average estimates of the known reserves from 1980 to 2009 are graphed in Fig. 2-9.

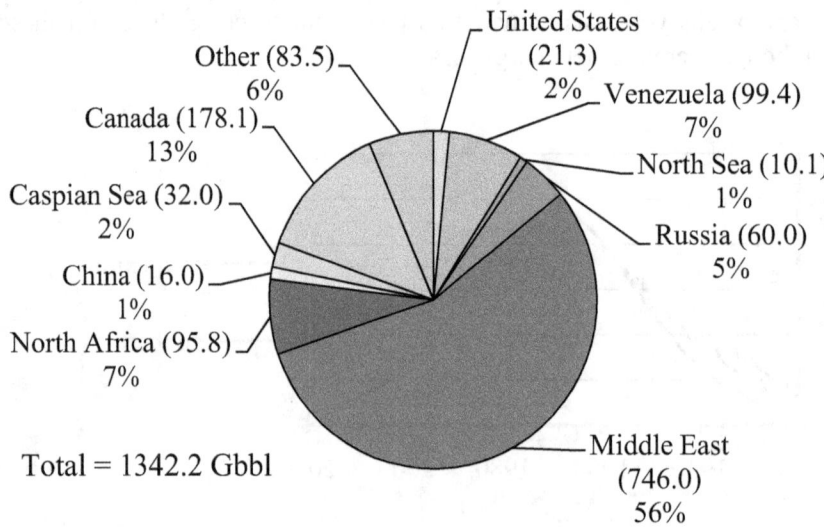

Fig. 2-8 2009 known world petroleum reserves [EIA].

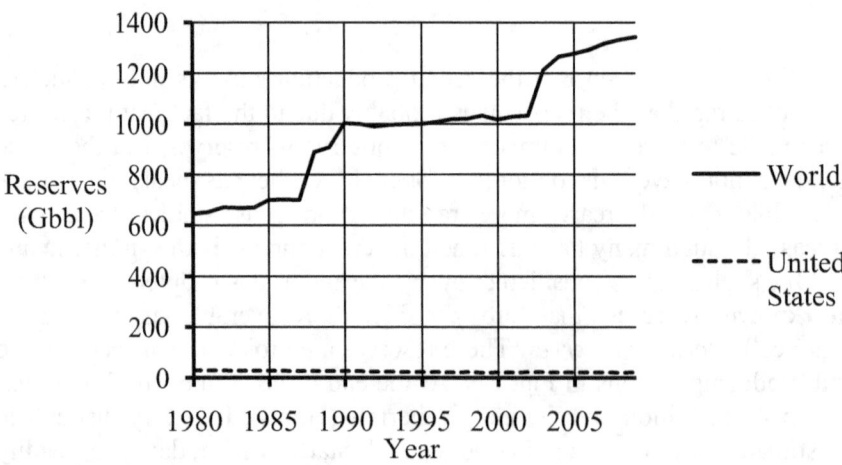

Fig. 2-9 Known petroleum reserves in the world and United States [EIA].

Not only are the proven reserves estimates, but also there is a question as to how much of the petroleum is recoverable even by tertiary methods. In addition, there are the unknown, or undiscovered reserves, and the uncertainty as to how much of it is recoverable. It is believed that the range of proven reserves is between 850 and 1600 Gbbl and there may be another 800 to 1200 Gbbl of unknown reserves that can be produced with the aid of future technologies. This means that by the most optimistic estimates the recoverable, economically viable reserves would be exhausted in less than 100 years at the 2009 rate of production. Considering that the rate of production is currently increasing, this time period could be significantly less. The most promising areas for expanding the world's production are the north slope of Alaska, Arctic Ocean, Gulf of Mexico using deepwater drilling, Caspian Sea and southern Sudan, all of which face significant technological, environmental and political challenges. If deepwater drilling proves to be practical, several other fields may be discovered throughout the world. But the 2010 explosion at the BP corporation Deepwater Horizon oil platform in the Gulf of Mexico and the resulting ecological disaster is likely to dampen the enthusiasm for drilling deep water wells. The cost of drilling such wells is in the billions of dollars, even if there are no accidents.

An important quantity that is often used to predict the future of oil is the reserves to production ratio (R/P), which is the number of years oil can be produced from known reserves at the current rate of production. The history of this ratio from 1980 to 2009 is illustrated in Fig. 2-10. It is seen from this graph that the discovery of new reserves has substantially out-stripped production since 1980 and it seems that the danger of running out of oil in the foreseeable future is unlikely. Such a conclusion is misleading, however, because we are approaching the point at which the cost of finding and extracting new oil is prohibitive relative to the cost of other sources of energy.

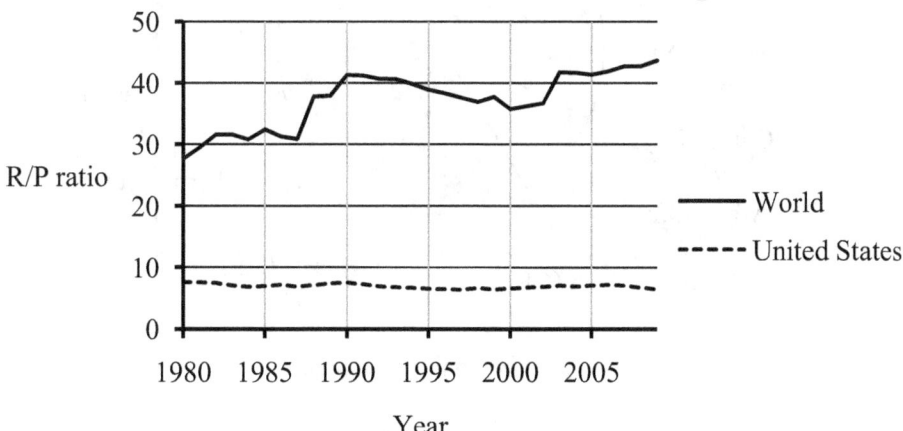

Fig. 2-10 Historical R/P ratio for the world and United States (EIA).

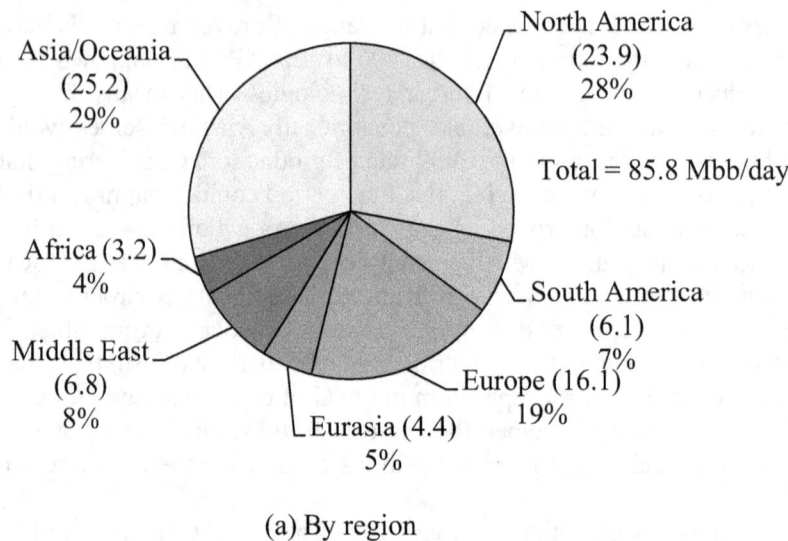

(a) By region

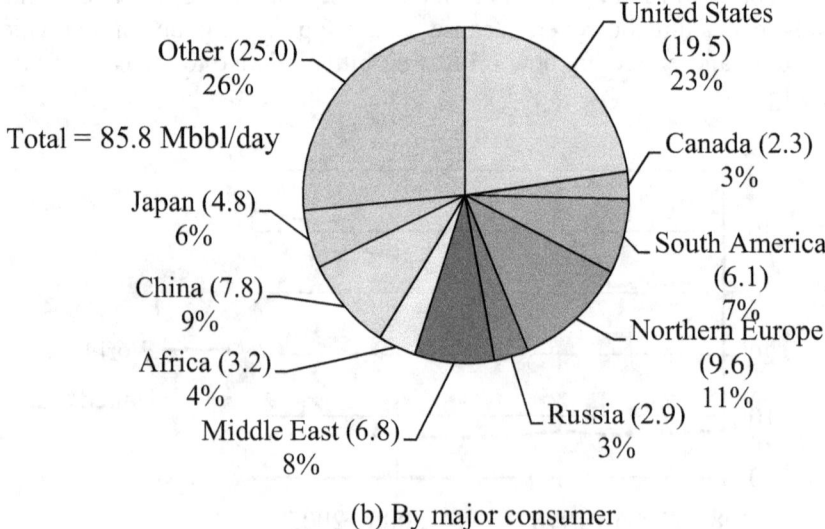

(b) By major consumer

Fig. 2-11 2008 consumption by region and major consumer [EIA].

The 2008 consumption of petroleum by major region and consuming area is shown in Fig. 2-11. The worldwide consumption that year was 85.8 Mbbl/day. The United States required 19.5 Mbbl/day, or the equivalent of 13 fully loaded ships the size of the Exxon Valdez. The differences between the areas of production shown in Fig. 2-6 and the consuming areas shown in Fig. 2-11 are abundantly clear. It is the industrialized nations with high standards of living that consume over half of the petroleum, even though they produce less than one-fifth of it. This discrepancy becomes even more apparent when one considers the consumptions per capita given in Fig. 2-12. The United States and Canada are particularly conspicuous, each using about five times the world average.

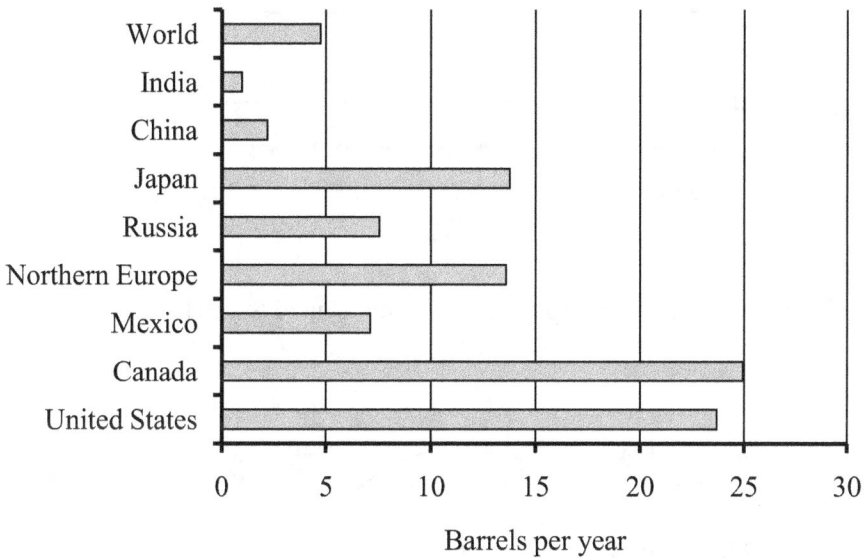

Fig. 2-12 2008 per capita consumption [EIA and USCB].

Another ratio of interest is the production to consumption ratio (P/C). A graph of this ratio for the United States from 1980 through 2009 is given in Fig. 2-13. It indicates the United States shortfall, which must be made up by imports. Where the 2008 imports came from is broken down in Fig. 2-14. In total, we imported about 13.7 Mbbl/day. Only about 28 percent of these imports are from our neighbors Canada and Mexico. Over half of our imports came from nations that are either unfriendly to the United States or are unstable or both. Nearly 10 Mbbl/day must be imported by ship every day. Counting the oil that is brought in from Alaska, each day roughly 8 oil tankers the size of the Exxon Valdez must land in the ports of the 48 contiguous states and that, at any given time, dozens of tankers may be destined for these ports. The United States economy very much depends on the reasonably smooth operation of these supply lines as well as the imports from Canada and Mexico.

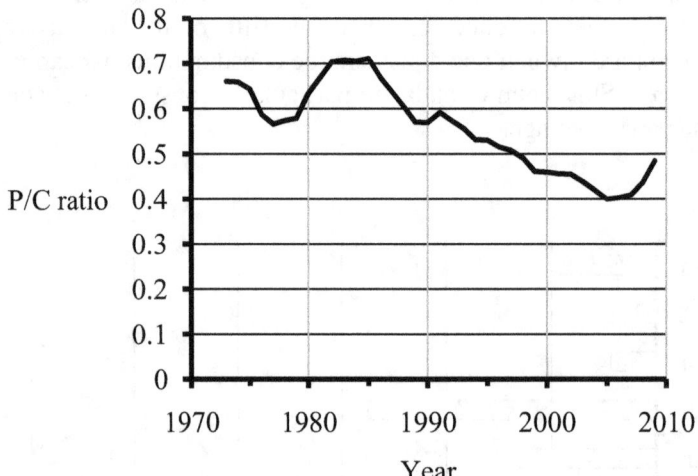

Fig. 2-13 Production to consumption ratio for the United States [EIA].

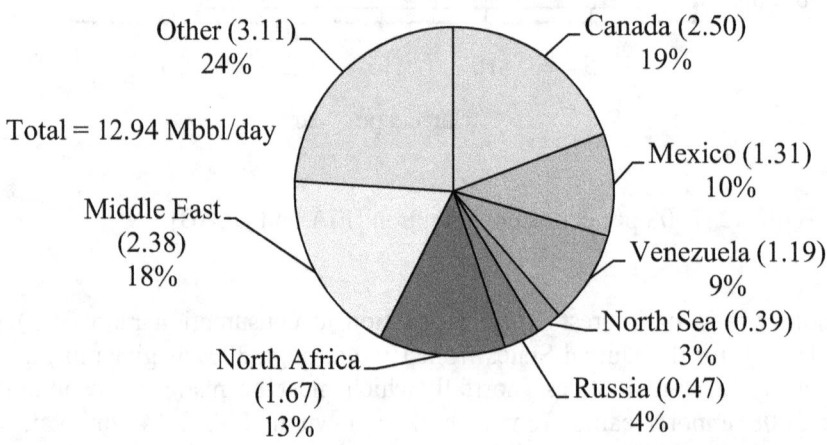

Fig. 2.14 2008 United States petroleum imports [EIA].

An important quantity for smoothing out the short-term effects of the supply of a good is, in general, the amount of the good at hand. *Petroleum stocks*, or *inventories*, are defined by the EIA to include all individual products held at refineries, in pipelines and bulk storage facilities with capacities of at least 50,000 bbls. Products stored by retailers or at any point of consumption are excluded. The stocks of petroleum products are a hedge against shortages in supply and are particularly important to investors in the petroleum futures market. Large corporations, especially those with heavy transportation costs, such as airlines whose fuel costs may be as much as 30 percent of their total costs, frequently use the futures market to lock in prices in advance.

Throughout this book, production and consumption have been expressed, and will continue to be expressed, as rates of extraction and usage as opposed to simply amounts. Therefore, for the world as a whole the

$$\text{Rate of change of stocks} = \text{Production} - \text{Consumption.}$$

If production exceeds consumption, then the worldwide stocks grow; otherwise they shrink. For an individual nation or collection of nations, imports and exports, which also are expressed as rates, must be considered as well, in which case the above relationship becomes

$$\text{Rate of change of stocks} = (\text{Production}+\text{Imports}) - (\text{Consumption}+\text{Exports})$$

By importing more oil than it exports, a nation or group of nations can maintain or increase its stocks even if its consumption exceeds its production. The United States, Japan and most European nations are able to keep their stocks at more or less a steady level only by importing large quantities of oil. If there were a serious interruption in supply these countries would have to rely on either their stocks or a reduction in consumption in order to preserve their vital petroleum requirements. The stocks held by the United States individually and the Organization for Economic Cooperation and Development (OCED) as a whole from 1973 to 2006 are graphed in Fig. 2-15. The OCED includes the United States and consists of the world's 30 most developed nations, most of which are oil-importing nations. At the end of 2009, the combined stocks in all OCED nations were 3.59 Gbbls and in the United States alone were 1.78 Gbbls.

The first time the United States considered establishing massive storage facilities under government control as an emergency supply was in 1944 during World War II. It also was proposed in 1952 during the Korean War and 1956 during the Suez crisis, but was not acted upon until 1975 after many of the Arab states cut off their exports to the United States during the1973-74 Oil Embargo. This embargo prompted Congress and President Ford to create the *Strategic Petroleum Reserve (SPR)*. The sites chosen for the storage facilities were more than 500 underground salt domes scattered along the coast of the Gulf of Mexico. The first oil was transferred into these domes in 1977 and at the end of 2009 they held 727 Mbbls, or

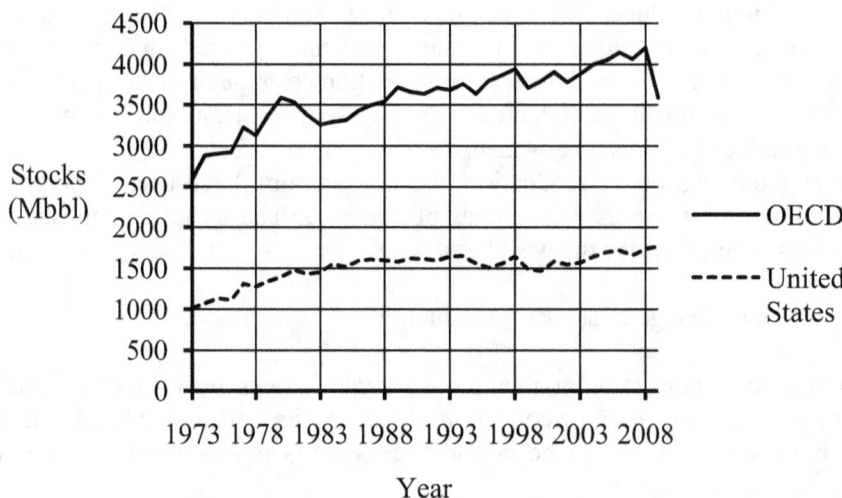

Fig. 2-15 2006 average OECD and United States stocks [EIA].

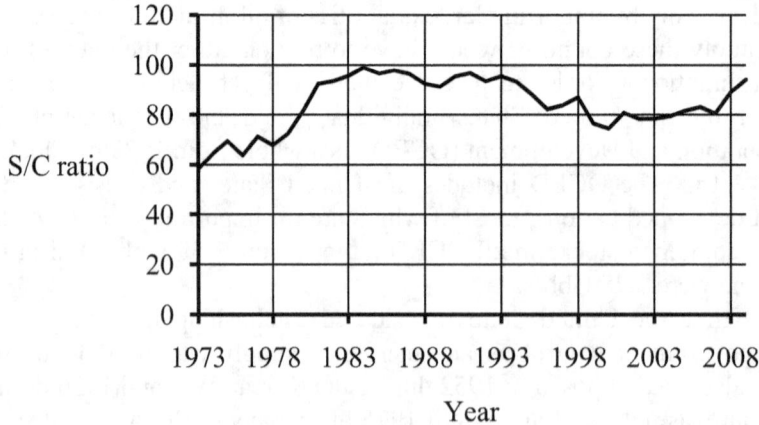

Fig. 2-16 Historical United States stocks to consumption ratio [EIA].

about a 38-day supply assuming our 2009 level of consumption. In 2005, the Emergency Policy Act authorized an increase in the capacity to one billion barrels. The power to tap the SPR was given to the President and it has been used twice, once in 1991, at the time of the First Gulf War, and again after the 2005 Hurricane Katrina severely impaired oil production in the Gulf of Mexico, thereby reducing production in the United States supply by 25 days assuming its 2009 level of consumption. The stocks-to-consumption (S/C) ratio, which indicates the number of days our stocks, including the SPR, would last in the absence of all production and imports, is given in Fig. 2-16.

Over the last 60 years, three important international organizations have been formed. Their memberships are listed in Table 2-3. Two of them, the Organization of Petroleum Exporting Countries (OPEC) and the Organization of Arab Petroleum Exporting Countries (OAPEC) are directly related to the production and marketing of oil. OPEC was founded in Baghdad in 1960 by five of its current twelve members. As its name indicates, it consists of the major oil exporting countries and its purpose is to stabilize the price of oil and to assure its members a reasonable price for their production. It accomplishes its control by limiting the production and marketing of oil. Each member is allotted a production quota based on its reserves and these quotas are updated by agreements among the member states.

Table 2-3 Memberships of three of the world's important organizations.

OPEC	OAPEC		OECD	
Algeria	Algeria	Australia	Hungary	Poland
Angola	Bahrain	Austria	Iceland	Portugal
Indonesia	Egypt	Belgium	Ireland	Slovakia
Iran	Iraq	Canada	Italy	South Korea
Iraq	Kuwait	Czech Rep.	Japan	Spain
Kuwait	Libya	Denmark	Luxembourg	Sweden
Libya	Qatar	Finland	Mexico	Switzerland
Nigeria	Saudi Arabia	France	Netherlands	Turkey
Qatar	Syria	Germany	New Zealand	UK
Saudi Arabia	UAE	Greece	Norway	United States
UAE				
Venezuela				

OAPEC was created in Beirut in 1968 by Kuwait, Libya and Saudi Arabia and now has ten members. The only countries in OAPEC that are not also OPEC members are Bahrain, Egypt and Syria. Although Egypt and Syria export very little oil, they do produce a significant amount for their own use. The OAPEC nations form an economic bloc based on their common cultural and Islamic traditions that go beyond the exportation of oil. The countries in OPEC that do not have an Arabic heritage, namely Angola, Iran, Indonesia, Nigeria and Venezuela, are not part of OAPEC.

The third organization, the Organization for Economic Cooperation and Development (OECD), originated with the Organization for European Economic Cooperation (OEEC). The OEEC was formed after World War II to administer the Marshall Plan aid provided by the United States and Canada for the purpose of reconstructing Europe. OECD replaced OEEC in 1961 and has grown to 30 nations from around the world. Although its stated purpose has never been to counter OPEC, it is made up of the world's most developed countries and, as a result, includes most of the world's chief consumers of petroleum.

When one considers the areas where most of the oil is located and produced, as opposed to where most of it is consumed, it takes little imagination to realize the possibility of conflict, especially when considering the differences in culture, economic development and military power. The first use of oil as a political weapon was, what is now known as the 1973-74 Oil Embargo. The embargo was instigated by the members of OAPEC to punish those who supported Israel during the Yom Kippur War. This war was initiated by threats from Syria and Egypt in an attempt to regain the territories lost during the Six-day War of 1967 when Israel made a preemptive strike against its Arab neighbors and gained possession of the West Bank, Gaza Strip and Jerusalem. Although the embargo was largely aimed at the United States and succeeded in causing substantial economic disruption there, it proved devastating to some of the world's poorer and weaker nations. In the long run, because the embargo more than quadrupled the price of oil, it caused non-OPEC countries to expand their exploration, production and conservation. Figure 2-17 shows both the monthly average price and inflation-adjusted average price of oil from 1973 through 2009. Because this graph is of annual averages, it does not indicate the daily highs. (Not seen in the graph are the 2008 short term prices of over $130 per barrel or the radical behavior that occurred thereafter.)

The second major economic recession related to oil occurred when the internal political situation in Iran resulted in the overthrow of the Shah and the installation of the Islamic regime headed by the Ayatollah Khomeini in early 1979. The disruption of the oil supply was exacerbated by the Iranian invasion of the American embassy in Tehran in November of that year and the Iran-Iraq War that began in September of 1980. Unlike the 1973-74 Oil Embargo, the shortage of oil was not caused by OAPEC, but by the loss of production in Iran and Iraq. In fact, most OAPEC countries increased their production during the 1979-'80 period. Worldwide production never dropped more than a few percent and the sharp rise in

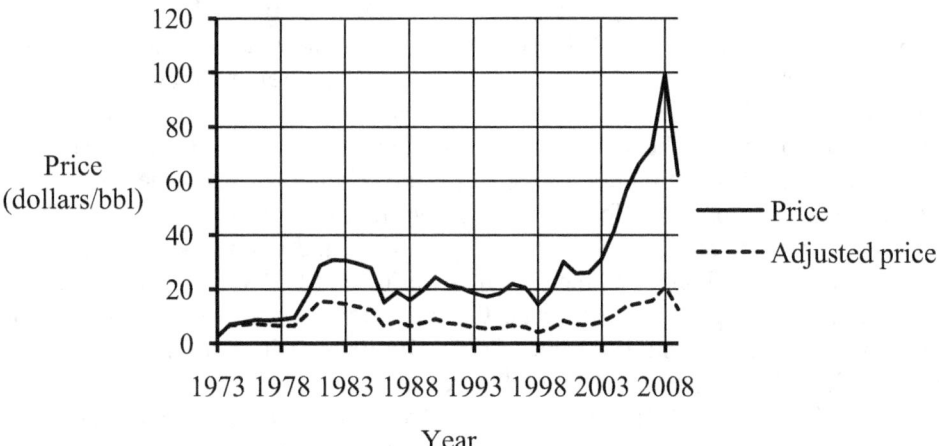

Fig. 2-17 Per barrel price and inflation adjusted price to 1973 dollars [EIA, BLS].

oil prices was partly psychological and was short-lived. After the price of oil shot up to almost $40 per barrel in 1980, the price began to drop and fell below $10 per barrel by July, 1986. By 1987, some of the OPEC nations had raised their estimates of reserves so that they could justify larger export quotas.

It was 1991 before the next spike occurred and was caused by the annexation of Kuwait and its vast oil fields by Iraq. This drove the price of oil to above $30 per barrel. The attack was condemned by the United Nations and in January of 1991, the sovereignty of Kuwait was reestablished after the First Gulf War by a coalition of forces from the United States, Saudi Arabia and several other countries. Even though the retreating Iraqis set fire to the Kuwaiti oil fields, within days of this very brief conflict the price of oil fell below $20. During the remainder of the 1990s the price of oil fluctuated over a considerable range and briefly dipped below $10 owing to the reduced demand caused by the severe Asian economic downturn in 1997-98. As Asia recovered, demand returned and in November of 2000, the price of a barrel of oil again averaged over $30.

Although oil peaked at $31 per barrel immediately following the September 11, 2001, Al-Qaeda attack on the World Trade Center and Pentagon, it quickly dropped to about $22 by September 24, and by November 6, it fell below $20. This decline, despite the terrorist attacks and the start of the Afghanistan War, was attributed to the American recession that began in 2000, and it was thought that the recession would worsen because of the attacks. However, as the Iraq War, or Second Gulf War, became a reality and the United States became mired in the Iraq and

Afganistan wars, the price of oil temporarily rose to over $75 per barrel in 2006 and for a time settled in the $50 to $80 range. But by mid-2008 it briefly had risen to over $140 per barrel.

The price of oil during this period has been also affected by the political upheaval and rise to power of Hugo Chavez in Venezuela. President Chavez specifically has shown his displeasure with United States foreign policy and publicly derided President Bush in an address to the United Nations. Although he has attempted to undermine the United States foreign policy in Latin America, it is unlikely he can afford to cut oil sales to the United States. The petroleum marketplace is too globalized for the trade between two countries to have much of an impact on the overall price of oil, even if one of the countries is the world's biggest consumer of oil and the other is a member of OPEC. Even considering all of the oil-related problems of recent years, the inflation-adjusted price of oil did not exceed the price reached in 1980 until 2007. By 2009, it had dropped below the inflation-adjusted 1980 price, but in 2010 it rebounded to the $75 to $90 dollars per barrel range. In February, 2011, oil again spiked at over $100 per barrel due political crises in Middle Eastern nations and the resulting uncertainties in oil supplies.

In addition to the major events that have affected the supply of petroleum there have been a number of minor events. For years there have been serious disagreements among the nations surrounding the Caspian Sea over the use of a pipeline connecting the oil fields of the Caspian Sea to ports on the Black Sea. Other routes are either under construction or are being considered, but have been equally met with political opposition of various sorts. In 2006 and early 2007 a political dispute between Russia and Belarus over tariffs and the price of natural gas resulted in a threat by Russia to shut off the Druzhba pipeline through Belarus. This caused considerable concern in central Europe because this pipeline also provided 1.2 Mbbl/day of oil to eastern and central Europe.

Figure 2-18 graphs the inflation-adjusted per capita gross domestic product (GDP) of the United States in terms of 1973 dollars. Superimposed on the graph is an unscaled graph of the inflation-adjusted price of oil. It is seen that associated with most rises in the price of oil there is a corresponding dip in the GDP. This association is particularly apparent for the 1973, 1979-80, 1991 and 2001-03 periods. There is little doubt about the relationship between the price of oil and overall productivity, and this relationship also is reflected in the world's stock markets. In contrast, after the sharp rise in the 2003 to 2007 period, the recession in 2008 caused the price of oil to drop.

While an oil price increase may temporarily cause a dip in productivity as measured by the GDP, it also prompts increases in oil exploration and production, conversions to other sources of energy and conservation. The oil shortage in 1973-'74 resulted in an oil glut that lessened the impact of the 1979-'80 shortage and ultimately helped the substantial expansion of our economy from 1982 to 2000. Figure 2-19 illustrates the inflation-adjusted GDP in 1973 dollars per barrel of oil consumed by the United States from 1973 through 2009. Although our consumption of oil increased by 9 percent during this period, the United States inflation-adjusted

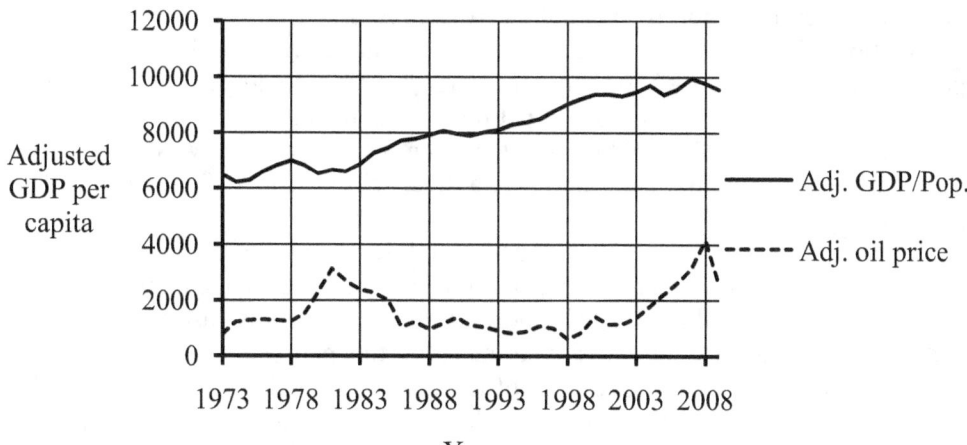

Fig. 2-18 United States GDP per capita in 1973 US dollars [EIA, USCB, BLS].

per capita GDP almost doubled. While petroleum is a major contributor to productivity, it alone does not control the productivity of a nation. There are other sources of energy and technological improvements allow us to use energy more efficiently. Energy consumption from all sources was up only 24 percent during the 1973-2009 period, indicating that some of the overall productivity gain was due to the better use of energy through technology.

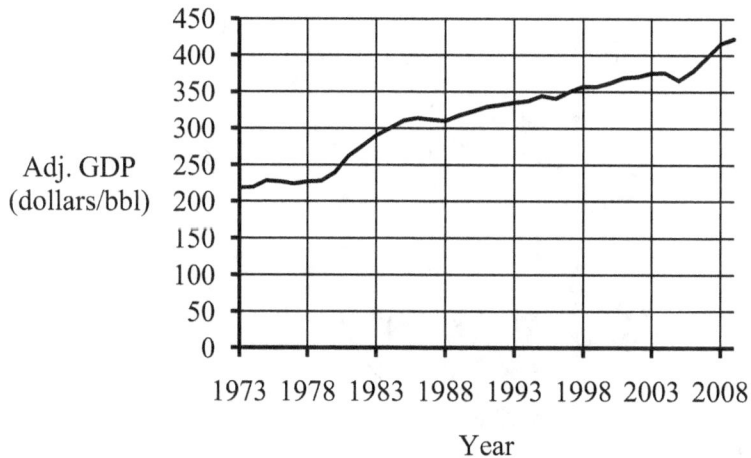

Fig. 2-19 GDP in 1973 dollars per barrel of oil consumed [EIA and BLS].

An indication of the United States vulnerability to the price of oil is its trade deficit. According to the United States Department of Commerce, since the year 2000 the United States trade deficit has varied between $366 billion and $760 billion per year. A large portion of this deficit, between 20 and 60 percent, is due to the importation of petroleum. In 2002, about 20 percent of the trade deficit was due to petroleum imports, but by 2007 this percentage had risen to 44 percent. In 2009, the difference between the United States petroleum imports and exports cost the United States approximately $215 billion and the trade deficit that year was $379 billion. Thus, about 57 percent of the deficit was due to the United States demand for petroleum.

Figure 2-20 gives the 2009 amounts and percentages of petroleum that were consumed by the four EIA energy sectors. When one considers that 82 percent of the petroleum consumed in the United States ultimately is used as gasoline, diesel fuel and jet fuel, it is not surprising that most of the petroleum is needed by the transportation sector. About one-third of the diesel fuel and essentially all gasoline and jet fuel are used for transportation. Therefore, approximately 72 percent of the petroleum we consume is for this purpose. The four EIA sectors are mutually exclusive, in that energy charged to one sector is excluded from the other sectors. The transportation sector includes all modes of transportation, cars, trucks, trains, airplanes and ships. The petroleum products used by personal cars are charged to the transportation sector, not the residential sector, and those used by trucks are charged to the transportation sector, not the commercial or industrial sector. There is no doubt as to which part of our lives is most affected by petroleum. There is also no doubt about where the changes must come from if we are to reduce our dependence on petroleum.   The ways in which the energy required to move people and goods is

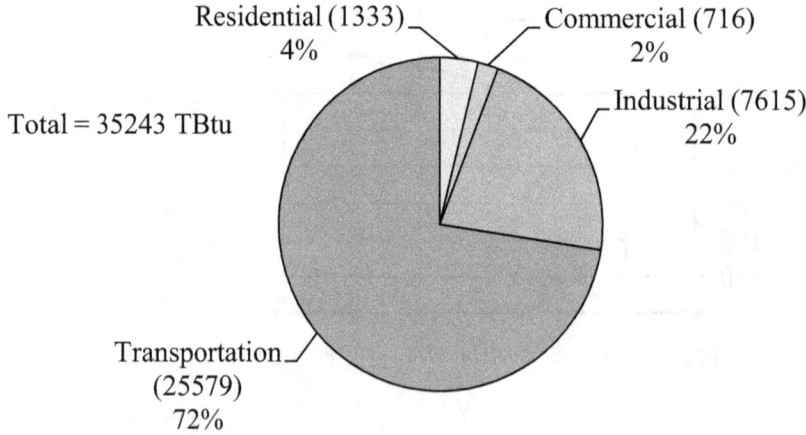

Fig. 2-20 2005 petroleum consumption by EIA sector [EIA].

supplied eventually will have to change drastically. These changes will most probably need to occur within this century, and perhaps within the next 50 years. Fortunately, there are several choices open to us.

One of the options is the production of synthetic fuels. Because hydrocarbons consist of only hydrogen and carbon, any hydrocarbon can be changed into any other hydrocarbon; it is only a matter of how much we are willing to spend in terms of both money and energy. In particular, coal can be converted into hydrocarbon gases and either coal or natural gas can be converted into gasoline and diesel fuel. Such conversions were first performed on a large scale by the Germans during World War II when they had very limited access to petroleum. They used a method known as the Fischer-Tropsch process, which is the basis of one of the steps in the synthetic processes used today. The energy cost is severe for the typical processes using coal as the feedstock. Additional discussions of synthetic conversions are presented in Chap. 6. Also, by adding oxygen, alternative fuels such as methanol and ethanol can be produced. Typical efficiencies are less than 60 percent (e.g., in converting coal to methanol, only about 58 percent of the coal's energy is retained in the energy of the methanol).

Worldwide changes in consumption habits and patterns take several years, and massive changes to the world's transportation infrastructure will take decades. Although the future supply of petroleum is uncertain, past data does allow us to make short-term projections. The petroleum consumption history of the world, OECD and United States is illustrated in Fig. 2-21. From the figure, it is seen that the world's consumption has increased by 41 percent since 1982, an average of 0.86 Mbbl/day per year. The increases by Europe and the United States have been 8 percent and 22 percent, respectively, but their consumptions have declined in recent years. In contrast, the increases in the developing countries have been quite large (e.g., Brazil 132 percent, India 304 percent and China 371 percent) and are increasing by a greater amount each year. The projected growth of worldwide consumption is shown in Fig. 2-22. Although Fig. 2-21 seems to indicate a linear growth rate of about one Mbbl/day per year, there is actually a slight upward trend. This upward trend indicates a slight exponential growth of approximately 1.5 percent.

In 2000, the United States Geological Survey (USGS) published the report of a study sponsored by the EIA. This report was updated in 2004. It examines twelve scenarios based on growth rates of 0, 1, 2 and 3 percent and three different amounts of ultimately recoverable oil, including the 700 to 800 Gbbl already produced. The three amounts are 2,248 Gbbl, 3,003 Gbbl (the mean estimate) and 3,896 Gbbl, with the latter amount estimated as having only one chance in twenty of being realized. Besides assuming an exponential growth, the report assumes that once peak production is reached, an immediate exponential decline will occur based on a constant R/P ratio equal to 10. The reason for choosing an R/P ratio of 10 is based on the observation of past large oil producing areas that have reached their peaks. From Fig. 2-10, note that since 1980 the United States R/P ratio has remained approximately constant at 10. Under this assumption, the world's oil production would drop to half its peak within seven years. Such a rapid shift away from oil would cause major disruptions in the world's economy and could cause a depression equal to or worse than that of the Great Depression of the 1930s.

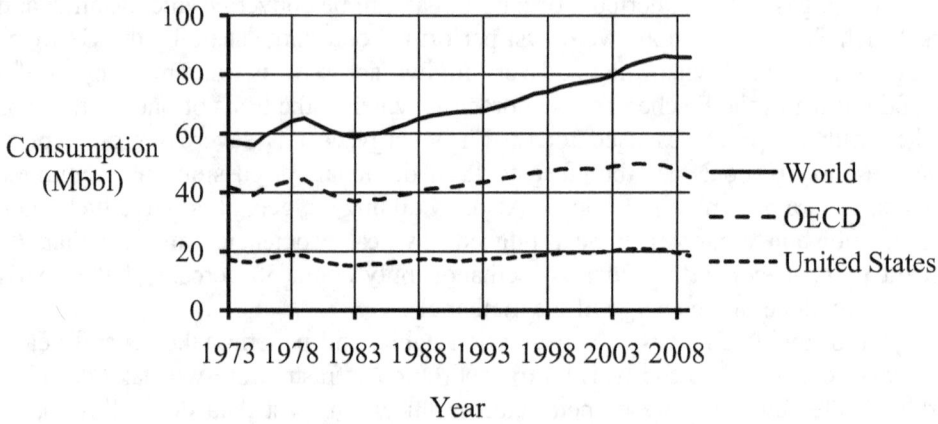

Fig. 2-21 Petroleum consumption by the world, OECD and the United States [EIA].

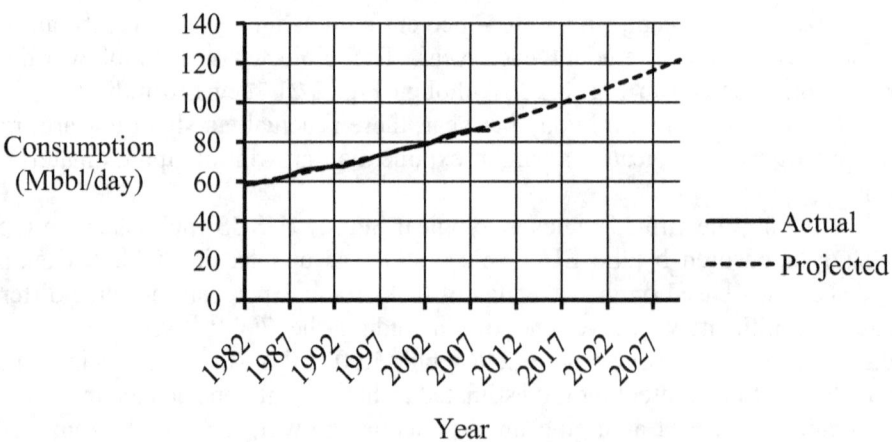

Fig. 2-22 Projected world consumption through 2030 [EIA].

Using the USGS model, Fig. 2-23 gives the possible future scenario based on a 1.5 percent rise in production per year and a total recovery of 3003 Gbbl. As seen from the figure, the peak would occur around 2050 and would fall back to the 2006 level of production by 2057. Unfortunately, it would continue to fall to half of the 2006 level by 2064 and the per capita consumption would be somewhat less than half, possibly less than one-third, of the 2006 consumption depending on the world's population growth. But it is doubtful the peak will be as spiked as the figure indicates. Rising oil prices relative to the prices of other sources of energy should smooth out the top of the curve and cause the descent to be less steep, although the descent could still be steep enough to be economically painful.

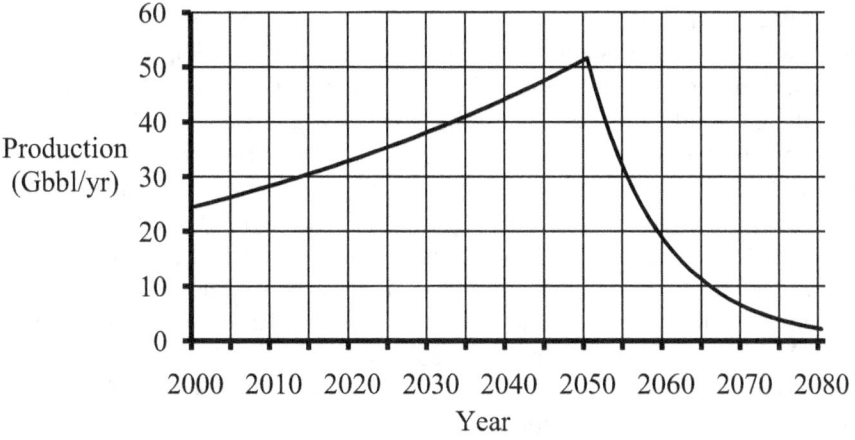

Fig. 2-23 Projected world production using USGS model.

One way of looking at a newly discovered field is to estimate how long it will extend the world's energy supply by dividing its assumed reserves by an estimated production rate. The USGS has approximated the total known and unknown reserves under all of the major areas of the north slope of Alaska to be between 25 and 30 Gbbls. If one assumes all of these areas, including the Arctic National Wildlife Reserve (ANWR), are opened for petroleum development and optimistically assume worldwide production will level off at 100 Mbbls/day, then the Alaskan oil would extend the world's supply less than ten months. If one assumes the Caspian Sea fields contain the extremely optimistic 200 Gbbls of reserves, then these reserves would extend the supply between five and six years at most.

Although most estimates of the world's reserves include the tar, or oil, sands in Canada and the heavy crude in Venezuela, they do not include the world's oil shale deposits, which are enormous. Most of these deposits are found within 1000 m (3300 ft) of the Earth's surface. *Tar sands, heavy crude and extra heavy crude* vary in quality from a mixture of soil with kerogen, heavy oil or tar. Much of the tar sands, heavy crude and extra heavy crude are classified as *bitumen*, a highly viscous mixture of hydrocarbons. Canada began counting tar sands among its reserves in 2003 and Venezuela began claiming its heavy crude in 1988 after other OPEC nations drastically increased their reserve estimates.

The world's largest oil shale deposits are centered on the point where the borders of Colorado, Utah and Wyoming meet. *Oil shale deposits* are rock formations permeated with kerogen, the precursor of natural gas and petroleum. Normally, natural gas and petroleum are the products of kerogen that has been subjected to high temperatures and pressures over millions of years, but the same result can be had by drilling holes in the oil shale deposits and using in-situ burning or electric heating. In 2005 Stephen Mut of the Shell Exploration and Production Company testified before the Senate Energy and Natural Resources Committee concerning the oil shale recovery tests the company had performed. These tests consisted of an in-situ conversion process (ICP) that electrically heated the shale over a period of three to four years. Mut claimed the tests showed that it took an energy equivalent of one barrel of oil to generate the electricity needed to retrieve 3.5 barrels of hydrocarbon products, a three and a half to one energy gain. A second method for extracting hydrocarbons from shale is to simply mine the shale and use an above ground facility to extract and refine the kerogen. Either way, it is estimated that 0.6 barrels of products could be extracted from each ton of shale and Fig. 2-24 gives an anticipated breakdown of the products that could be produced. (*Naphtha* is hydrocarbon with a medium number of carbon atoms.) These products could, of

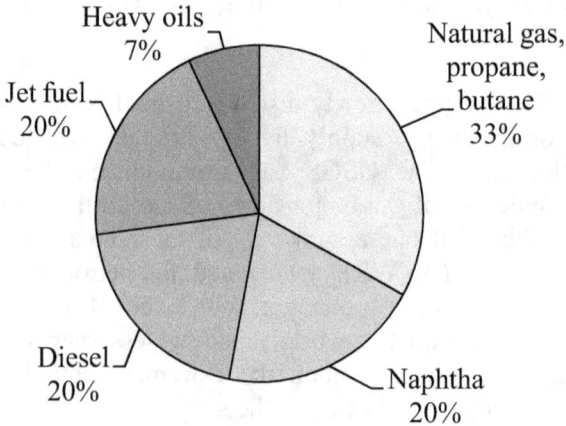

Fig. 2-24 Relative amounts of products obtained from oil shale.

course, be further refined into other products, including gasoline. The DOE predicts that about 1,000 Gbbls of hydrocarbon products could be recovered from the shale in the Colorado-Utah-Wyoming area alone.

If this were the end of the story, these deposits could enable the United States to be petroleum independent for well over 100 years. However, there are significant environmental and technical problems to be overcome. With in-situ processing, measures must be taken to keep the hydrocarbons from contaminating the groundwater. The Shell tests protected the groundwater by freezing the ground around the area being heated, thus creating a barrier of frozen earth around the kerogen. Clearly, providing a frozen container around an area being heated would require a considerable amount of additional energy. Mining the shale would not only disfigure the land, but would require huge amounts of water in an area that has very little to spare.

While the average energy efficiency of extracting and refining oil from wells is about 90 percent, extracting and refining hydrocarbons from tar sands, heavy crude, extra heavy crude and oil shale may vary from 50 to 80 percent depending on the quality of the source. The uncertainties surrounding these energy sources are too great to accurately predict average efficiencies or even the quantities of useful products that could be recovered from them.

There have certainly been advantages to using oil as a major energy source. It has been relatively cheap to produce, store, transport, refine and use and, even at over a hundred dollars a barrel, it is still cheap when compared to sources other than natural gas and coal. In its unrefined form it is flammable, but not volatile. Although storage tanks do occasionally catch fire, the damage is minimal when compared to a natural gas explosion or nuclear accident. Its non-volatility and its high energy density in terms of both volume and weight (see Table 1-1) easily make it the best choice for storing, transporting and using it to power moving vehicles. A single barrel of oil occupies 0.159 m^3 (5.615 ft^3), weighs 140 kg (308 lb) and contains 5.8 MBtu of useable energy. Even freight trains and ships now use diesel instead of coal. Most cars now may go 500 miles on a single tankful of gasoline or diesel. A ship the size of the Exxon Valdez carries 1.48 million barrels of oil and some ships transport as much as 3.82 million barrels.

A railroad tank car typically holds 550 barrels. Petroleum may also be moved by pipelines, which range from 0.1 to 1.2 m (4 to 48 in) in diameter and have flow rates from 1 to 6 m/s (3.28 to 19.68 ft/s), and a large pipeline may transport over a million barrels per day. At its peak, the Trans-Alaska Pipeline transported over two million barrels per day. Some pipelines are capable of sequentially transporting a variety of petroleum products. Figure 2-25 illustrates the important components of a pipeline. Pump stations are positioned at various points along the pipeline wherever they are needed to keep the pressure at the desired level. Block valve stations are typically spaced approximately 48 km (30 mi) apart and are for isolating sections of the pipeline for maintenance purposes and for halting the flow when a rupture occurs. In addition to the components shown, there may be intermediate delivery stations. A serious problem with pipelines is that they may extend over very long distances and are therefore vulnerable to vandalism and sabotage.

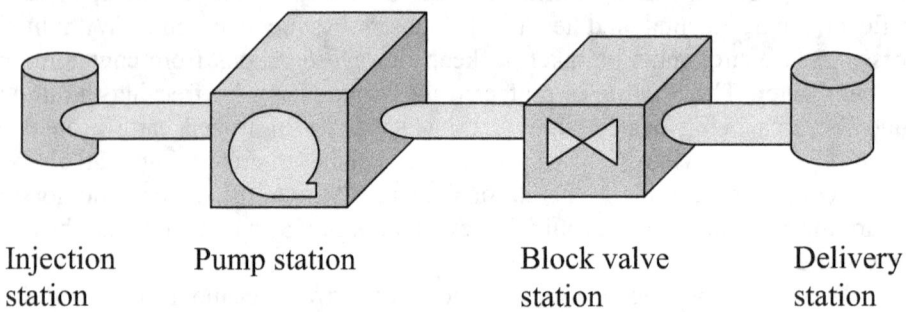

Injection   Pump station   Block valve   Delivery
station   station   station

Fig. 2-25 Major components of a petroleum pipeline.

On the other hand, there are certainly serious downsides to using oil as the principal energy source. Besides the uncertainty of supply and the fact that there is a limited amount of petroleum that can be extracted, there is, of course, the matter of pollution and global warming. It is appearing more and more likely that global warming is the overriding problem related to our consumption of all fossil fuels, not just petroleum. In fact, global warming may render our concerns over finiteness a moot point. It is becoming increasingly clear that the Earth's atmosphere cannot hold the increased carbon dioxide produced from the carbon locked up in the Earth's fossil fuels and still support life as we know it. It is not suggested that life will be destroyed, but much of our land mass would be returned to the oceans and the changes in weather patterns is uncertain and could be severe. Ironically, actions taken to lessen the effects of global warming could reduce the economic impact and political tensions related to petroleum. Any reduction in petroleum usage would tend to delay and smooth the anticipated peak in production and lessen the steepness of the subsequent decline. Chapter 10 has been reserved for examining the effects of all energy sources on pollution and global warming. With regard to international politics and economics, the United States problems will be exacerbated as our dependency on foreign oil increases.

3

NATURAL GAS

The first known use of natural gas was by the Chinese, who used it in an evaporation process to obtain salt from brine. The Native Americans were discovered using natural gas that leaked from the ground as early as the 17th century and Americans and Europeans used it for lighting throughout the 19th century. It began to be used more extensively after we began to drill oil and gas wells in the latter half of the 19th century. In the early 20th century when the oil industry was booming, the natural gas that accompanied oil production was often considered a useless byproduct and simply burned, or flared, off. Today, it is a valuable commodity and is primarily used for heating, generating electricity and producing fertilizers and synthetic fuels, including gasoline. As the use of coal for generating electricity becomes more unacceptable because of the resultant pollution, power companies are increasingly turning to natural gas to satisfy their energy needs.

Natural gas is hydrocarbon gases that originate in the ground. Although gas wells and oil wells are separate, some natural gas accompanies oil production and some liquid petroleum products are obtained from gas wells. The fact that both petroleum and natural gas came from kerogen and were created by the same processes over millions of years means that the locations where they are found tend to be in the same areas. As a consequence, although not so much of a problem now, in the future natural gas is likely to present the same geopolitical problems as oil does today. Natural gas differs from petroleum in that its constituents, particularly methane, can be more readily obtained by other means, such as decaying biomass, than most petroleum products.

Natural gas also differs from petroleum in that its volumetric energy density is less than that of petroleum, even if it is liquefied. As indicated in Table 1-1, its uncompressed volumetric energy density is 0.036 MBtu/m^3 and in liquefied form is 22 MBtu/m^3 (0.623 MBtu/cf). In reality, its energy density varies widely depending on its quality, but 1,025 Btu/cf was its average energy density in 2007 and is the estimate used in this book. The weight density of natural gas varies around 48 MBtu/t (21,800 Btu/lb), which is about ten percent more than gasoline or diesel. To liquefy natural gas its temperature must be lowered to − 162°C (-260°F) and then kept in a high-strength, pressurized vessel. The extra weight of the tank and associated high-pressure equipment is likely to make the overall energy weight density of an entire system less than that of gasoline or diesel.

Alternatively, natural gas may be compressed, in which case its volumetric density is proportional to its pressure if the temperature is held constant. According to the laws of Boyle and Charles, the amount of a gas in a given volume depends on both pressure and temperature—for a constant temperature, it is proportional to the pressure and, for a constant pressure, it is inversely proportional to temperature. Normally, natural gas is compressed to between ten million and forty million Pascal [100 to 400 atmospheres—one *atmosphere* being the atmospheric pressure at sea level when the temperature is 0°C (32°F)]. In the United States, the standard method for stating the amount of natural gas is to give the amount in cubic feet at a temperature of 15°C (59°F) and a pressure of 75,000 Pascal (0.74 atmosphere) The unit of measurement most often used by the gas industry in the United States when billing residential customers is hundreds of cubic feet, which is abbreviated Ccf with C indicating a hundred and cf being used instead of ft^3. When referring to millions of cubic feet, the abbreviation MMcf is used instead of Mcf, which includes the more familiar symbol M for a million. (In fact, a single "M" in the gas industry indicates a thousand!) One MMcf of natural gas has the energy equivalent of 176 barrels of petroleum. Sometimes statistics are given in terms of energy instead of volume and millions, billions, trillions or quadrillions of Btu are indicated by the normal MBtu, GBtu, TBtu or PBtu, respectively.

When pricing natural gas, its quality, energy density and volume all must be considered. The quality of natural gas depends on the pollutants it contains and gas sold to residential customers must meet the standards set by the Federal Energy Regulatory Commission (FERC). Because the amount of gas in a volume depends on pressure and the value of natural gas lies in its energy content, which is proportional to amount, gas meters must take into account both the flow and pressure. When usage in an area is heavy, the gas pressure tends to drop and meters must adjust the amount accordingly.

The processing of natural gas is done in the two stages depicted in Fig. 3-1. From a well or group of wells the raw, or *wet*, natural gas is piped to a condensate facility where the lease condensates, i.e., liquid hydrocarbons and waste water, are extracted. The lease condensates are then sent to an oil refinery and most of the gases are sent to a central natural gas processing plant. Between five and fifteen percent of the gas leaving the condensate facilities may be used without further processing. There is considerable variation in the composition of the natural gas sent to the processing plant, but after the liquids are removed, it is more or less as indicated in Fig. 3-2. The share of methane, CH_4, may vary from 70 to 90 percent. The other constituents are carbon dioxide, hydrogen, nitrogen, hydrogen sulfide, and helium, with traces of mercury present in some locations. During processing, the pollutants and useful gaseous byproducts are removed. What is left is *dry natural gas*, which is almost entirely methane and is output for consumption. The output of the processing plant must meet FERC standards. Because methane is odorless and deadly to humans, the processing plant also adds a gas with a distinct odor to its output so that leaks can be readily detected.

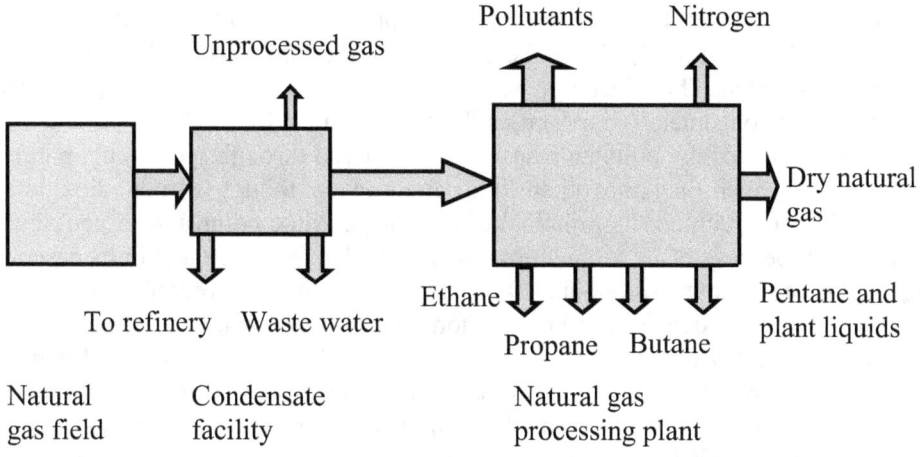

Fig. 3-1　Natural gas processing.

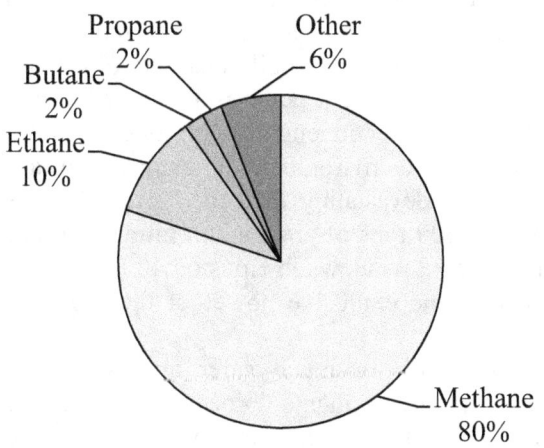

Fig. 3-2　Typical composition of wellhead natural gas.

At the processing plant, sequentially applied methods for separating the gaseous byproducts from the natural gas are used. The first is to pass the natural gas through an absorption tower where it is absorbed by special oil, and then the gaseous byproducts are boiled off using their different boiling points in manner similar to the way the petroleum hydrocarbons are separated in a distillation tower. Even more of these byproducts can be extracted by using a cryogenic expansion process in which the temperature of the gas output from the absorption process is reduced to -120°C (-158°F). Then the different compounds are boiled off as the temperature is raised. The efficiency with which natural gas is produced, processed and delivered to consumers is approximately 90 percent.

After most of the pollutants have been removed through processing, natural gas is much cleaner than the other fossil fuels, even though carbon dioxide is released when it is burned. Its principal disadvantage when compared to petroleum or coal is that, because of its low volumetric energy density when it is in its gaseous state, it is more difficult to store and transport. As indicated above, the two ways of increasing its energy density are liquefaction and compression. Liquid natural gas (LNG) has an energy density that is over 600 times that of uncompressed natural gas. Compressed natural gas (CNG) normally has a density that is no more than half that of LNG. The fact that LNG must be liquefied and then stored in high-strength tanks makes it difficult to store in large quantities or transport in small quantities. LNG can be transported by ships, but storing LNG in cars or ordinary trucks is impractical. Transporting LNG overseas is done using specialized ships referred to as *LNG carriers*. Some current LNG carriers have capacities as much as 150,000 m³, but there are carriers on order that will transport up to 260,000 m³. These larger carriers will transport about 6 TBtu of energy as compared to the largest oil tankers that can carry over 22 TBtu. Therefore, it would take more than three and a half times as many large LNG carriers as large oil tankers to transport the same amount of energy. All LNG tanks have some leakage from *boil-off*. Ships either reliquefy the boil-off or use it to help power its engines. Although an oil spill from a large tanker would be very bad for the environment, an explosion of a large LNG carrier while in port could be more devastating than the Texas City disaster in 1947 in which a ship loaded with 2,300 tons of ammonium nitrate exploded and killed 581 people. The explosion shattered windows in Houston 40 miles away and was judged to be as much as one fifth the explosive power of the atomic blast that leveled Hiroshima.

On land, pipelines offer a much safer and less expensive means of transporting natural gas. Pipeline accidents do occur, but seldom cause fatalities or serious loss of property. The major components of a long-distance gas pipeline are the same as those for the petroleum pipeline shown in Fig. 2-25, except compressors are used in place of pumps. Also, there are regulator stations for relieving excess pressure and more intermediate delivery stations where gas is diverted to local pipelines. There are generally three types of pipelines. *Gathering pipelines* are those that move gas from wells to the condensation facilities and then on to the processing plants. *Transportation pipelines* are large, high-capacity pipelines for moving gas from the processing plants to major distribution points and normally extend over long distances. The third type, *distribution pipelines*, are those that transport gas to individual residences, businesses and industries.

Although there are undersea pipelines, they are very expensive and technologically challenging to construct and maintain. The deepest undersea pipeline lies 2,150 m (7,054 ft) below the surface of the Black Sea. Not only is it difficult to lay a pipeline at such depths, but a transportation gas pipeline must have compressor and block valve stations that need to be brought to the surface for maintenance. Islands such as Japan and Taiwan must rely on LNG carriers to satisfy their import requirements. The United States received 584 Gcf, 2.7 percent, of its 2006 consumption and 17 percent of its imports by LNG carriers.

To minimize the cost of designing adequate pipelines, particularly transportation pipelines, while still satisfying peak demands, large storage facilities are needed close to where large demands exist. The most important attributes of natural gas storage facilities are their capacities and losses, the speed with which they can be filled and discharged and their construction, startup and operating costs. Large, high-strength LNG tanks can be used to level out short term demands. But even though LNG is more than 600 times denser than uncompressed natural gas and can be rapidly *cycled* (i.e., charged and discharged), tanks do not have enough capacity to hold the huge quantities that are sometimes required. LNG storage facilities tend to be expensive to construct and maintain if boil-off and other losses are to be minimized. Approximately one fifth of the gas used in winter must first be stored.

To store huge quantities of natural gas, large caverns or other underground reservoirs are normally used to store compressed natural gas. Unfortunately, compressing a large amount of gas requires a significant amount of energy and this lowers the efficiency of delivering natural gas. Three types of reservoirs are employed: depleted gas wells, aquifers and salt caverns. Depleted gas wells have the advantages of already having known geological structures, and much of the required equipment, including dehydration equipment, is already present. Also, one major startup cost is filling a reservoir with enough gas, called *base gas*, to provide a sufficient discharge rate. Depleted wells already contain some gas that can be used as base gas. How fast a depleted well can be cycled is dependent on its porosity and permeability.

Salt caverns and aquifer reservoirs require the injection of base gas and this gas may never be recovered. They also require new equipment and a study to determine their capacities. Salt caverns tend to be smaller, but have relatively fast cycle times. In the East, the depleted gas wells in Pennsylvania tend to be the most available storage sites. The Southwest is where most salt caverns are used for gas storage and aquifers are used in the Midwest. A large amount of natural gas is annually transported from the productive areas in the Southwest and Gulf of Mexico to the Midwest and East where it is stored until winter when it is most needed. Natural gas that is stored and later recovered is referred to as *working gas*. According to the EIA, the total United States underground gas storage capacity in

2009 was 8,494 billion cubic feet (Gcf) at 397 sites and the total maximum working gas capacity was 3,835 Gcf. The 2007 cost of storing and later recovering working gas was estimated to be 64 cents per million cubic feet (MMcf). Figure 3-3 gives the amounts and percentages of the underground gas storage capacities of each of the three types of storage facilities.

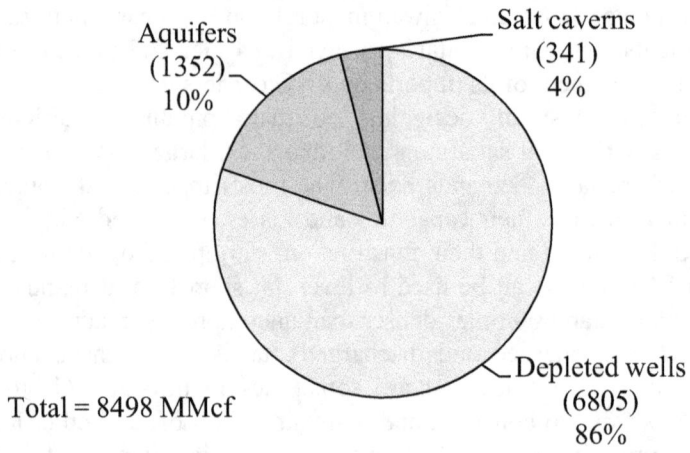

Fig. 3-3 2009 underground gas storage capacities in the United States.

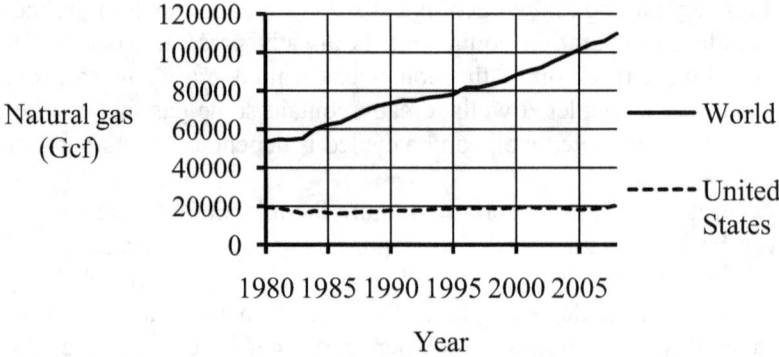

Fig. 3-4 Production of dry natural gas in Gcf [EIA)].

The production of dry natural gas from 1980 through 2008 is given in Fig. 3-4. While the production in the United States has remained almost constant at around 20,000 Gcf per year, production worldwide has more than doubled from 53,351 to 109,789 Gcf. The primary reason for this, of course, is the increased energy demand of the rest of the world. Europe has experienced a moderate 22 percent rise from 9,144 to 12,853 Gcf, probably due to the discovery of oil and gas deposits in the North Sea and the availability of gas from Russia.

Figure 3-5 provides the 2008 production of natural gas by the world's major areas. Over 94 percent of the gas produced in North America was produced in Canada and the United States. In Europe, 83 percent of the production was in the North Sea nations of the United Kingdom, Norway and the Netherlands. Russia produced 76 percent of Eurasia's natural gas and Egypt and Algeria produced 64 percent of Africa's natural gas. In the Middle East and Asia, there are no nations that dominate the production, and production seems to be more or less dictated by national consumption.

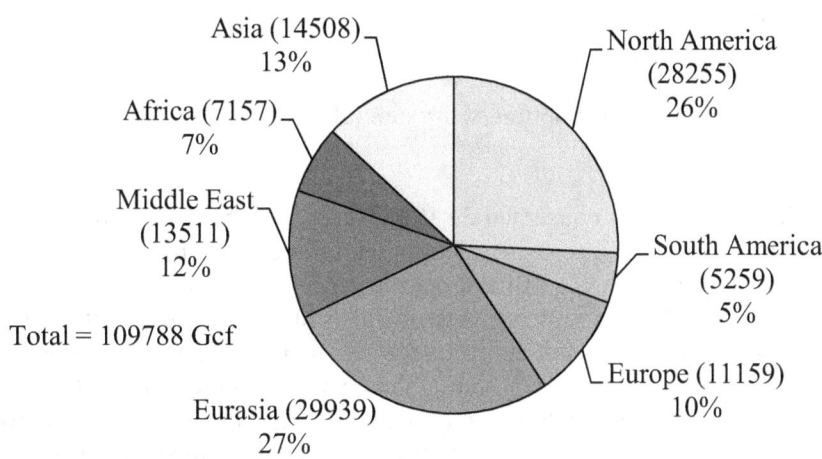

Fig.3-5 2008 production of dry natural gas by area in Gcf [EIA].

The 2008 consumption of dry natural gas by the world's major areas is shown in Fig. 3-6. Note the correspondence of production and consumption worldwide. This was not true for petroleum because it is easily and inexpensively transported by ship. But for natural gas, there is a need to consume it in areas that are adjacent to those where it is produced so that it can be shipped by pipeline. The United States shortfall of 3,734 Gcf was almost entirely met by imports from

Canada. The biggest shortfall was in the European nations unable to draw gas from the North Sea. These nations had to make up most of their deficit by importing gas from Russia. Only a small percent of the world's natural gas was transported by ship, although Japan had to import 94 percent of the 3,572 Gcf it consumed in 2008 by ship.

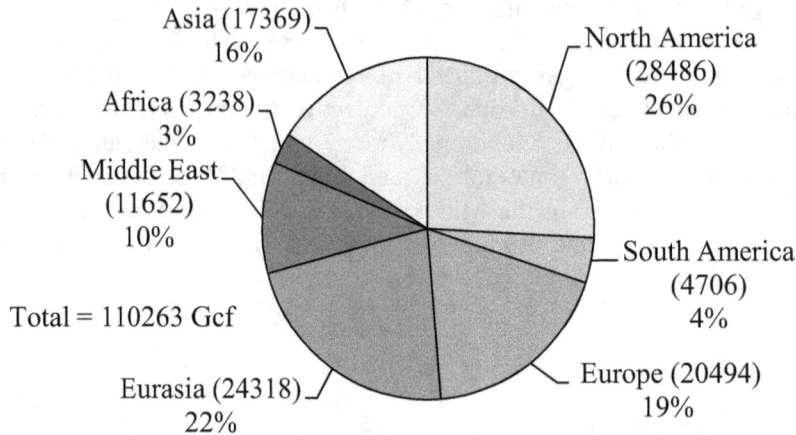

Fig. 3-6 2008 consumption of dry natural gas in Gcf [EIA].

The United States consumed 23,195 Gcf, or 23,825 TBtu, of dry natural gas in 2008 versus 37,280 TBtu of petroleum. The world consumed 110,263 Gcf, or 113,258 TBtu, versus 173,874 TBtu of petroleum. The United States got 56 percent more of its energy from petroleum than it did from natural gas while worldwide petroleum provided 54 percent more energy than natural gas. This is important because petroleum is much more polluting than natural gas (see Chap. 10). Figure 3-7 gives the 2008 daily per capita usage of dry natural gas in Ccf. As with petroleum, each person in the United States and Canada consumes four to six times as much as the average person in the world, but unlike petroleum, Russia is an even greater consumer of natural gas than either the United States or Canada. As noted in Chap. 8, a sizable portion of natural gas is used for space heating, so countries that have colder climates are heavy consumers of natural gas.

Figures 3-8 and 3-9 indicate the 2009 EIA sector consumption of dry natural gas in the United States. To show how much natural gas is used by first converting its energy to electricity, Fig. 3-8 shows the percentage and amount used by electrical power plants. Figure 3-9 assigns this electrical energy to the sectors according to how much electricity they use. Because a good is being transported, the transportation sector includes the natural gas siphoned off by compressor stations to provide the energy they need to move the gas through their pipelines. In fact, almost all of the natural gas assigned to this sector is for this purpose. Less than one percent of the natural gas consumed is used in vehicles. As discussed in Chap. 7, in 2009 natural gas supplied 17 percent of the energy needed to generate the electricity consumed in the United States; most of the remainder of this energy, 52 percent, was provided by coal.

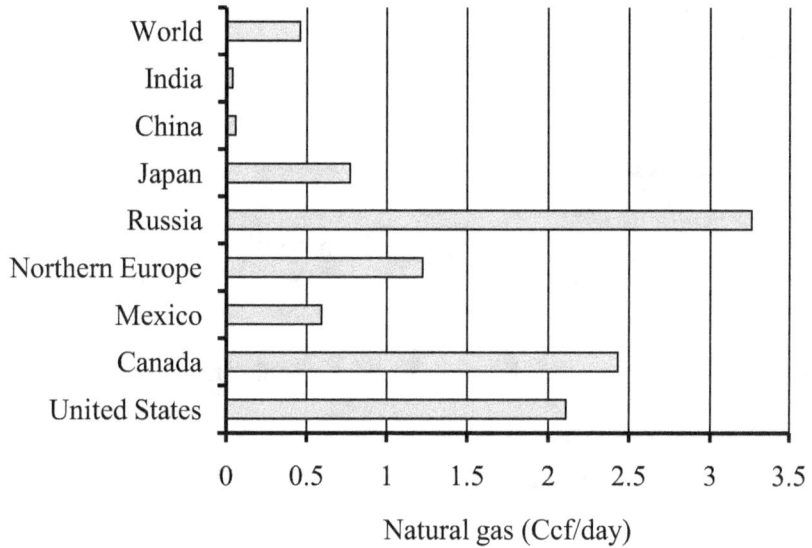

Fig. 3-7 2008 daily per capita consumption of dry natural gas [EIA, USCB].

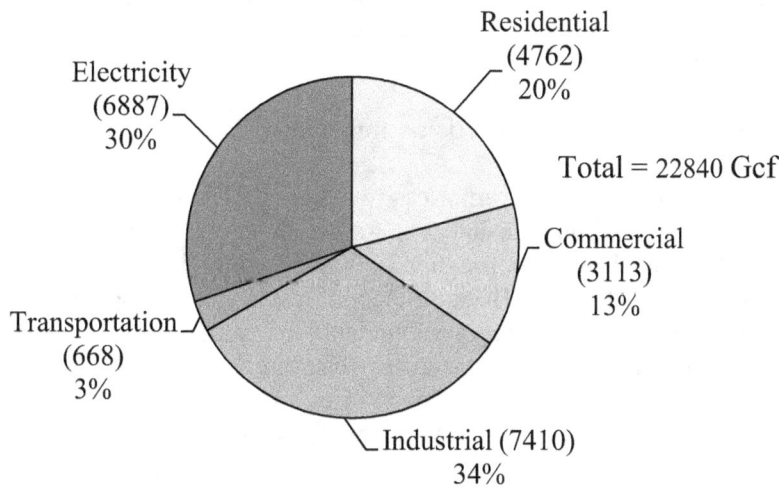

Fig. 3-8 2009 natural gas usage by sector and electrical power plants in Gcf [EIA].

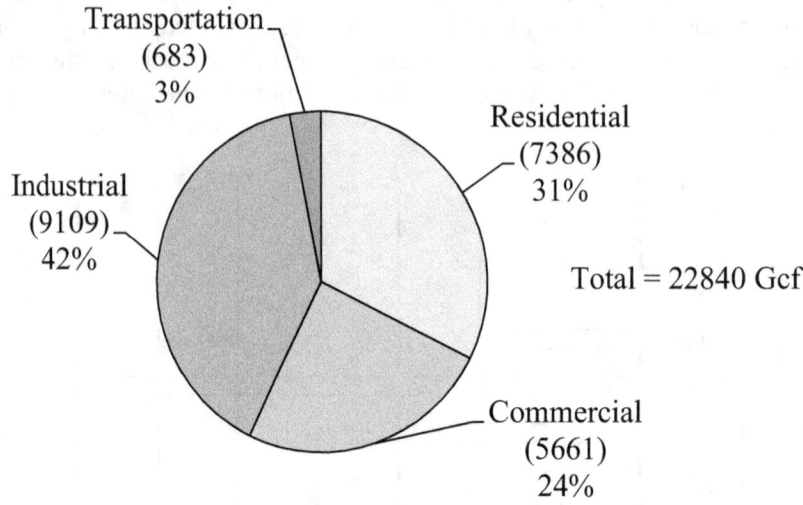

Fig. 3-9 2009 natural gas usage by end-user sectors only [EIA].

Inventory, or stocks, is computed in the same way it is for petroleum, which for the world is:

Stocks' rate of change = Production – Consumption

As with petroleum, if production exceeds consumption, then the worldwide stocks grow; otherwise they shrink. For an individual nation or collection of nations the equation becomes:

Stocks' rate of change = (Production + Imports) – (Consumption + Exports)

Almost all of the natural gas inventories are held in the underground reservoirs used to compensate for seasonal fluctuations. As indicated above the total working gas capacity in the United States is roughly 3,800 Gcf, which is about one-fifth to one-sixth our annual consumption. Because the various areas of the world as well as the world as a whole presently have an abundance of natural gas, the tendency is to produce natural gas as it is consumed. Therefore, except for storing gas to accommodate seasonal fluctuations, inventories are small. The United States does not have a strategic reserve of natural gas as it does for petroleum, but depends on the readily available supply in North America. The 2009 imports to the United States are shown in Fig. 3-10. The total imports were 4,602 Gcf, of which 3,831 Gcf was by pipeline from Canada and Mexico. About 88 percent of the United States imports were from our friendly neighbors to the north and south and were brought in by pipeline. All of the other 451 Gcf was brought by LNG carriers, mostly from the friendly nation of Trinidad/Tobago. Our total imports were 21 percent of our consumption. Approximately 16 percent was from offshore wells within the United States ocean boundaries.

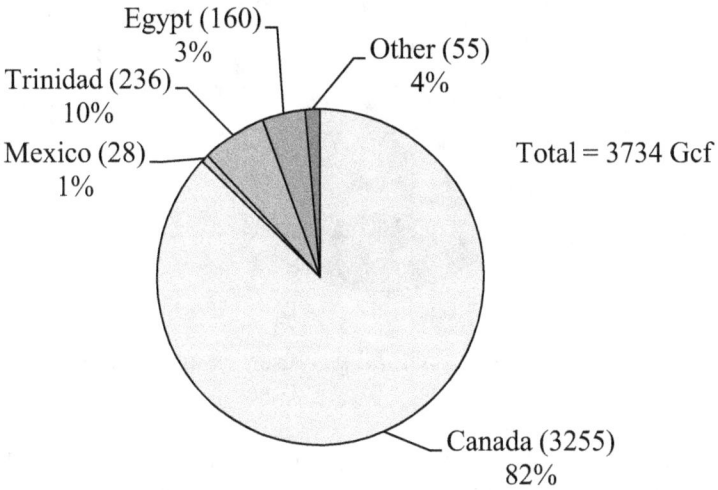

Egypt (160)
3%

Other (55)
4%

Trinidad (236)
10%

Mexico (28)
1%

Total = 3734 Gcf

Canada (3255)
82%

Fig. 3-10 2009 imports to the United States in Gcf [EIA].

Figure 3-11 gives the proven natural gas reserves for the various areas at the end of 2009. Although these data were obtained from the EIA website, they were originally gotten from sources outside the agency and, as noted in Chap. 1, are not certified by the EIA. The data shown was originally taken from the *Oil and Gas Journal*. The EIA website also gives the petroleum and natural gas data provided by *World Oil*. The unreliability of these international data is made worse by the fact that it is for proven reserves. Natural gas exploration historically has not been as extensive as petroleum exploration because its value was not fully realized until about 1970. Eighty years ago it was treated more as a byproduct of petroleum mining than as a valuable product in its own right and was often wasted. As a result, there are likely many more unknown gas reserves than unknown petroleum reserves. On the other hand, these data do serve as a rough gauge of the world's natural gas resources.

By comparing Fig. 3-11 with Fig. 3-5 it is apparent that Fig. 3-5 is misleading in that the Middle East's 2006 production does not reflect its capacity to produce. The Middle East's 2009 production was 11,822 Gcf while its proven reserves were 2,566 Tcf, a reserves-to-production ratio (R/P) of 219. This ratio

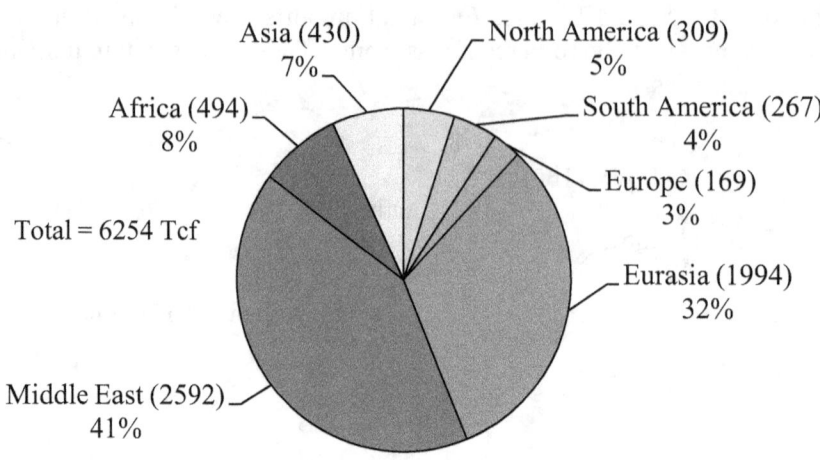

Fig. 3-11 2009 proven natural gas reserves in Tcf [EIA].

indicates that at its present rate of production its proven reserves would last the area 219 years. Of course, as future world demand increases and petroleum reserves are depleted, the Middle East will export more natural gas. Over one-third of the Middle East's proven reserves are in Iran, a statistic that is likely to change as exports increase and other nations in the area concentrate more on natural gas and less on petroleum. Also, Eurasia's production is 24 percent of the world total and its proven reserves are 33 percent of the total. Its R/P ratio is 81. Russia, which has 1,680 Tcf of Eurasia's 2,014 Tcf proven reserves, has an R/P ratio of 73. Despite Russia's heavy per capita consumption indicated in Fig. 3-7, it will continue to be a major exporter of natural gas for some time. The United States had 204 Tcf of the 276 Tcf in North America and a R/P ratio of 11. The world had 6,184 Tcf in proven reserves and an R/P ratio of 59. If natural gas is to be exploited by the industrialized nations, it is clear that much of it will have to be transported by lengthy pipelines through areas of instability or by LNG carriers.

      The histories of proven reserves and R/P ratios for both the world and the United States are given in Figs. 3-12 and 3-13, respectively. For the United States, the R/P ratio has remained almost constant near ten, indicating that the proven reserves have kept pace with production. For the world, the R/P ratio has grown by 21 percent, indicating that the discovery of new reserves has actually outstripped production. This increase in the R/P ratio is probably due to the interest in natural gas exploration that has grown substantially over the last 27 years. Whatever the reason, the rate of discovery of new natural gas fields is expected to continue for some time to come. Because there is a finite amount of natural gas, its global production must follow a Hubbert curve, but because of the uncertainties surrounding this production, it is too early to develop such a curve. As petroleum becomes depleted, the consumption of natural gas will accelerate and its abundance could decrease rapidly. However, dry natural gas is almost entirely methane, and methane can be synthesized from coal and occurs as a natural byproduct of decaying biomass. Creating methane in these ways would tend to slow down the depletion of natural gas.

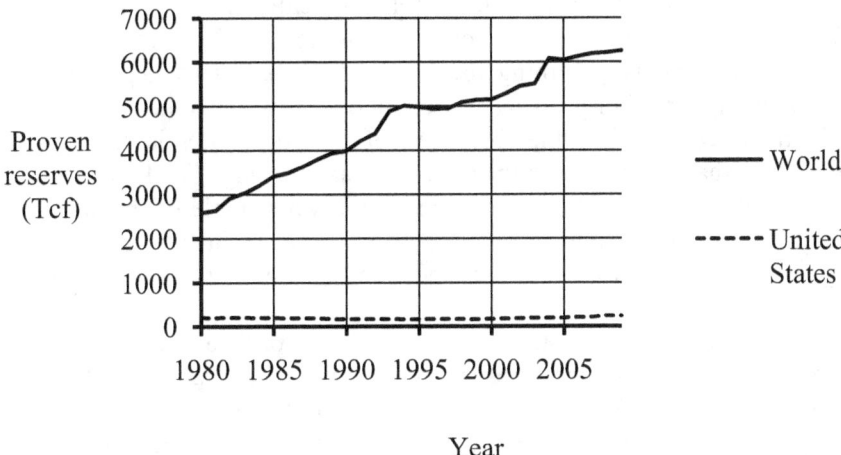

Fig. 3-12 History of proven reserves for the world and United States [EIA].

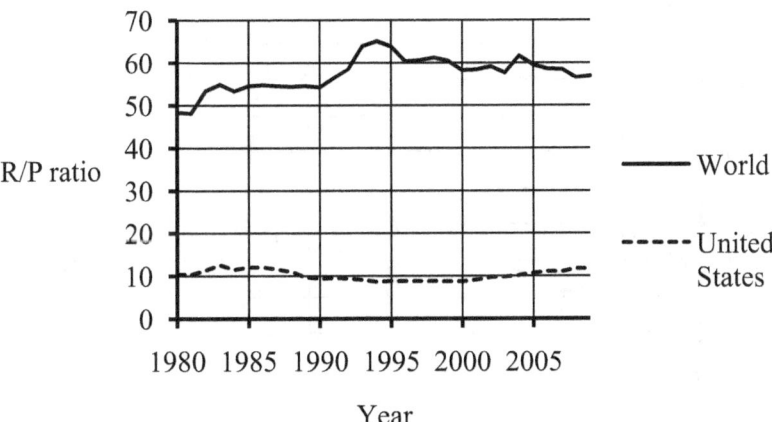

Fig. 3-13 History of R/P ratios for the world and United States [EIA].

The Potential Gas Committee estimates that shale deposits in Pennsylvania, Michigan, West Virginia, New York, Ohio and Texas may contain as much as 1,836 trillion cubic feet of natural gas that may eventually be recovered. However, as indicated in the article "Natural Gas Changes the Energy Map" that appeared in a November/December, 2009, issue of the *Technical Review* the recoverable amount may be only about 616 trillion cubic feet. The amount recovered depends on the cost of extraction. To extract gas from shale is technologically challenging. The largest known field lies about two kilometers below the surface and is accessed by drilling vertically and then gradually turning the pipe until it is horizontal. The shale is fractured by pumping high-pressure water into the well. When the shale is sufficiently fractured to allow the free flow of gas, the water is removed. The fracturing of the shale is colloquially referred to as *fracking*. Fracking could be a serious source of pollution if the fissures reach the water table or the surface and allow the natural gas to seep upward into the water or air. To justify full exploitation of shale gas the price of natural gas would need to be at least six (year 2009) dollars per thousand cubic feet.

Figure 3-14 gives the histories of the wellhead price and inflation adjusted wellhead price to the 1980 dollar. These prices reached their peaks in 2005 at which time the price was 4.6 times as much as it was in 1980 and the adjusted price was doubled. By the end of 2009 these prices had dropped back to 2.33 and 0.89 times those of 1980, respectively. To pay for the processing and distribution, customers must pay more than the wellhead price. Typically, the gas producer supplies dry natural gas to a transportation pipeline company that transports the gas to a local

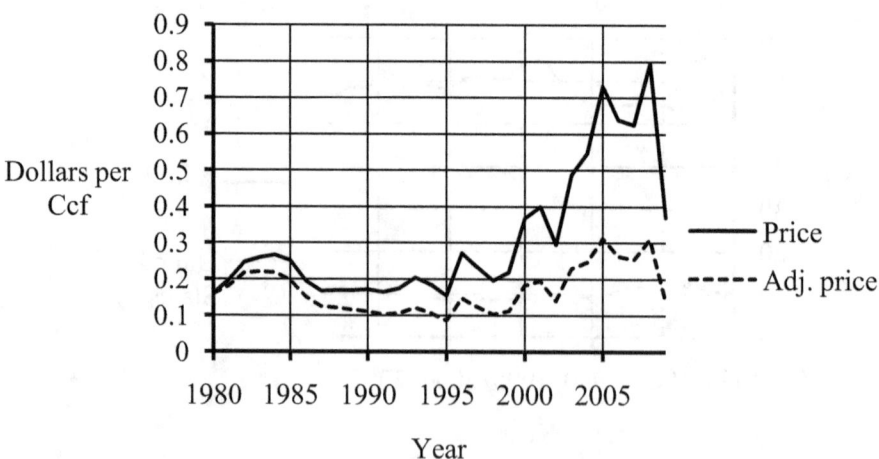

Fig. 3-14 Wellhead and inflation adjusted 1980 wellhead prices of natural gas [EIA,BLS].

distributor. Over the 2002 to 2007 period the average cost of delivering dry natural gas to customers was 46 percent of the total cost and the wellhead cost of the gas was 54 percent. The expense of delivering dry natural gas to the various sectors depends on the sector and those that consume large quantities are the least expensive to serve. Residences require the most infrastructure and overhead and must pay the greatest price. In 2007, the average wellhead price was 64 cents per Ccf. On average, the residential sector paid 131 cents per Ccf, the commercial sector paid 113 cents, the industrial sector paid 76 cents and electrical power plants paid 73 cents. Historically the ups and downs of the average annual price of petroleum have been loosely matched by those of natural gas, but in the areas with cold climates the demand for natural gas is highly seasonal. Due to its use for space heating, the natural gas demand in the United States is 60 to 70 percent more in the winter than in the summer. Although gasoline tends to be used more in the summer because of increased driving, petroleum consumption as a whole is not noticeably seasonal.

For purposes of comparison, for the United States Fig. 3-15 gives the natural gas consumption in Tcf, the per capita GDP adjusted to 1980 dollars and the price of natural gas adjusted to 1980 dollars. There are four periods of interest. The 1980 to 1986 period saw an increase and then a drop in the price. The 39 percent rise in price between 1980 and 1983 triggered a 15 percent drop in consumption. From Fig. 1-8 it is seen that the total energy consumption increased very slowly indicating a shift away from natural gas to a cheaper energy source. The shift mainly occurred in the electrical power plants where coal replaced natural gas as the energy source for

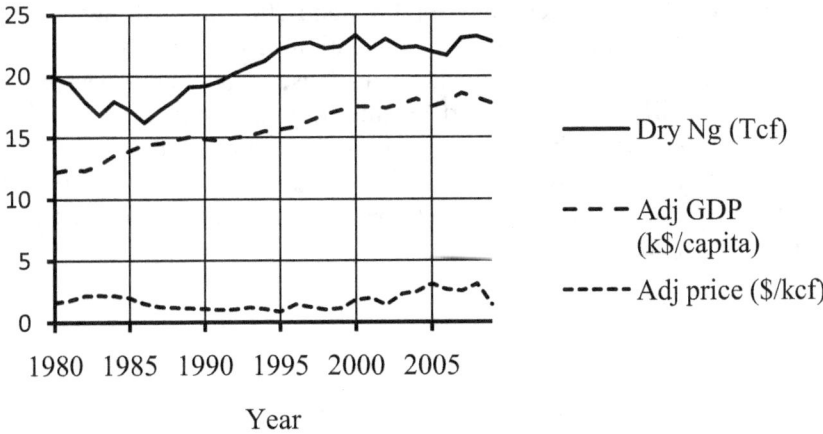

Fig. 3-15 Natural gas usage and adj. price and adj. per capita GDP [EIA, USCB, BLS].

generating electricity. Because the overall cost of energy did not change very much, the growth in adjusted per capita GDP between 1980 and 1986, which was 18 percent, was not noticeably affected by the average cost of energy. From 1986 to 1995, the price of natural gas plummeted 42 percent while consumption increased 37 percent. The per capita GDP increased by 8.5 percent and the total energy consumption increased by 19 percent, primarily due to the increased consumption of cheaper petroleum. From 1995 to 2005, the price of natural gas in 1980 dollars was up 257 percent and the consumption of natural gas dropped one percent despite the environmental pressure to use less coal. During this period, the per capita economy grew by 12 percent and total energy consumption increased 11 percent. There was a rise in energy consumption even though less natural gas was consumed because 12 percent more coal and 16 percent more nuclear energy were used to generate our electric power. From 2005 to 2009, the price dropped 55 percent, the consumption increased 3.7 percent and the per capita GDP increased by a meager 1.5 percent. The annual figures used to plot this graph do not reflect the radical swings of the economy caused by the deep recession that occurred during this period, but do show a small decline in consumption and a steep decline in price during the latter part of the period.

Although burning natural gas also produces the greenhouse gas carbon dioxide, it gives off much less carbon dioxide than burning petroleum or coal. For this reason, there will be increasing environmental and political incentive to use less coal and the past cost trade-offs between natural gas and coal in generating electricity will, in the future, favor natural gas. However, both nuclear and renewable energy, which produce very small amounts of greenhouse gases, may play an even greater role in generating electricity than natural gas. The use of solar and wind energy more than doubled between 2003 and 2007 and will undoubtedly be an important part of our future electricity generation as the price of natural gas increases and research improves the efficiencies of these technologies. Despite the fact that natural gas is difficult to transport by ship, it will become an increasingly important resource. However, like petroleum, it is finite and sooner or later will become difficult and expensive to find. When this happens, its use for manufacturing fertilizer and other products will become more financially viable than its use as an energy source.

4

COAL

Although man has probably used coal for hundreds of thousands of years, the first evidence of its early use is a 120,000 year old open pit mine discovered in Germany. The coal from this mine is believed to have been for cooking. Coal is known to have been a common fuel source in China for at least 10,000 years and by 2000 BC its use had become widespread. The Chinese smelted metals with coal as early as the 5th century BC, but it wasn't until the late 18th century AD, when the industrial revolution began, that coal became an extensively used and commercially valuable commodity. Throughout the 19th century it powered industry, steam locomotives and steamships. Around the beginning of the 20th century coal consumption took another leap when it began to be used to drive steam-powered electrical generators.

Coal is a black or brownish-black sedimentary rock that is capable of being ignited and is mostly composed of carbon and hydrocarbons, but also contains sulfur and other compounds that are considered impurities by its users. It was formed from decayed plant life over millions of years in a manner similar to that of petroleum and natural gas, but local environmental conditions and compounds caused it to be pressed into a solid form. It is normally found in layers called *seams*. Although it is quite abundant in many parts of the world, including the United States, it is finite and will eventually be depleted. The unit of measure most commonly used by the coal industry in the United States is the *short ton*, or *ton*, which is equivalent to 2,000 pounds. The metric tonne (t), which is 1,000 kg (2205 pounds), is sometimes used, but the statistics given in this chapter use short tons to indicate weight. In this chapter, the term short ton is often used in place of ton to clearly distinguish it from the metric tonne.

Coal is produced by either surface or underground mining with roughly two-thirds of our coal coming from surface mining. Surface mining has the advantages of being safer and less costly than underground mining. It is accomplished by using giant mechanically-driven shovels to first remove the topsoil covering the seam of coal and then to remove the coal. Recent laws require that once the coal has been extracted the topsoil must be restored so that the land is again usable. Underground mining is accomplished by tunneling into mountain sides or digging vertical shafts or both until the seam of coal is reached, at which point men and machines are used to extract and transport the coal to the surface. It is dangerous work for the miners because of cave-ins, fire and the possible presence of methane gas. Historically, miners have been exploited and even enslaved, and in the 19th and early 20th centuries, they often rioted and were the cause of political unrest. After they formed unions there was considerable conflict between the unions and mine owners over wages and working conditions. Many of the leftist leanings in the United Kingdom in the middle of the 20th century were initiated by the coal miners.

In the United States, there were strikes as late as 1922 that resulted in people being killed. In 1921, unrest among the miners culminated in what is known as the Battle of Blair Mountain in which 13,000 miners staged a demonstration in Logan, West Virginia, and 30 people were killed. In 1922, 21 were killed in Herrin, Illinois, in what is known as the Herrin Massacre. The federal government has instituted several laws to make the mines safer, but even today the enforcement of these laws is somewhat spotty. There are also laws that regulate when and how strikes may take place and unionization practices. These laws were designed to protect both the workers and their employers. To reduce the dangers of underground mining, surface mining has recently become more prevalent and the use of modern techniques and machinery has increased. According to the EIA website, the amount of coal produced in the United States per miner per hour has more than tripled since 1978, partly due to surface mining being much less labor intensive than underground mining.

It should be noted that, despite the considerable publicity given to mine accidents, mining is not the most dangerous occupation. In the United States, the Bureau of Labor Statistics (BLS) reported that in 2006 the logging industry was the most dangerous with 87.1 fatalities per 100,000 employees, fishing was second with 84.1 fatalities per 100,000. Coal mining had 27 fatalities and a fatality rate of 28.4. Private industry as a whole had 4,956 fatalities and a fatality rate of 4.0. These statistics do not include long term health-related deaths due to environmental conditions.

Coal is transported by trains, trucks, barges, ships and pipelines, with trains being the preferred land-based method for transporting coal long distances. But barges are often used where rivers are large enough to allow sufficient barge traffic, and ships are used between seaports. Trucks are used primarily for local distribution. To transport coal by pipeline it must first be crushed and mixed with water to create a slurry. The problems with moving coal by pipeline include the lack of an adequate source of water at the beginning of the pipeline and the fact that water must be removed at the destination. Figure 4-1 gives the average 1997 distances coal was transported in the United States by the various modes of transportation. The percentages of the 1997 means of transporting coal are given in Fig. 4-2. "Other" includes conveyors, tramways, slurry pipelines and ships between American seaports. "Multimodal" indicates that two or more means of transportation contributed a significant percentage of the distance.

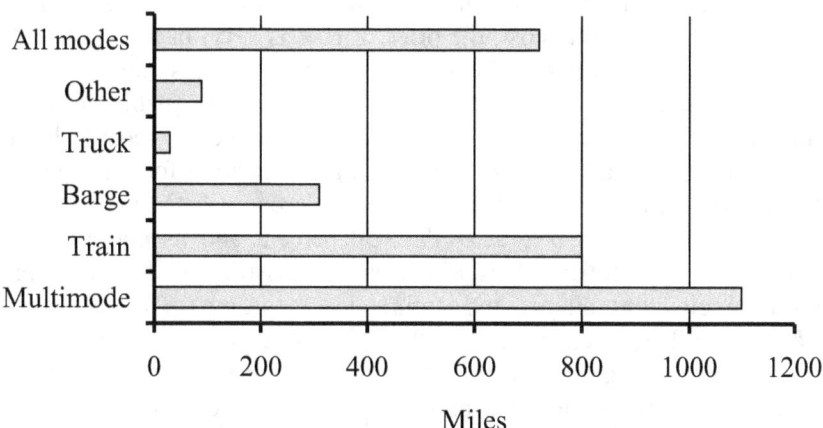

Fig. 4-1 1997 average distances coal was transported in the United States[EIA].

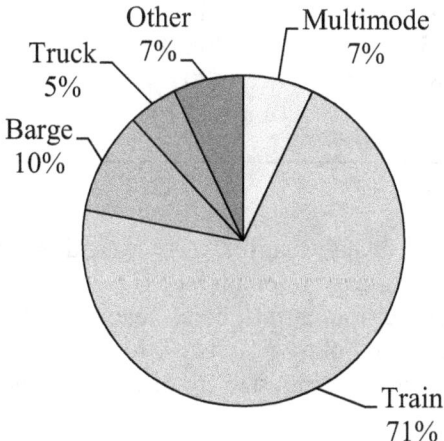

Fig. 4-2 1997 tonnage percentages of transportation modes in the United States [EIA].

The cost of transporting coal is a significant portion of its delivered cost. In 1997 the cost was 35 percent of the delivered cost and the cost per short ton was approximately $11 or 1.34 cents per short ton-mile. At that time the cost of shipping coal by train was 1.36 cents per short ton-mile, by barge was 0.93 cents per short ton-mile and by truck was 14 cents per short ton-mile. Because shipping costs are a significant percentage of the total cost, the price of coal is very much dependent on the location of the end user relative to the location where the coal is mined.

There is a wide range of soil that contains carbon and is capable of burning, with peat containing the least carbon and anthracite containing the most. Although peat, after being air-dried, has been used for cooking and heating for centuries, it is found on the surface in marshy areas and is not considered coal. Even after drying, its moisture content by weight is greater than 50 percent and its energy content typically is about nine MBtu per short ton. Coal is classified in four categories, lignite, sub-bituminous, bituminous and anthracite, according to the amount of carbon it contains. These categories and their principal characteristics are summarized in Table 4-1. Lignite and sub-bituminous coal are sometimes referred to as *soft*, or *brown*, *coal* while bituminous and anthracite coal is called *hard coal*. The amounts and percentages of the various types of coal produced in the United States in 2006 are given in Fig. 4-3. The total coal production in the United States that year was 1,163 million short tons.

Table 4-1 Characteristics of coal types [EIA].

Type	Percent carbon	Percent moisture by wgt.	MBtu/short ton
Lignite	25-35	<45	9-17
Sub-bituminous	35-45	20-30	17-24
Bituminous	45-86	<20	21-30
Anthracite	86-97	<15	22-32

Coal of all types also contains sulfur, mercury and other impurities. When it is burned, these impurities form compounds such as sulfur dioxide and nitrogen oxides which, upon entering the atmosphere, react with the moisture in the atmosphere to form acid rain and other pollutants. Mercury often finds its way into rivers, lakes and oceans where it is ingested by marine life and ends up in animals that prey on marine life.

The 2008 amounts and percentages of coal produced in the different major world areas are indicated in Fig. 4-4. The United States provided 93 percent of North America's coal that year. Canada produced six percent and Mexico one percent. In Europe, Germany and Poland were responsible for almost half of the production, in Eurasia nearly two-thirds of the coal was produced in Russia, and in

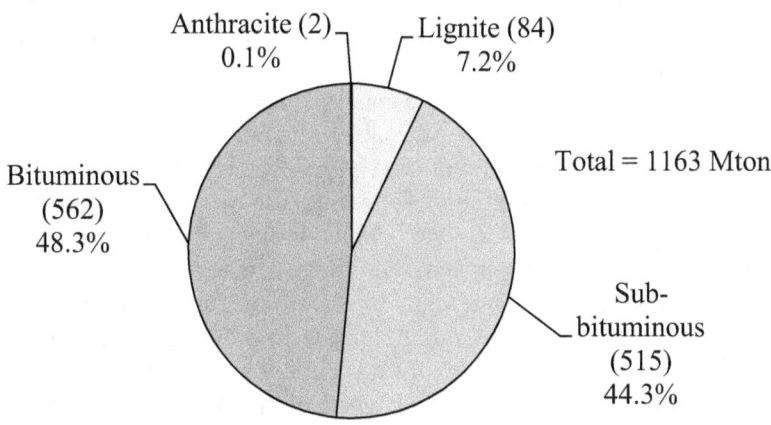

Fig. 4-3 2006 coal production in the United States in millions of short tons [EIA].

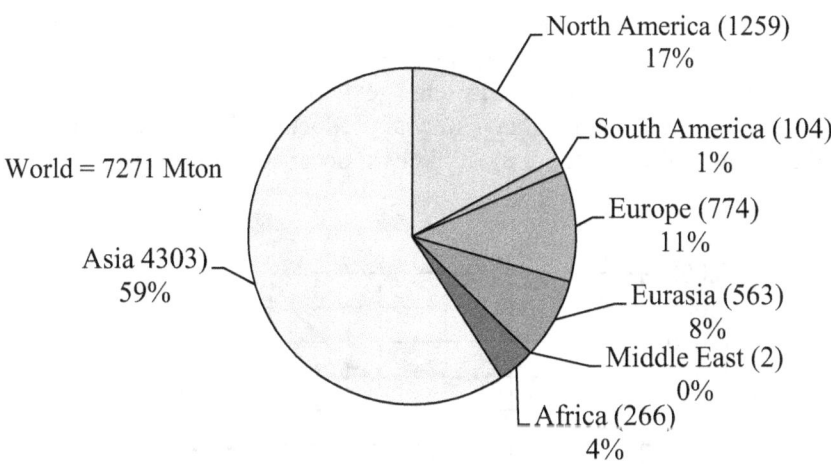

Fig. 4-4 2008 total coal production in major areas in millions of short tons [EIA].

Asia two-thirds was produced in China. China has historically gotten much of its energy from coal. India and Australia accounted for most of the remaining Asian production. Japan no longer produces enough coal to be measured using the EIA's criteria. All Middle East coal was mined in Iran and five-sixths of South America's coal was produced in Colombia. As indicated below, these figures do not necessarily reflect where the coal reserves are located, but, with the exception of Colombia, more accurately point to where most of the coal is in demand.

Figure 4-5 shows the history of coal production in the world and United States from 1980 through 2008. For the ten years from 1998 to 2008 the United States production remained essentially flat while the world production rose 44 percent. Most of this increase is due to China, India and Australia where their combined output went up by 86 percent. The rate of increase of world production seems to be lessening, but the past data is too erratic and there are too many uncertainties to make a reasonably accurate long term prediction. Because coal is a finite commodity, its production eventually will follow some sort of Hubbert curve, but it is too early to predict its shape and when the peak production of coal will occur.

Figure 4-6 gives the percentages and amounts of the 2005 known recoverable coal reserves in the major areas of the world. Of the 272 billion short tons of the North American reserves, 264 are in the United States. Of this amount, 141 billion short tons are classified as lignite and sub-bituminous and 123 billion short tons as bituminous and anthracite. For the world as a whole, 438 billion short tons are lignite and sub-bituminous and 493 are bituminous and anthracite. Russia has 173 billion short tons of reserves, the second greatest amount of reserves. No other country had even half the 2005 known recoverable reserves as the United States, which had 28 percent of the world's 931 billion short tons of reserves. Even though these figures will undoubtedly change with time, the United States will almost certainly maintain an advantage in coal production for the foreseeable future. Increased estimates of reserves are most likely to occur in South America and Africa

Fig. 4-5 Total coal production in millions of short tons [EIA].

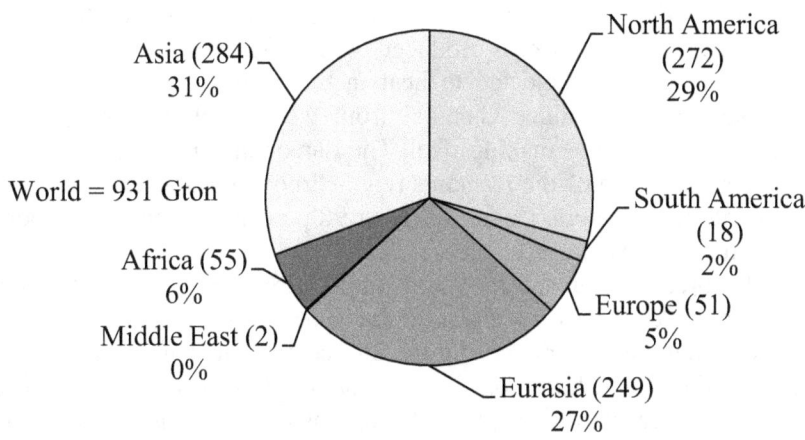

Fig. 4-6 2005 known recoverable coal reserves in billions of short tons [EIA].

where the geology has not been examined as closely as in the other regions of the world.

Using 2007 figures, the EIA predicts that, at the 2007 level of production, the recoverable coal would last the United States 286 years. Also, the 2007 United States demonstrated reserve base, which includes coal that is currently economically unattractive, but technologically capable of being extracted, is 491 billion short tons. On the other hand, if one views the world as a whole, its known recoverable reserves may not last more than 160 years. China and India, the greatest and third greatest consumers of coal, have known recoverable reserves-to-production, R/P, ratios of 54 and 200, respectively. China's consumption is increasing and it is clear that they will need to find new sources of energy within the next half century. The question is not whether the United States (or the world) has enough coal to sustain it for the next few centuries; the question is whether it can be consumed without irreparably damaging our atmosphere (see Chap. 10). In addition, the demand for coal could change if electricity demand is increased due to electric vehicles, the energy sources for generating electricity change appreciably or coal becomes a significant source of synthetic fuels. These uncertainties make the future of coal very unpredictable. It is emphasized that the R/P ratios discussed here use data for known recoverable reserves and the reserves that are eventually recovered may be more than twice those used in computing these ratios.

Coal primarily is used to generate electricity, smelt metals and make coke, which, in turn, provides industrial heat for high temperature applications. Indeed coal is sometimes classified according to its prevalent application. Although steam can be generated by burning any coal and even peat, coal having 24 to 30 MBtu/ton (normally bituminous coal) is more economically suitable for this purpose and is called *steam coal*. On the other hand, very soft bituminous coal is used for making coke and is called *coking coal*. Coke is produced by *destructive distillation* of coal, a process in which the coal is subjected to heat in the absence of air. Destructive distillation is also used to produce charcoal from wood. Coke is used in high-temperature applications such as making steel. The flame temperature of a solid fuel depends on the concentration of the reactants (e.g., carbon and oxygen), the surface area of the fuel and other factors. Coke is at least 98 percent carbon and its porous texture increases its surface area.

Figure 4-7 breaks down the 2009 consumption of coal in the United States by end use. Although coal was widely used for space and water heating by the residential and commercial sectors early in the 20th century, its smoke caused cities to become very polluted and little coal is used for these purposes today. Because trains and ships now use petroleum products for power, the transportation sector consumes virtually no coal directly. Steam boilers in electrical power plants consumed 92 percent of the coal used and the industrial sector consumed only 6 percent. The total consumption of 1,001 million short tons amounted to 2.74 million short tons per day, enough to fill 265 coal trains, each having 100 coal cars. A 2006 breakdown of the 44,977 million short tons consumed by those industries classified as "Other industrial" is given in Fig.4-8.

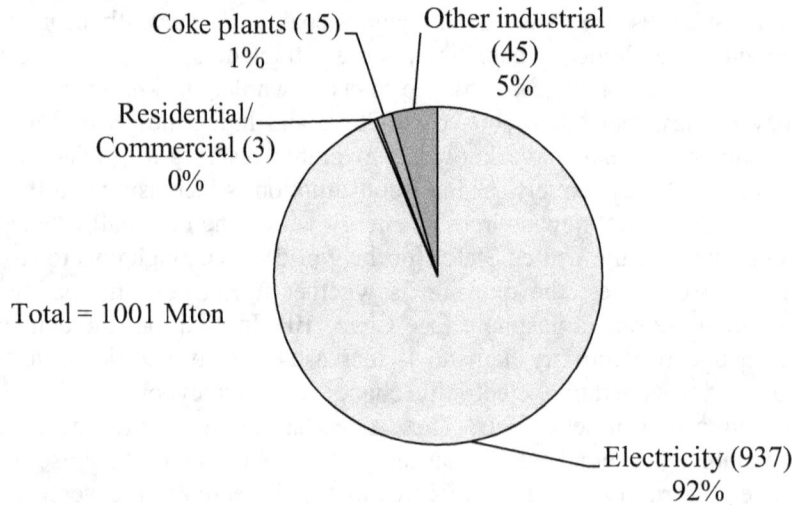

Fig.4-7 2009 coal consumption in the United States in millions of short
        tons [EIA].

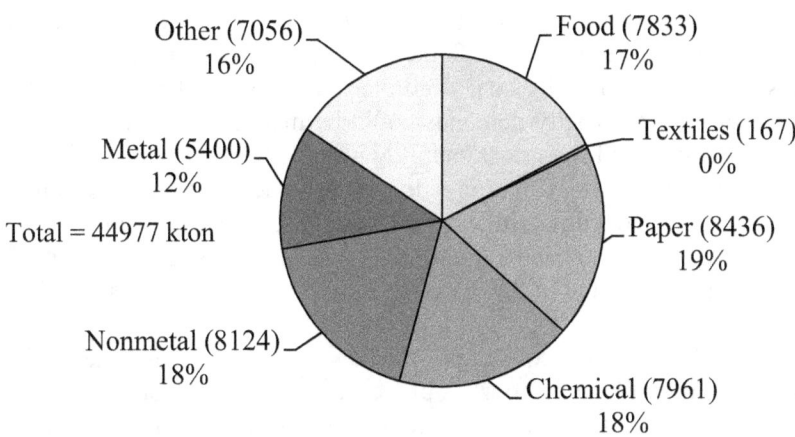

Fig. 4-8 2006 coal consumption by "Other industrial" in thousands of short tons [EIA].

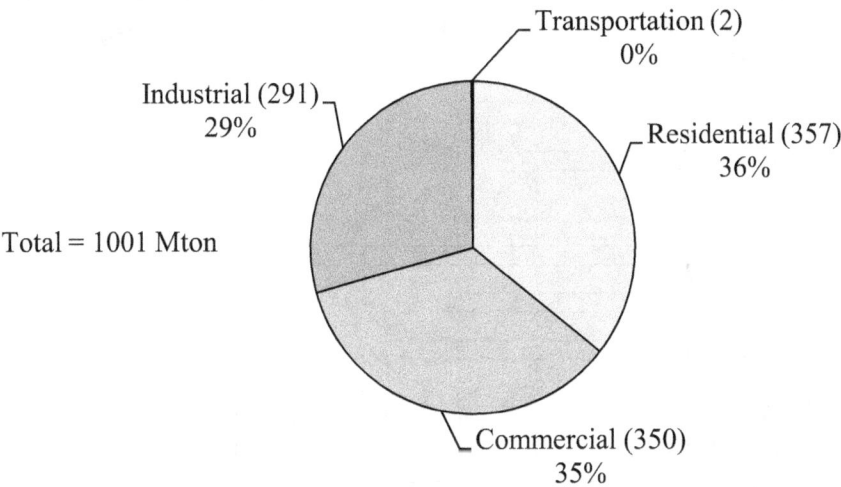

Fig. 4-9 2009 coal usage in millions of short tons including electricity consumption [EIA].

There are considerable changes when the coal used for generating electricity is divided among the four end-user sectors as shown in Fig. 4-9. This figure shows that, when both direct and indirect usage is taken into account, the residential sector was the greatest user of coal, the commercial sector was second and the industrial sector was third. Even transportation is responsible for some coal consumption because of the electricity used by light rail.

Coal stocks in the United States at the end of 2006 were 187 million short tons, approximately one sixth of its 2006 consumption—a two months supply. Most of these stocks were stored by coal-fired electrical power plants so that they could guarantee that their electricity demands could be met. Long term stocks of coal are not important because of the ready supply within the United States. The most likely long term supply problem is a nationwide strike by the mine workers' unions. When lengthy strikes have occurred in the past, the federal government has intervened before a shortage of coal became a national crisis. In 1943, during World War II, a coal industry strike resulted in the government taking the extreme step of seizing the mines.

The 2005 per capita consumption of several countries and the world is given in Fig. 4-10. Northern Europe consists of Germany and all European countries to the north and west of Germany. Germany consumes 68 percent of this area's coal and, despite its size, France consumes less than six percent because 80 percent of its electricity is generated using nuclear power. Germany, however, is currently subsidizing the construction of wind and solar facilities to ease their reliance on coal and other fossil fuels. Although Canada uses more energy per capita than the United States, on average a Canadian uses only 54 percent as much coal as an American. The average American uses over three and a half times as much as the average person in the world.

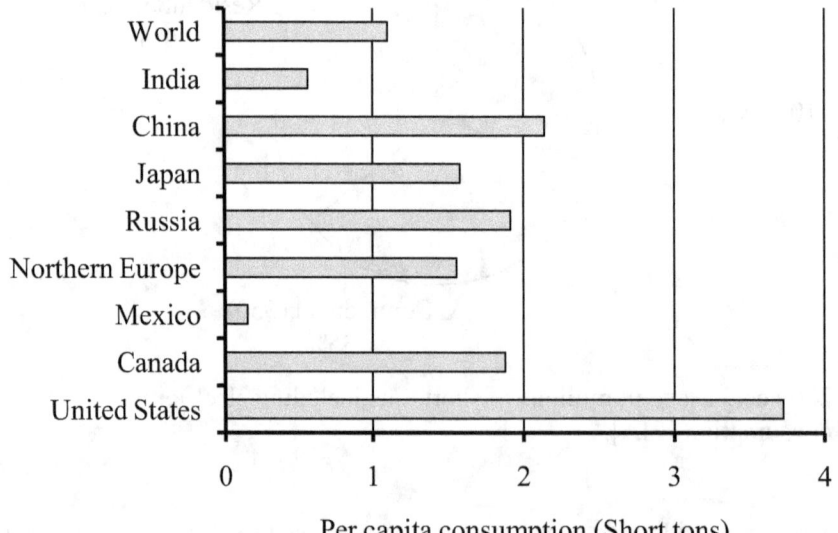

Per capita consumption (Short tons)

Fig. 4-10 2008 per capita consumption of coal in short tons [EIA, USCB].

Figure 4-11 shows both the United States 2009 imports from and exports to the world's major regions. The United States exported about 36 million short tons more than it imported and most of its 59 million short tons of exports, 69 percent, were bound for Europe and Canada. Almost 79 percent of its imports were from Colombia. Because the United States has an abundance of coal, there is a question as to why it imports any coal. The answer apparently has to do with politics and trade agreements.

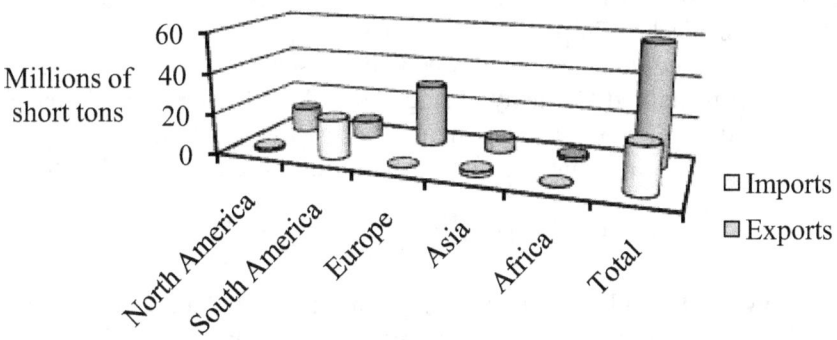

Fig. 4-11 2009 United States coal imports and exports in millions of short tons [EIA].

The price of delivered coal varies according to the end-user and depends not only on the quality of the coal the end user tends to order, but also on the volume of the orders and the infrastructure arrangements for its delivery. Figure 4-12 gives the average prices paid by the three major users in the United States at the end of 2008.

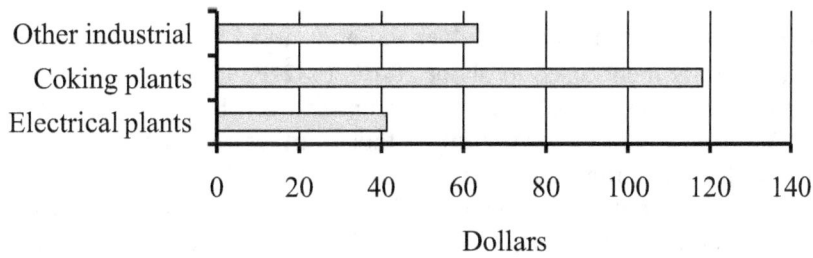

Fig. 4-12 2008 delivered price per short ton of coal to major users [EIA].

Electricity generating plants pay much less, partly because their orders are regular and large and they have special arrangements with the railroads. Except for Canada, countries that receive the United States exports pay even more because of large shipping expenses. For generating steam, the cost of fuel per MBtu of steam output is almost entirely due to the recoverable energy in the fuel. At the end of 2008, the oil distillate used by generating plants cost $19.83 per MBtu, natural gas cost $7.82 per MBtu and coal cost $1.02 per MBtu. It is clear why so many electrical generating plants that use fossil fuel are designed to use coal.

A use for coal that was briefly introduced in Chap. 2 is that of synthesizing hydrocarbon fuels from coal. There are basically three processes for accomplishing this synthesis, the Fischer-Tropsch process, the Bergius process and the Karrick process. The best known of these processes is the Fischer-Tropsch process, which was used by energy starved Germany during World War II to provide liquid fuels for its war machine. The basic reaction in this process combines hydrogen and carbon monoxide to produce chainlike petroleum products (*alkanes*) and water:

$$(2n+1)H_2 + nCO \rightarrow C_nH_{(2n+2)} + nH_2O.$$

Further refining of the alkanes may then be used to produce other petroleum products of the desired quality. The required reactants are in a mixture called *syngas*, or *water gas*. Syngas is about 50 percent carbon monoxide, 40 percent hydrogen and 10 percent carbon dioxide and nitrogen, and is obtained by producing a char from coal or biomass and passing high-temperature steam over the char. The residue from syngas production also contains useful energy in what is sometimes called semi-coke. Unfortunately, another byproduct is carbon dioxide and other impurities that must be removed from the syngas. The process employs a catalyst and the catalyst, as well as the temperatures and pressures employed, determine the end products. The entire process is depicted in Fig. 4-13.

The fundamental reaction underlying the Bergius process converts carbon and hydrogen into alkanes:

$$nC + (n+1)H_2 \rightarrow C_nH_{(2n+2)} .$$

Lignite or sub-bituminous coal is first pulverized and mixed with heavy oil and a catalyst. The mixture is then raised to about 500°C (932°F) in a reaction chamber filled with hydrogen. The pressure in the chamber may be as much as 650 atmospheres. As with the Fischer-Tropsch method, further refining may then be applied to obtain the desired products. An advantage of the Bergius process is that liquid hydrocarbons are obtained directly and, as a result, the process is often referred to as a *direct liquefaction* process. Over 97 percent of the input coal can be converted to synthetic fuels. One disadvantage is that it requires a large quantity of water.

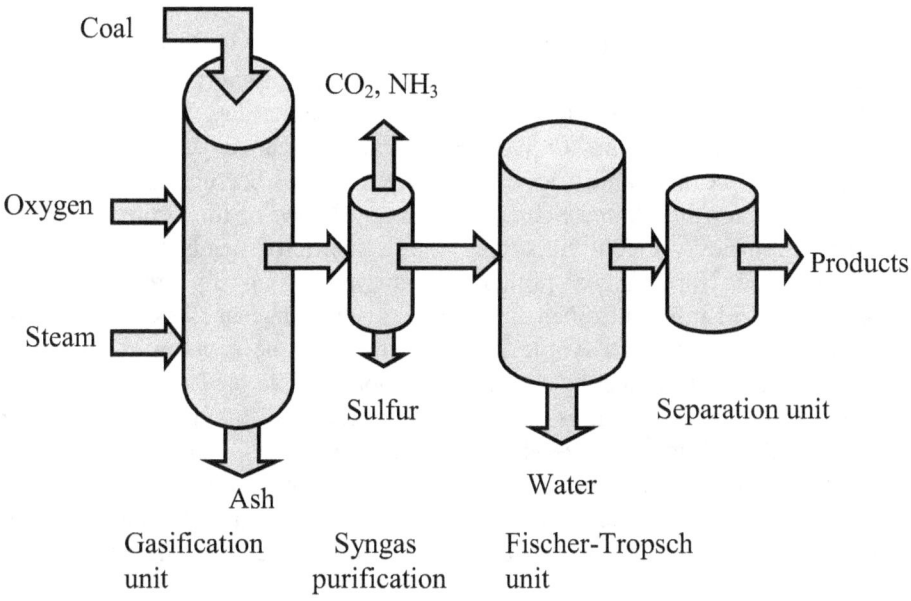

Fig. 4-13 Gasification and Fischer-Tropsch process.

The input to a Karrick process may be either crushed, but not pulverized, oil shale or lignite. Superheated steam between 540°C and 760°C (1000°F and 1400°F) is passed through the crushed input and the solids, liquids and gases are separated and further processed. Although the fraction of liquids produced is relatively low, very little of the total energy in the input feedstock is lost and both the output gases and solids have high energy content that can be used for general heating purposes.

How much liquid hydrocarbon is produced by these processes varies from one-half barrel to two barrels per input short ton and depends on the quality of the input and the process parameters. The Karrick process typically yields about two-thirds of a barrel per ton and the other two processes tend to yield at least one barrel per ton. Generally, the total energy in the useful products divided by the total energy input is less than 60 percent. South Africa, the nation that produces the greatest portion of its consumption of petroleum products by synthesizing coal, has an abundance of coal and very limited petroleum reserves. Its largest synthesis plant uses a type of Fischer-Tropsch process and is capable of outputting 150,000 barrels of liquid petroleum products per day, 30 percent of South Africa's needs. The plant yields approximately 1.2 barrels per short ton of coal and the energy content of its liquid products is about 27 percent of that of the input coal. The DOE's National Energy Technological Laboratory (NETL) has designed a Fischer-Tropsch facility that it predicts would input 24,533 short tons of coal per day and output 50,000 barrels of diesel and naphtha. It would yield over two barrels per short ton and generate 125 MW of electricity from the otherwise wasted process heat. When operating at full capacity, it would also expel 32 thousand short tons, 15.8 million cubic meters (560 million cubic feet), of carbon dioxide per day. The estimated cost of the plant is 4.5 billion dollars, or $90,000 per barrel of capacity.

Although the availability of coal is not likely to become an international political problem for the United States, the heavy reliance of the United States and other large countries on coal to supply their electricity needs has become an international issue. Burning coal pollutes the air more than any other fuel and is responsible for acid rain and much of the carbon dioxide that is expelled into the Earth's atmosphere. Although synthesizing coal has become economically viable in some cases, it does not reduce the amount of carbon dioxide produced. Synthesizing processes create large quantities of carbon dioxide and the products produced release as much carbon dioxide when they are burned as their non-synthesized counterparts. The carbon does not disappear and, if it is oxidized, the products of the chemical reaction are primarily carbon dioxide and water. The basic reaction involved when a hydrocarbon is burned is demonstrated by the oxidation of propane:

$$C_3H_8 + 5O_2 \rightarrow 3CO_2 + 4H_2O .$$

For coal, the chemical reactions are very complex and some of the carbon bearing reactants do not burn, but remain in the ash and other residue. However, some of the carbon dioxide generated in a synthesizing process can be captured and sequestered (see Chap. 10).

Regardless of whether coal is used for generating electricity, to provide heat for industrial processes or to produce other hydrocarbon products, the resulting byproducts include considerable amounts of carbon dioxide and other pollutants. In addition, mining tends to pollute nearby streams and surface mining disfigures the land. It takes both time and money to return the land to something resembling its original appearance. Although mining is not the most dangerous occupation, mining is dangerous and underground mines are, by nature, unhealthy environments. Despite these downsides, coal must remain an important component of our energy future for many years. For many industrial applications, such as smelting, there is no reasonable substitute for coal. For generating electricity, natural gas, nuclear energy, hydroelectric energy, geothermal energy, solar energy and wind energy are the alternatives. Natural gas could be used to fire newly constructed electricity generating facilities and even replace some of older the generating facilities, but could not appreciably reduce the use of coal for decades. After all, natural gas is also a finite commodity and also produces carbon dioxide when it is burned, albeit much less than coal. Also, both natural gas and fuel oil are much more expensive than coal. As indicated in Chap. 5, nuclear plants, although they are competitive with coal fired plants, take a long time to build, produce radioactive waste and have serious political implications. There are a limited number of locations where hydroelectric or geothermal facilities are feasible and solar energy is, for now, quite expensive and can be exploited only when the sun is shining. Wind energy is becoming competitive, but is variable and can supply no more than 50 percent of our electricity needs without substantial facilities for storing the energy they produce. (See Chaps. 6 and 7 for discussions of renewable energy sources and electricity generation.) More efficient use of heat and conservation offer the best hope of reducing the increasing utilization of coal in the near future. A meaningful worldwide energy policy could possibly begin to reduce the consumption of coal within two or three decades.

5

NUCLEAR

There is some disagreement as to when the nuclear age began. Did it begin in 1905 with Albert Einstein's publication of his special theory of relativity and its implied equation $E=mc^2$; or in 1939 when Otto Hahn, Fritz Strassmann, Lise Meitner and Otto Frisch discovered that when a uranium atom decays it splits into fragments with a total mass less than that of the uranium atom; or in 1942 when Enrico Fermi achieved the first known chain reaction from a pile of uranium and graphite; or in 1945 when the Manhattan Project exploded an atomic bomb at the Trinity Site north of Alamogordo, New Mexico? By any account the nuclear age was born in the first half of the 20th century. World War II ended abruptly after the United States dropped atomic bombs on Hiroshima and Nagasaki, Japan, in early August, 1945. The United States, under the supervision of Edward Teller, constructed the first hydrogen bomb in 1952 and realized the fusion of hydrogen into helium.

After the war the United States funded Project Plowshare which was charged with finding peaceful applications for atomic energy. In the early 1950s the United States, partly due to the insistence of Admiral Hyman Rickover, used piles similar to the one built by Enrico Fermi to develop controllable uranium reactors capable of generating steam and the first electricity was generated using nuclear energy in 1951 at the nuclear experimental laboratory near Arco, Idaho. The first nuclear submarine, the Nautilus, was launched in January, 1954, at the Groton, Connecticut, shipyard. Also in 1954, the first nuclear power plant connected to a power grid became operational at the Union of Soviet Socialist Republics' Chernobyl Power Plant. The first commercial power plant became operational at Sellafield, England, in 1956, and in 1961 the first large commercial nuclear power station went into service at the Yankee Atomic Electric Power Plant in Rowe, Massachusetts. The world's capacity to use nuclear power to generate electricity grew rapidly from 1,000 MW in 1960 to more than 385 times that amount by 2007. There are now over 30 countries using nuclear energy to generate electricity. Although Project Plowshare and the Department of Defense experimented with other uses for nuclear energy, including excavating a new Panama Canal and developing a nuclear powered airplane, the only current applications of nuclear energy are to make bombs, generate electricity and power military submarines and surface ships.

There are two fundamental types of nuclear reactions, fission and fusion. In a fission reaction, an atom splits into fragments such that the total mass of the fragments is less than the mass of the atom. The difference in the masses becomes the energy released by the reaction according to the equation energy equals the mass times the velocity of light squared, $E=mc^2$. One of the fission reactions used to generate electricity is that which splits uranium 235, ^{235}U, and a typical fission of this uranium isotope is depicted in Fig. 5-1. This fission of ^{235}U is one of many possibilities and is the one that led to the energy-producing fission discovery of Hahn, Meitner and Strassmann. Other common elements used as sources for fission are thorium and plutonium. In contrast, nuclear fusion combines two or more fragments such that the products of the reaction have less mass than the reactants. The most common fusion reaction combines two hydrogen atoms to produce a helium atom and is illustrated in Fig. 5-2. There are fusion reactions that combine elements having higher atomic numbers, but they occur only under extreme temperatures and pressures. Because fusion requires temperatures of hundreds of millions of degrees, it is difficult to control. The sole use of fusion on Earth is presently to create tremendous explosions, but it is hoped that someday it can be tamed and used to generate electricity. Until then we must be content to rely on fission to supply our nuclear power generation.

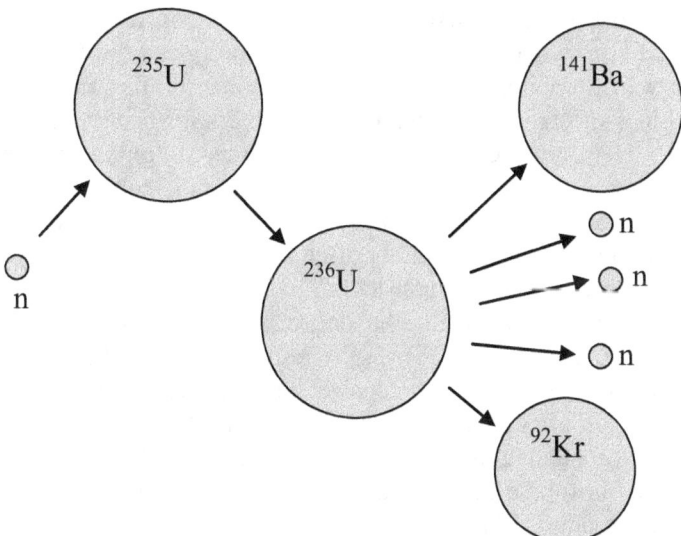

Fig. 5-1 The fission of ^{235}U into krypton and barium.

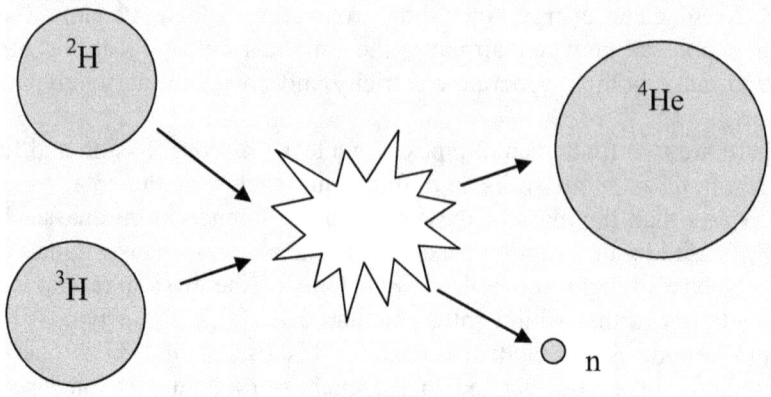

Fig. 5-2 The fusion of hydrogen into helium.

From Fig. 5-1 it is seen that a neutron entering a ^{235}U atom initiates a temporary conversion to a ^{236}U isotope that immediately fractures into fragments that may include up to three neutrons. The neutron fragments may in turn cause fissions of other ^{235}U atoms and if on average at least one of the neutron fragments causes a fission, then a chain reaction occurs. If a high percentage of neutrons cause fissions and there is sufficient mass, then there is a violent explosion. Oddly enough, not only is the chance of fission reduced if a neutron is too slow, but, if the concentration of ^{235}U in a substance is low, it may also be reduced if a neutron is too fast. As a result most nuclear reactors include a moderator in the reactor's core. For the pile used by Fermi when he and his colleagues produced the first chain reaction, the moderator was graphite. In addition, the number of fissions depends on the density of ^{235}U in a substance. In most nuclear reactors for generating electricity, the percentage of ^{235}U is between three and five percent. For an atomic bomb, the percentage of ^{235}U must be at least 85 percent. To provide control over the speed of reaction, a reactor core also includes an absorber substance that absorbs some of the neutrons and thereby reduces the number of neutrons that cause fissions.

Uranium ore requires considerable processing before an adequate concentration of ^{235}U is produced. The entire procedure for obtaining a suitable concentration of ^{235}U is shown in Fig. 5-3. Uranium ore mined in the United States typically contains only 0.05 to 0.3 percent of the uranium oxide, U_3O_8, required to begin the process of obtaining the needed three to five percent of ^{235}U. Some uranium deposits in other countries contain a significantly higher percentage of U_3O_8. Also, some phosphate deposits are sufficiently rich in uranium that U_3O_8 can be economically produced as a byproduct by plants that produce phosphate fertilizers. The milling process involves grinding the ore and then chemically leaching it to extract the U_3O_8, which is sold as *natural uranium*, or *yellowcake*.

Yellowcake is about 0.7 percent ^{235}U and 99.3 percent ^{238}U, an isotope of uranium that cannot sustain a chain reaction and must undergo an enrichment process. Ore mined in the United States yields about 3.5 kilograms (7.7 lbs) of U_3O_8 per tonne (1.1 short ton).

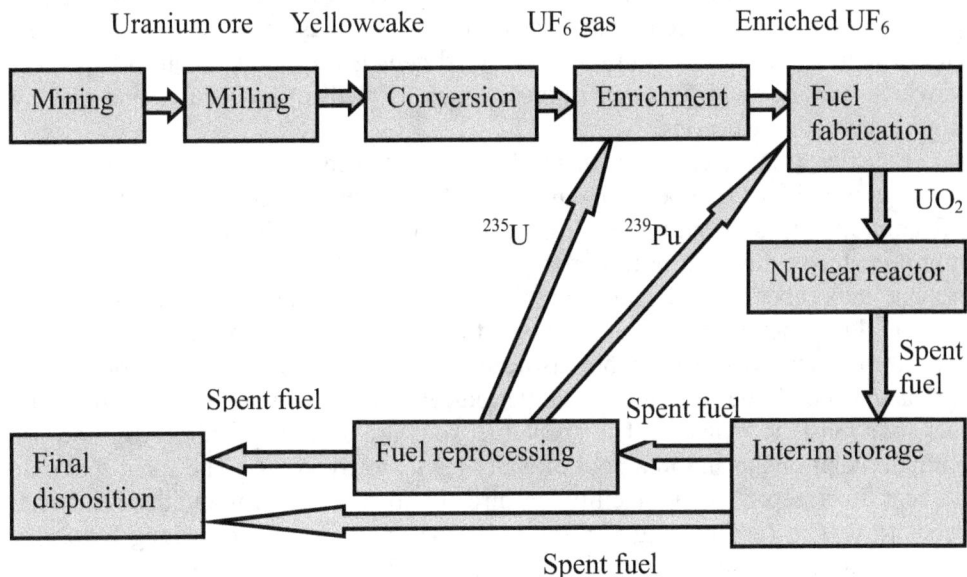

Fig. 5-3 Process undergone by uranium used as reactor fuel [EIA].

Enrichment is accomplished by first converting yellowcake to uranium hexafluoride, UF_6, which is then heated to its gaseous state. There are a number of methods for separating out the UF_6 molecules containing ^{235}U atoms from those containing ^{238}U. Because the differences between ^{235}U and ^{238}U are slight, all methods involve a cascade of stages in which each stage increases the percentage of ^{235}U relative to that of ^{238}U. The thermal diffusion, gaseous diffusion and gas centrifuge methods all depend on the fact that the ^{235}U isotope is 1.26 percent lighter than the ^{238}U isotope. The earliest methods were thermal diffusion and gaseous diffusion. Thermal diffusion is possible because lighter molecules tend toward hotter surfaces more than heavier molecules and gaseous diffusion uses high pressure to force the UF_6 gas through a series of semi-permeable membranes that allow the lighter gas to pass more readily than the heavier gas. Thermal diffusion was soon abandoned in favor of gaseous diffusion. Although gaseous diffusion still accounts for about one third of enriched uranium production, it has been largely replaced by the gas centrifuge method that uses centrifugal force to separate the lighter ^{235}U from the heavier ^{238}U. Although a gas diffusion plant is less expensive to construct, it is more expensive to operate than a gas centrifuge plant. An improvement to the gas centrifuge is the Zippe centrifuge that uses both heat and centrifugal force to separate the isotopes. The most recently developed technique is to use lasers tuned to frequencies that ionize ^{235}U but not ^{238}U, thereby allowing the positively charged ^{235}U to be separated out by using an electric field. This method is expected to be employed commercially within the next few years. Other methods being investigated include aerodynamic, electromagnetic, chemical and plasma separation processes. That which is not fabricated into fuel is called *depleted uranium*, about five percent of which may be used for other purposes such as medical procedures and making armor penetrating weapons.

After a sufficient percentage of ^{235}U is attained, the UF_6 is processed into uranium dioxide, UO_2, a powdery substance that is then fabricated into pellets about 1.5 cm (0.6 in) long and 1 cm (0.4 in) in diameter. Each pellet has the same approximately equivalent useful energy content as 150 barrels of oil. These pellets are put into tubes, called *fuel rods*, that serve as the energy source for nuclear reactors. In a reactor, the ^{235}U atoms begin to fission. These fissions produce the heat needed to generate electricity, but also cause the percentage of ^{235}U to drop. As the ^{235}U is expended, the reactor must be periodically shut down and some or all of the fuel rods must be replaced. The spent fuel rods are still quite radioactive and are normally kept onsite and in a pool of water for up to five years. The spent fuel may be sent to a reprocessing facility or directly to a long term radioactive waste depository. The reprocessing facility waste may also be sent to a long term, dry storage waste depository. The radioactivity of most of the waste stored in such depositories is relatively low, but remains dangerous to humans and animals for centuries or, for some materials, millennia.

The spent fuel not only still contains a useful amount of ^{235}U, but also contains plutonium 239, ^{239}Pu, which is also a fissile material. (In fact, the bomb exploded over Nagasaki was a plutonium bomb.) The reason the spent fuel contains ^{239}Pu is that several chain reactions occur within a reactor and one of them is the reaction shown in Fig. 5-4. When a ^{238}U atom in a fuel rod is struck by a neutron, it normally becomes a ^{239}U atom that quickly and naturally decays into a neptunium 239, ^{239}Np, atom by means of a beta minus, β^-, decay. This decay converts a neutron into a proton, an electron and an antineutrino. After a short time, a second β^- decay converts the ^{239}Np atom into a ^{239}Pu atom. Some of the ^{239}Pu helps supply heat to the reactor as it fissions into fragments including particles consisting of two neutrons and two protons. Such particles are helium nuclei and are referred to as *alpha particles*. As a fissile material, the ^{239}Pu from the spent fuel may be separated out and fabricated into new fuel pellets.

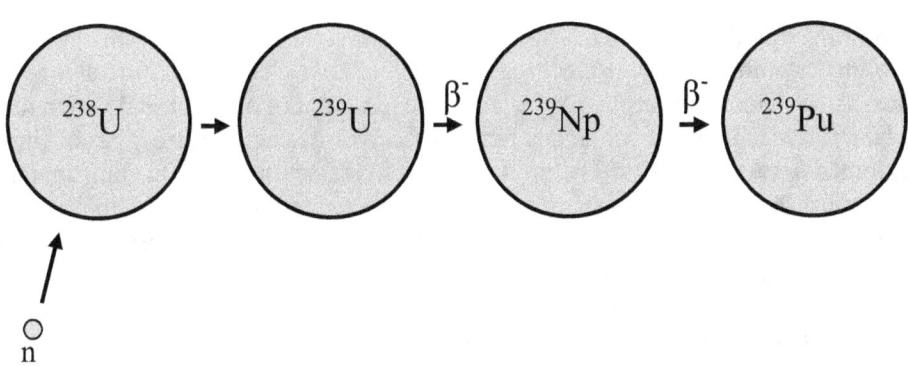

Fig. 5-4 Process by which ^{238}U is converted into ^{239}Pu.

The worldwide organization for promoting safe, secure and peaceful nuclear technologies and monitoring activities related to radioactive materials is the International Atomic Energy Agency (IAEA). Within the United States it is the Nuclear Regulatory Commission (NRC) that is charged with licensing and inspecting all commercial fuel facilities that process or fabricate uranium ore into nuclear fuel. It was created in 1974 and replaced the Atomic Energy Commission that had existed since 1947. The NRC must report to Congress on a regular basis regarding the status of all such facilities within the United States. It also has the responsibility of certifying new reactors and enforcing the laws concerning civilian nuclear facilities.

Except for a reactor replacing a boiler, most nuclear electrical generating plants are similar to fossil fuel powered generating plants. A simplified diagram of a nuclear power plant is given in Fig. 5-5. Basically, heat from the reactor (or boiler in

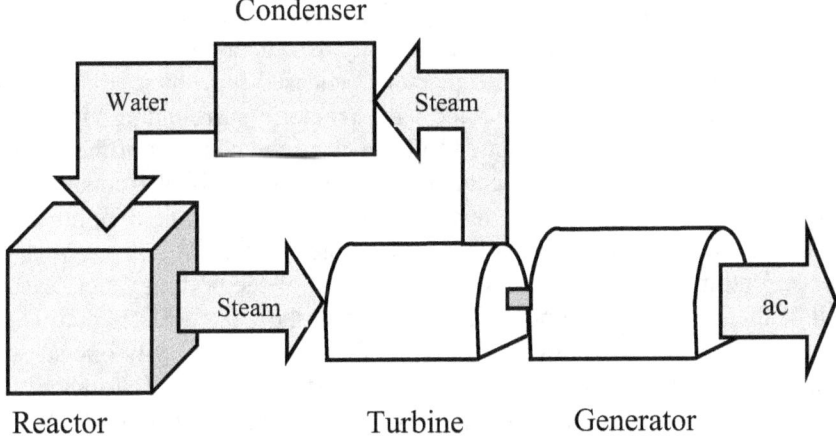

Fig. 5-5 Simplified diagram of a nuclear power plant.

the case of a fossil fuel plant) is used to generate steam and the steam drives a turbine that, in turn, drives an electrical generator. Most electricity is created by a turbine that provides the mechanical energy needed to drive an electrical generator. A *turbine* is a machine that contains a fan that receives its energy from a gas or fluid being forced across its blades. The high pressure fluid, steam or gas pushing against the blades causes the turbine's rotor, which is connected to the generator, to revolve. The resulting rotation of the generator's rotor creates a voltage across the output terminals of the generator. As discussed in Chaps. 6 and 7, the driving force may be from either steam or an exploding gas (as is the case in a turbo-jet engine). In order to force the steam through the turbine there must be a pressure difference between the turbine's input and output steam. This difference is created by a condenser that condenses the steam back into water. The condenser is a large heat exchanger that uses a cooling tower to vent the steam's heat into the atmosphere. A *heat exchanger* is a piece of equipment for transferring heat from one medium to another without mixing the two media. It is normally constructed by enclosing an array of pipes inside a vessel so that one medium, a gas or liquid, can be passed through the pipes as the other medium is circulated through the surrounding vessel. The heat is transferred through the walls of the pipes to the other medium by conduction.

There are numerous reactor designs, but the most common is the pressurized water reactor (PWR) design shown in Fig. 5-6. The water is raised to approximately 325°C (617°F) causing the pressure to rise to 150 atmospheres. Of the 104 licensed reactors in the United States in 2007, 69 were PWRs, and in the world 269 of the 439 reactors were PWRs. The heart of the reactor is its core and the fission energy generated in the core is released as heat which heats the pressurized water. The water is circulated out of the container surrounding the core through a heat exchange that turns a separate stream of water into the steam that drives the turbine. The medium that transfers heat from the core, in this case the pressurized water, is referred to as the reactor's *coolant*. The steam output by the turbine is condensed back into water and is circulated back through the heat exchanger. The reason the pressurized water around the core is not used directly to generate the required steam is that it becomes radioactive. The entire reactor is encased in a thick containment wall made of concrete and steel that protects the reactor's surroundings from the radiation within the reactor and is designed to keep all of the contents of the reactor inside the reactor in the event of an accident. The central part of the containment wall is forged as a single piece of very strong steel by a limited number of manufacturers, all of which are located outside the United States.

The second most common reactor design is that of the boiling water reactor (BWR). All of the 35 remaining operational reactors in the United States and 94 of the reactors in the world are BWRs. In this type of reactor there is only one stream consisting of both water and steam. The water is pressurized, but only to about 75 atm, half as much as in a PWR, and at this pressure the water boils at about 285°C (545°F). As shown in Fig. 5-7, the coolant stream is converted to steam and this steam drives the turbine directly. Although the design is much simpler, because the

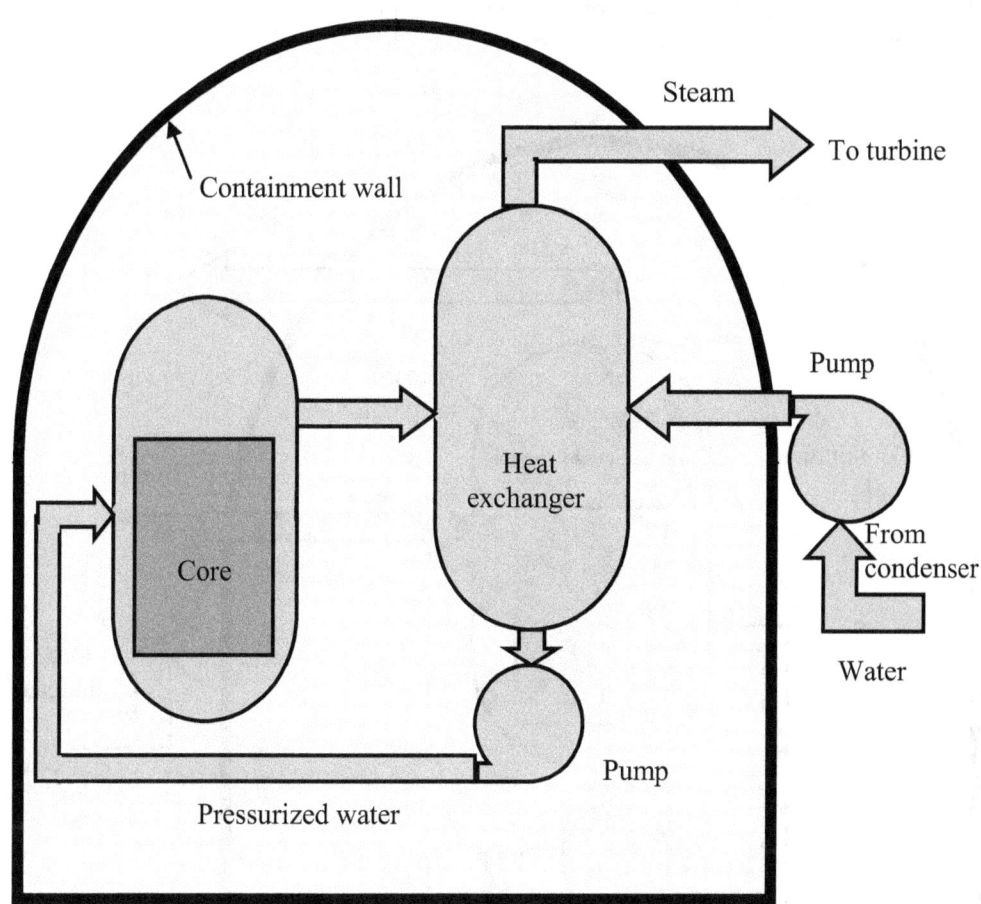

Fig. 5-6 Simplified pressurized water reactor design.

coolant is radioactive, the turbine, the heat exchanger in the condenser and associated piping must be shielded as well as the reactor. Fortunately, the radioactivity in the coolant is short-lived and maintenance on the equipment outside of the reactor may be performed shortly after being shut down.

     The core of a reactor is where the fuel rods are located and the coolant flows around and through the core. The fuel rods consist of Zircaloy tubes that encase the UO_2 pellets. Zircaloy is a hard, corrosive-resistant substance that is primarily zirconium, Zr, and is permeable to neutrons. Fuel rods are normally three to four meters (9.3 to 13.1 ft) in length and are grouped in assemblies of 300 to 400 rods each. There may be over a hundred such assemblies giving a total of several thousand rods, and the total amount of uranium in the fuel rods typically weighs between 90 and 100 short tons (81 to 90 tonnes).

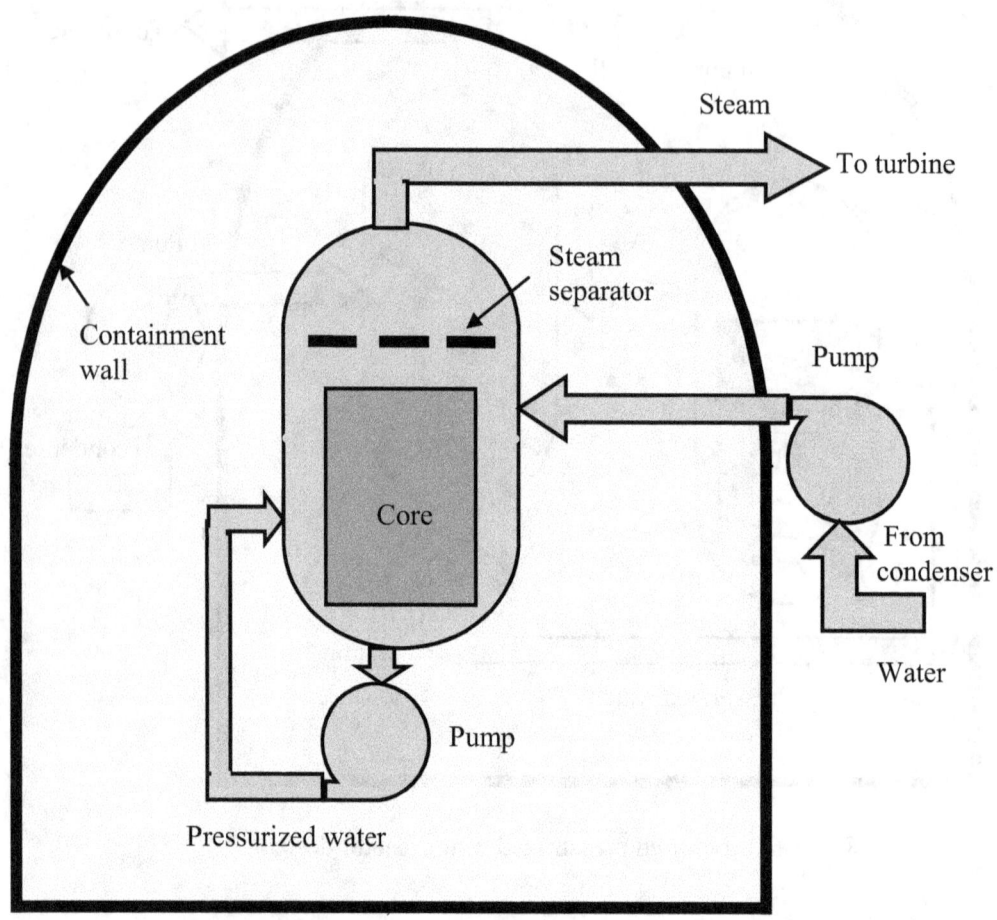

Fig. 5-7 Simplified boiling water reactor design.

As shown in Fig. 5-8, the core also contains control rods and a moderator for slowing down the neutrons. As indicated above, fissions are more likely to occur if the neutrons are not traveling too fast. The most common moderators are water and graphite. If the moderator in a reactor is ordinary water, then the reactor is called a *light water reactor*. If the hydrogen in the water has a neutron in each atom (i.e., it is deuterium), then the water is said to be *heavy water* and the reactor is correspondingly called a *heavy water reactor*. Most heavy water reactors are PWRs.

Even when heavy water is used as the moderator, light water may be the coolant. An important class of pressurized heavy water reactors (PHWRs), commonly known

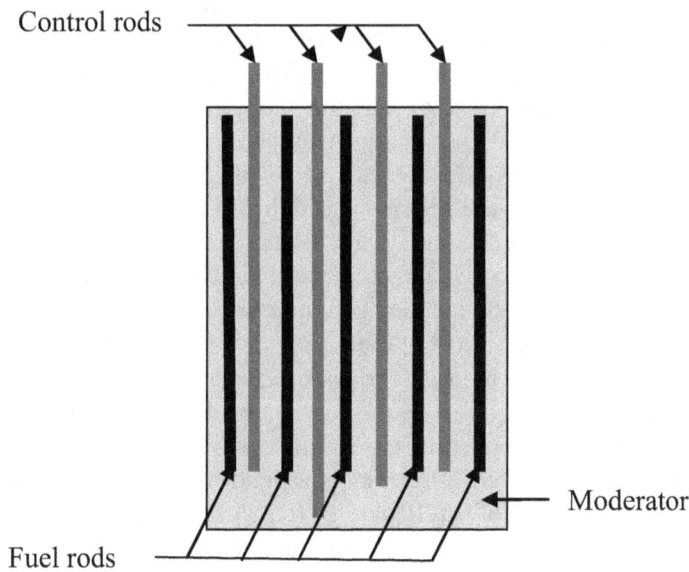

Control rods

Moderator

Fuel rods

Fig. 5-8 Major components of a reactor's core.

as CANDU reactors, was developed in Canada. CANDU reactors are best known for using yellowcake as fuel, thus eliminating the enrichment process. Pressurized heavy water is a more effective modulator and this allows the fuel to contain less ^{235}U. Worldwide, there are 43 CANDU reactors. They have become popular in some countries because the fuel is cheaper, but they include a continuous refueling process that has raised nuclear proliferation concerns because the spent fuel contains a relatively high percentage of highly radioactive ^{239}Pu.

The control rods contain a *neutron absorber* that reduces the number of neutrons available for fissions. They may be inserted into or withdrawn from the core and thereby control the number of fissions and resulting heat generated by those fissions. Typically, the neutron absorber is cadmium, hafnium or boron. In addition, some fuel assemblies may include *burnable absorber rods*. They tend to even out the rate at which fissions occur and are particularly useful in new assemblies that produce more than the average number of fissions. Also, by absorbing neutrons the absorber becomes fissile material that lengthens the useful life of the assembly as its nuclear material decays. The most common substance used for this purpose is gadolinium, Gd.

In addition to the reactor types discussed above, there are gas-cooled reactors (AGRs), light water graphite reactors (RBKMs) and fast breeder reactors

(FBRs). If a reactor is cooled by gas, then the gas may be passed through a heat exchanger to generate steam for a steam turbine or passed directly through a gas turbine. The AGRs are mainly located in the United Kingdom, the RBKMs are mainly in Russia and the FBRs are in Japan, France and Russia. Several countries, including the United States, have researched a new type of AGR in which the core contains over 300,000 billard ball sized spheres, called *pebbles*, instead of the usual rod oriented construction. Each pebble contains thousands of small particles of fissionable material coated with high-density materials designed to withstand high temperatures. The core is encased in a metal surrounded by graphite blocks. In addition to the fuel pebbles, graphite spheres are included in the core to control the temperature. The pebbles are continually circulated through the core and this circulation allows spent pebbles to be easily replaced. The spent pebbles are handled the same way spent fuel rods are handled. Helium or carbon dioxide is passed through the core and is heated to 900°C (1650°F) and fed into the gas turbine that drives the generator. If helium, which is inert, is used there is very little corrosion. The elaborate cooling system required in BWRs and PWRs is not needed and the simpler design lends itself to constructing smaller, more flexible designs with capacities as low as 124 MW. Their efficiency is around 40 percent, a considerable improvement over the 33 percent of BWRs and PWRs. A more complete description of gas-cooled designs based on spherical fuel modules is given in a January, 2002, article in *Scientific American* by J. A. Lake, R. G. Bennett and J. F. Kotek.

An FBR does not utilize a moderator, but requires faster neutrons and a more highly enriched fuel consisting of up to 20 percent plutonium. They are called breeder reactors because they include breeder blankets containing ^{238}U in their cores that create more fuel than is consumed. They do not use water as a coolant because it is a moderator. Instead, they use a liquid metal, most often molten sodium, but mercury and molten lead have also been employed. The plutonium in the fuel may be from spent fuel (see Fig. 5-3) or dismantled nuclear weapons. There are several methods for reprocessing spent fuel and breeder blankets to remove the fuel gain, and the content of the fissile material extracted varies according to the method. The fuel gain is mainly ^{239}Pu and results from a sequence of reactions, such as the one depicted in Fig. 5-4, that converts the ^{238}U in the breeder blanket into ^{239}Pu. Some FBR methods are especially worrisome from a proliferation standpoint because they are capable of producing relatively large amounts of weapons grade plutonium.

The twin problems of producing weapons grade plutonium and disposing of radioactive waste are largely overcome by a design described in the December, 2005, issue of *Scientific American* written by W. H. Hannum, G. E. Marsh and G.S. Stanford. Not all fast-neutron reactors are breeder reactors and this design uses a fast-neutron reactor to keep recycling spent fuel until ultimately almost all of the energy is extracted. This continual recycling could extend uranium reserves indefinitely. As seen in Fig. 5-3, spent fuel may or may not be reprocessed. If it is not reprocessed then there is a lot of radioactive waste that still contains much of the energy and must be stored for a very long time. A system that does not recycle the waste is called a *once-through system* or *open system*. In order to avoid the availability of weapons grade plutonium, the United States decided to build once-through systems only, but is now reconsidering this decision. Designs that include recycling are called *closed systems*. Some closed systems have been built by other countries including Russia, Japan, France and the United Kingdom. Most closed systems use the *plutonium uranium extraction* (*PUREX*) method. PUREX produces high grade plutonium and is the same method used for extracting the plutonium for bombs. The design described by Hannum, Marsh and Stanford is referred to as *full recycling* because the fuel is recycled until it is almost entirely depleted of energy.

The fission products output by a nuclear reactor may be divided into completely useless waste known as *true waste*, uranium waste that is mostly ^{238}U, and *transuranic waste* consisting of elements heavier than uranium. True waste is about five percent of the spent fuel and is made up of elements that are lighter than uranium and decay to safe levels in about three hundred years. Uranium waste makes up approximately 94 percent of the spent fuel, and is only mildly radioactive because it contains a reduced amount of ^{235}U and can be safely stored. Although the transuranic waste is only one percent of the waste, it is highly radioactive and, if stored, it must be stored for tens of thousands of years. But it is the waste component that can be recycled. It can be separated from the true waste by the pyrometallurgical process depicted in Fig. 5-9. The spent fuel may come from either fast-neutron reactors that utilize high energy neutrons or from thermal reactors. If it is from a thermal reactor it must first pass through an oxide reduction unit to convert it into a metal. Spent fuel from a fast-neutron reactor need not pass through this step, but can go directly to the electrorefiner. The electrorefiner uses electroplating to draw the metal ions onto the cathode from a chemical solution that contains the dissolved spent fuel or output of the oxide reduction unit. Most of the transuranic matter and much of the uranium is attracted to the cathode. The cathode processor removes this material and the casting system turns it into fuel rods that can be used in fast-neutron reactors. The solution in the cathode processor is periodically cleansed of the true

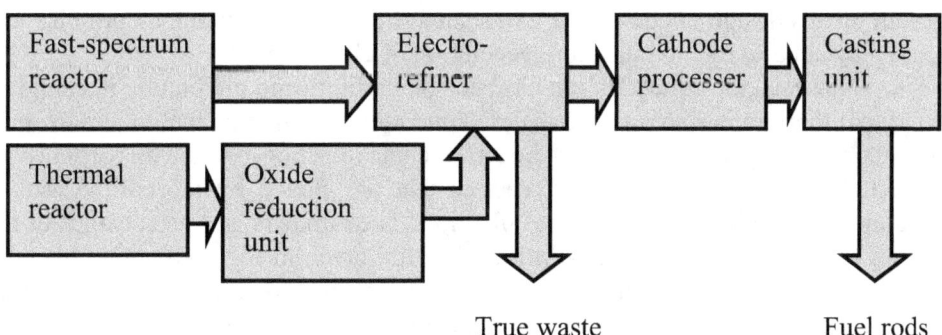

True waste   Fuel rods

Fig. 5-9 Pyrometallugical processing of spent fuel.

waste, which is sent away for disposal. Some of the material collected on the cathode may need to be returned to the electro-refiner for further processing.

Radioactive matter is broken into three categories. Matter, such as ^{238}U, that can occasionally split into fragments, but cannot sustain a chain reaction, is said to be *fissionable*, while matter that can sustain a chain reaction, such as ^{235}U and ^{239}Pu, is said to be *fissile*. A fissionable isotope, such as ^{238}U, is also said to be *fertile* because it may transmute into a fissile isotope, such as ^{239}Pu as shown in Fig. 5-4. The true waste consists mainly of strontium and cesium and is not fertile. The uranium attracted to the cathode is mostly fertile, but may include some fissile ^{235}U. Because the pyrometallurgical process outputs a mixture of fertile and fissile matter (and perhaps some matter that is not fertile), the output fuel is not at any point weapons grade plutonium as is the output of a PUREX system. This significantly reduces the chance that enriched plutonium will fall into the wrong hands.

The other half of the system indicated by Hannum, Marsh and Stanford, a fast-neutron reactor, must, as indicated for FBRs, use a coolant that does not act as a moderator. The United States has developed an advanced liquid metal reactor (ALMR) that uses a metallic liquid such as sodium or lead as its coolant. The metallic liquid is used to heat a secondary coolant that is the coolant that drives the turbine. A liquid metal has an advantage over water in that it can be kept at a low pressure and, thereby, lessens the chance of a radioactive leak. However, the core of an ALMR must operate at very high temperatures and sodium may catch fire. The pyrometallugical unit could be located at the same site as the fast-neutron reactor, further reducing the possibility of a black market in plutonium.

It is estimated that a once-through system extracts only five percent of the energy in thermal reactor fuel and PUREX recycling captures about six percent of the available energy. By repeated recycling fuel using the proposed system, it is believed that up to 99 percent of the available energy can be used. This means that the current uranium reserves could last for thousands of years. Although the proposed system is technologically possible, no commercial-sized plant has been built and there are still unanswered questions. For one, there is uncertainty as to how much such systems would cost. At best, it will be several years before a commercial system of this design is completed, but such systems hold tremendous promise and could play a major role in our future generation of electricity.

Although the concentration has been on uranium and plutonium, it should be noted that thorium is also a fissile material and could be used as fuel in a reactor. In fact, thorium is part of several radioactive chain reactions. India in particular is interested in constructing nuclear reactors that are fueled by thorium. Although uranium is scarce in India, thorium is plentiful. The numbers and percentages of the various types of reactors in the world in 2007 are given in Fig. 5-10. Also given in the figure following the numbers of reactors are the total electrical generating capacities of the corresponding reactor types in gigawatts. In 2007, the total number of reactors was 439 and their combined capacity was 385 GW.

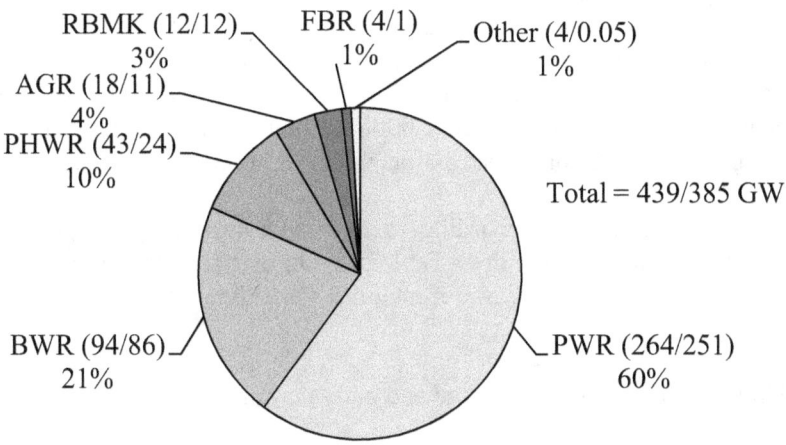

Fig. 5-10 Numbers of nuclear reactors in the world in 2007 and their combined capacities.

The growth of nuclear generated electrical power from 1980 to 2007 for both the world and the United States in billions of kilowatt-hours (GkWh—one kilowatt-hour is 3412 Btu) is shown in Fig. 5-11. In 1980, the United States produced 287 GkWh of nuclear generated power, or about 42 percent of the world total of 684 GkWh. By 1990 that percentage was reduced to about 30 percent and has remained there ever since. In 2005 the nuclear plants in the United States produced 782 GkWh and the world's plants produced 2,626 GkWh. If the cost of fossil fuels remains high, the contribution of using nuclear energy for generating

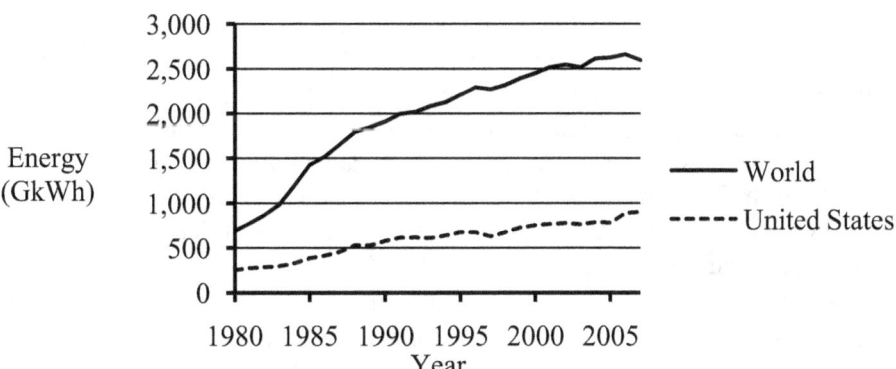

Fig. 5-11 Growth of nuclear power generation [EIA].

electricity is likely to increase dramatically, not just in the United States, but in the world as a whole. A pound of uranium fuel on average generates 164,000 kWh of electricity. A 1,000 MW nuclear plant typically requires 33 short tons (29.7 tonnes) of low-enriched uranium per year as opposed to 2.9 million short tons (2.61 million tonnes) of coal per year for a coal-fired 1,000MW plant. But, obtaining this much uranium fuel may require the processing of as much as 300,000 short tons (270,000 tonnes) of ore.

Uranium ore is graded according to the percentage or parts per million (ppm) of uranium contained in the ore. Table 5-1 summarizes the major categories of uranium ore. All soil and even seawater includes some uranium.

Table 5-1 Uranium content in percentage and parts per million.

Grade	Percentage	Parts per million
Very high grade	20	200,000
High grade	2	20,000
Low grade	0.1	1,000
Earth's crust		2.8
Seawater		0.003

The percentage of yellowcake (U_3O_8) in uranium ore varies widely from less than 0.1 percent to 20 percent with the richest ore being mined in Canada. Much of the rest of the high grade ore is found in Australia. The amounts and percentages of yellowcake produced from the mines of the various nations are given in Fig. 5-12. As the figure indicates, in 2007 the United States was a minor player in the production of yellowcake from its mines. Of the 53,775 short tons (48,398 tonnes) of yellowcake produced in the world, only 2,266 (2039) came from mines located in the United States. Sixty percent came from mines in Canada, Australia and Kazakhstan. The reason is not that the United States could not obtain more of its yellowcake from its own mines, the reason is that the ore from those countries is much richer and less expensive to process. This international data was not obtained from the EIA, but was obtained from the TradeTech website, *www.uranium. info/index*. (Their data for the United States matches that of the EIA.) The EIA website makes available only a limited amount of international uranium data. Although the IAEA tries to monitor nuclear activity, the accuracy of international data is in question because of the aura of secrecy surrounding the production and reserves of uranium.

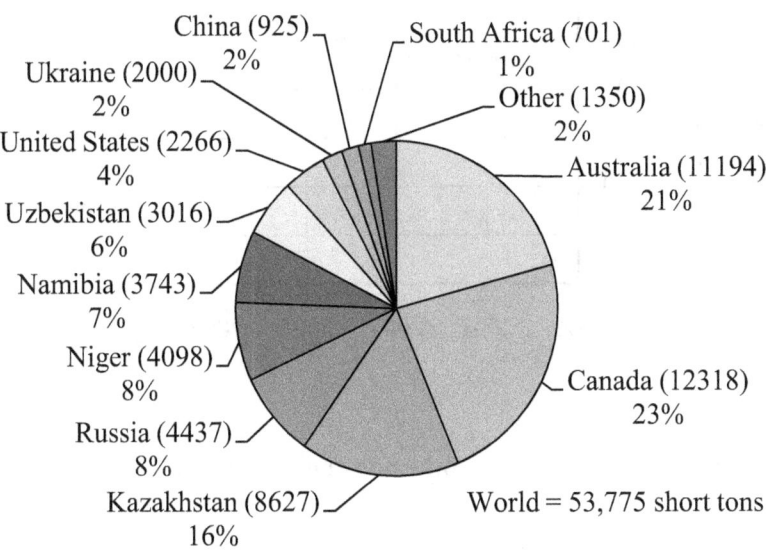

Fig. 5-12 2007 yellowcake produced from mines in various nations in short tons.

Figures 5-13 and 5-14 give the past United States and world production of yellowcake, respectively. The production from the United States mines varies while

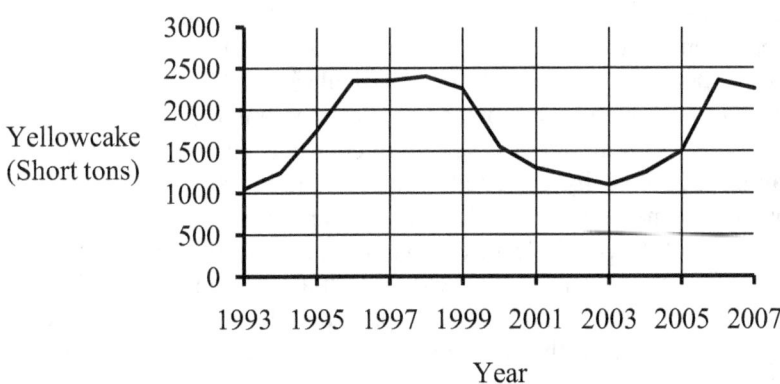

Fig. 5-13 Yellowcake from United States mines in short tons [EIA].

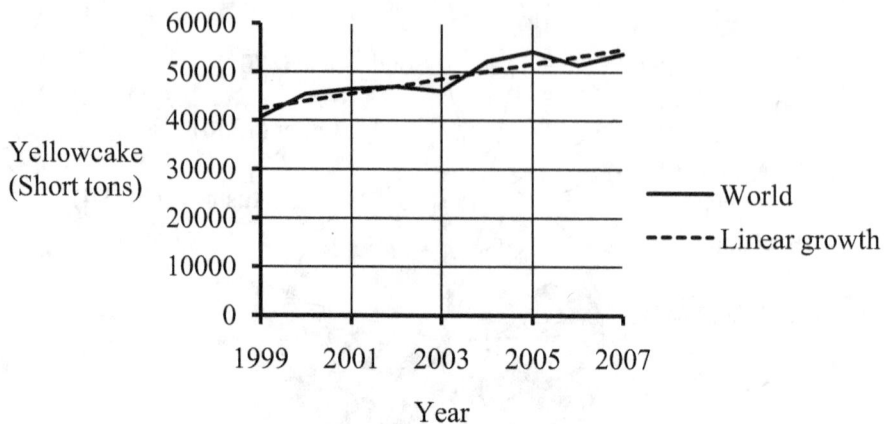

Fig. 5-14 Yellowcake from worldwide mines in short tons.

that of the mines worldwide has a definite up-trend. The United States demands are satisfied according to where they can be met most economically. Most of its present demands are met by purchasing uranium from places such as Canada and Australia that have higher-yield ore. The world's yellowcake production linear growth curve is superimposed on the world's actual production curve and indicates a growth of about 1,500 short tons (1,350 tonnes) per year. However, the exploration for new uranium deposits, which remained almost constant from 1985 to 2005, has increased dramatically since 2005. Just between 2005 and 2007 the amount of drilling for new deposits in the United States went from 1.7 million feet to 5.1 million feet. The primary reasons for this renewed interest in the United States are the need to become energy independent and to reduce the carbon dioxide put into the atmosphere by electrical power plants that use fossil fuels, particularly coal.

Uranium is about as common as tin or germanium and is at least 35 times as common as silver. Uranium ore reserves are rated according to the price at which the uranium can be economically extracted. The $60/lb ($130/kg) reserves in thousands of short tons as estimated by the IAEA in 2007 are shown in Fig. 5-15. The total estimated reserves at the $60/lb level was 6,016 thousand short tons (5,414 thousand tonnes). The total production in 2007 was 54.7 thousand short tons (49.2 thousand tonnes) giving an R/P ratio of 110, which implies that the current reserves would last only 110 years. However, this amount is very misleading for several reasons including the following:

- It has been only in the last three years that serious exploration has renewed after a twenty-year lapse. It is expected that new exploration will result in

$60/lb reserves being at least doubled in the next few years. This is particularly true in the United States, South Africa and Nigeria.

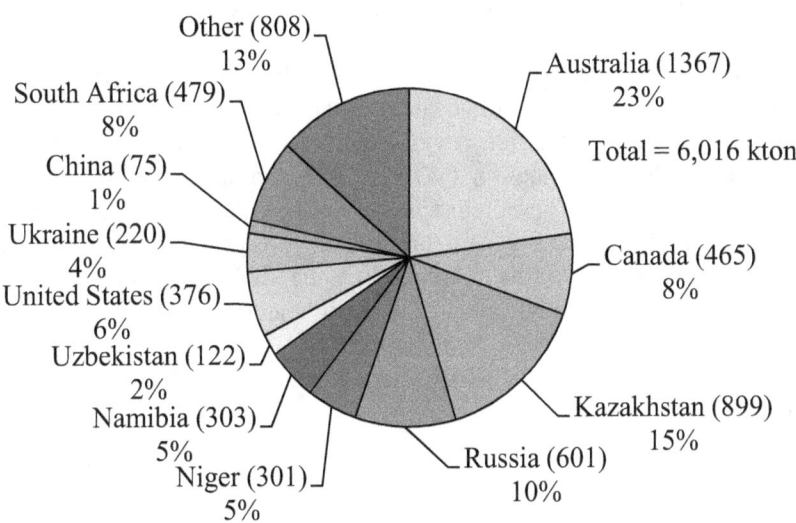

Fig. 5-15 2007 known yellowcake reserves in thousands of short tons at $60/lb [IAEA].

- Because of the secrecy surrounding uranium, the current reserve figures may be inaccurate.
- Spent fuel can be reprocessed and, in the future, breeder reactors and pyrometallurgical processing may become more widely used, thereby requiring less fuel to be mined. Recycling nuclear fuel becomes increasingly economically feasible as the price of yellowcake rises and could potentially extend the use of nuclear power plants indefinitely.
- Nuclear plants that are fueled by thorium instead of uranium may be built, especially in places like India where thorium is more plentiful. It is estimated that there is three times as much thorium in the world as uranium.
- The use of nuclear fission to generate electricity may increase considerably as there is an attempt to reduce the carbon dioxide released into the atmosphere by fossil fuels.

- Unlike the generation of electricity by fossil fuels, the principal costs do not include the cost of fuel. The principal costs are the capital costs of constructing nuclear plants with adequate safeguards, operating and maintaining these plants and disposing of spent fuel. Therefore, more expensive processed uranium is not a serious impediment to using uranium to generate electricity.

In 2007, 51 million pounds (23 million kilograms) of yellowcake was purchased by power plants in the United States at an average price of about $33 per pound, and 787 GkWh of electricity was generated by those plants. A slightly modified World Nuclear Association (WNA) 2007 example cost breakdown for obtaining one pound of fuel from nine pounds of yellowcake costing $33/lb is given in Fig. 5-16. The cost of each stage of processing is given in parentheses and is followed by the percentage of the total cost of the fuel. It is assumed that the conversion produces 7.5 lb and costs $5.45/lb, enrichment produces 7.3 lb and costs $61.40/lb and fabrication produces 1 lb and costs $109. The total cost is $895. If one pound of fuel generates 164,000 kWh, then the cost of fuel is 0.55 cents per kilowatt-hour.

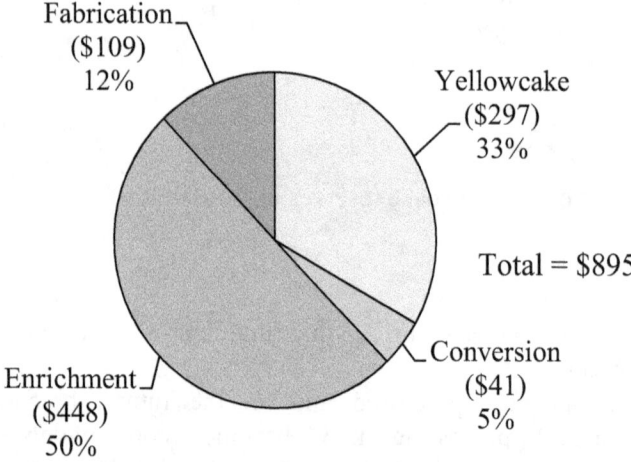

Fig. 5-16 Example cost breakdown of uranium processing [WNA].

As one might guess, determining the total cost per kilowatt-hour of generating electricity using uranium depends on the many assumptions one must make regarding capital costs, waste disposal costs and decommissioning costs as well as the cost of fuel. All of these factors vary considerably with time. There have been several studies conducted by different organizations in a variety of countries. Some of these studies have been summarized by organizations such as the WNA and the National Energy Institute (NEI). Typical assumptions are to use a $1500/kW construction cost, a 40-year plant lifetime, an 85 percent load factor (i.e., the plant will operate, on average, at 85 percent of capacity), a five percent discount rate on investment, and decommissioning costs of nine to fifteen percent of initial capital costs. In the United States, 0.1 cent per kilowatt-hour is levied on the generated power to pay for disposal. A representative estimate of the division of the 2007 total cost of generating power from uranium in cents per kilowatt-hour in the United States is shown in Fig. 5-17. In this figure, capital cost includes decommissioning and waste disposal. The total is 3.0 cents per kilowatt-hour, which is comparable to electricity generated by coal. Nuclear plants tend to operate at more than 85 percent capacity because their fuel is so much cheaper than that of a fossil fuel plant, 0.55 cents per kilowatt-hour versus 2.3 cents per kilowatt-hour. For comparison, the average charge for electricity in the United States in 2007, including distribution, was 8.77 cents per kilowatt-hour.

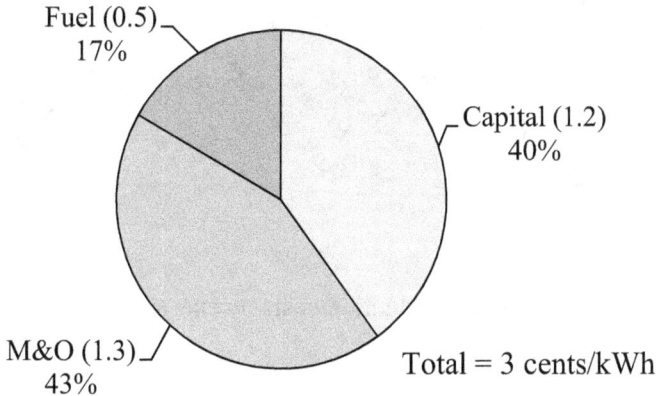

Fig. 5-17 Example cost division of nuclear power in the United States in cents/kWh [WNA].

The article "Giant Holes in the Ground" by Matthew L. Wald in the November/December, 2010, issue of *Technology Review* discusses the 2,234 megawatt nuclear generating plant being constructed in Georgia and the costs associated with producing electricity from various energy sources in general. Figure 5-18 summarizes the total costs in dollars per megawatt-hour quoted in the article for the different energy sources. The origin of the estimates is the EIA. Coal is divided into three categories, conventional, advanced and advanced with carbon capture and storage (CCS). Natural gas is divided into conventional combined cycle

(CCC), advanced combined cycle (ACC) and ACC with CCS. Included in the estimates are transmission, operation, maintenance, fuel and capital costs. However, these estimates depend on the future costs of the sources relative to those of fossil fuels, the disposal of radioactive waste and the laws regarding carbon dioxide emissions (see Chap. 9), all of which would have a significant impact on the actual costs.

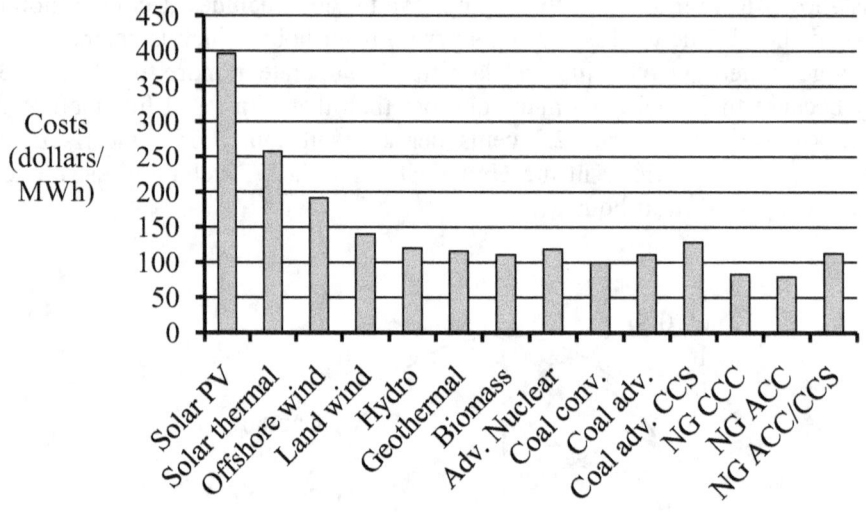

Fig. 5-18 Total electricity production costs by energy source [EIA].

The *Technology Review* article also gives the 2009 nuclear generating amounts and total generating amounts in gigakilowatt-hours for several countries. These capacities are shown in Fig. 5-19. The total worldwide electrical energy generated during 2009 was 18,286 GkWh, 2,560 GkWh of which was generated by nuclear power. Although the percentage of China's electricity generated by nuclear power is low, China has plans to build 33 new nuclear reactors versus nine new nuclear reactors planned in the United States. Twenty-four of China's 33 reactors are currently under construction while only two of the planned nine reactors in the United States are presently being constructed. Japan has plans to build 12 new reactors and Russia has plans to build 14 with 12 currently under construction.

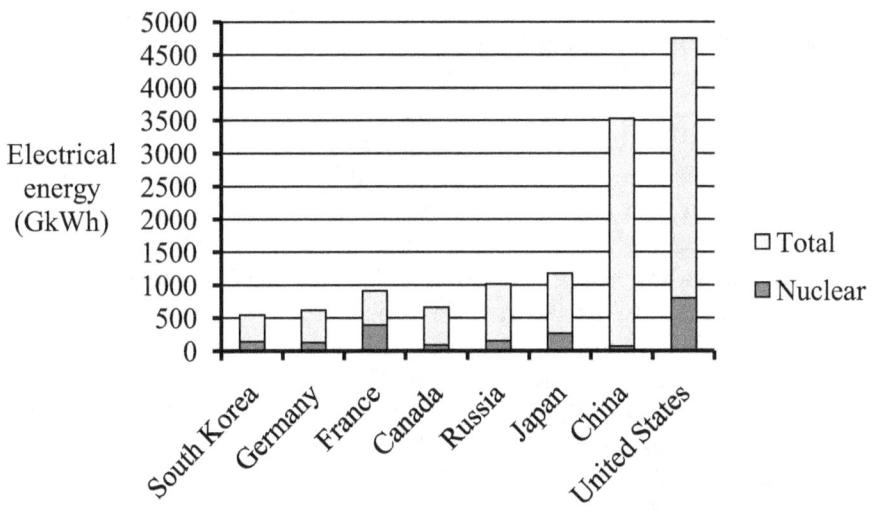

Fig. 5-19 Electrical energy generated in GkWh [EIA].

Eventually a nuclear reactor must be permanently shut down. Since January 1, 1999, there have been 24 closures with a total generating capacity of 6,364 MW, and another 15 closures with a total capacity of 9,323 MW are anticipated by the end of 2010. There are three major reasons for shutdowns: policies/laws, international agreements and economics. Several countries simply have laws forcing the shutdown of reactors after a certain time has elapsed since they began operation. For example, in Germany the period of operation is typically 32 years. Four closures were mandated by an agreement between the European Union (EU) and some of the former Soviet dominated countries as a condition for entry into the EU. Five more reactors are to be shut down before 2010 for the same reason. These reactors were deemed unsafe by the EU following the accident at Chernobyl. Most reactors, however, are shut down because their need for repairs have rendered them uneconomical they have gotten too old. When a reactor is decommissioned, not only does its fuel need to be disposed of safely, but much of the equipment and materials associated with the reactor are also radioactive to varying degrees and must be quarantined. Although it is not nearly as radioactive as spent fuel, it must still be transported and stored with care.

The WNA lists four classifications of nuclear waste, very low-level waste (VLLW), low-level waste (LLW), intermediate-level waste (ILW) and high-level waste (HLW). VLLW consists of materials that have so little radiation that it is considered not harmful to animals or the environment and special means are not required for its disposal. It is mainly construction materials used to build nuclear reactor enclosures and is no more toxic than most other industrial waste. About 90 percent of the waste that is radioactive enough to require special treatment is LLW and consists of clothing, wiping rags, mops, filters, tools, residues and other materials that had incidental contact with radioactive substances. Some LLW is produced by medical and other uses of nuclear substances not related to electricity generation. LLW is responsible for about one percent of the radiation from nuclear waste and its radiation generally decays quickly. ILW is about seven percent of the total radioactive waste and accounts for around four percent of the radiation. It is materials that have had close contact with radioactive substances, such as moderating and shielding materials. Both LLW and ILW waste sites must be licensed by either the NRC or the state in which they are located. Both must be isolated and some shielding may be required for ILW.

Only three percent of nuclear waste is HLW, but HLW produces 95 percent of the radiation. According to the Waste Policy Act of 1982, its disposal is strictly controlled by the Environmental Protection Agency (EPA), which is responsible for setting the disposal standards; the DOE, which must construct, operate and maintain the disposal sites; and NRC, which must license the sites. The WNA estimates that, on average, a 1,000 MW plant produces 200 cubic meters (7,063 ft^3) of LLW per year, 70 cubic meters (2,472 ft^3) of ILW per year and 10 cubic meters (353 ft^3) of HLW per year. This implies that the 385 billion watts of nuclear capacity in the world produces over one hundred thousand cubic meters (three and a half million cubic feet) of LLW, ILW and HLW that must be disposed of every year. Newer plants produce much less nuclear waste than older plants and new designs should produce even less waste.

Every 12 to 18 months one fourth to third of the fuel assemblies in a reactor have decayed to the point that they must be replaced. When a reactor is shut down there is passive decay heat that could build up in the core and extreme care must be taken to remove this heat. Fortunately, the decay is short lived and the decay heat rapidly drops from six percent of full power heat to less than one percent. The replaced assemblies must be stored in pools of water that are at least 6 meters (20 ft) deep for at least a year. Then they, along with other HLW, are put in dry concrete and steel casks for permanent storage. The government has been attempting to complete and license an underground permanent disposal site in Yucca Mountain, Nevada, for several years, but their attempts have met with substantial resistance from environmental groups. As a consequence, disposal of HLW has become a serious problem. The primary dangers related to HLW are that it may be spilled while it is being transported and, over time, it may accidentally seep into a widespread underground aquifer.

How long a material remains too radioactive to be considered safe depends on its constituents and their concentrations. The rate of natural decay of an isotope is a percentage of the amount of the element and is constant (i.e., it is a constant exponential decline). The rate of decay is usually denoted by the length of time it takes half of the isotope's atoms to fission into fragments. For example, if it takes a million years for 200 atoms to be reduced to 100, then it will take another million years for the remaining 100 atoms to be reduced to 50 and so on. This length of time is called the *half-life* of the isotope. The half-life of thorium 231 is 25.5 hours, of neptunium 235 is 400 days, of thorium 228 is 1.9 years, of plutonium 239 is 24,200 years, of uranium 235 is 704 million years and of uranium 238 is 4.47 billion years.

Except for the radioactive waste, there is virtually no pollution emanating from a nuclear plant unless there is an accident. Although numerous lesser accidents have occurred, there have been only two that have been of serious concern. They were the accidents at Three Mile Island in Pennsylvania in March, 1979, and at Chernobyl in Pripyat, Ukraine, in April, 1986. The accident at Three Mile Island caused considerable alarm, but in the end it was determined that little radioactivity escaped the reactor's containment structure. It was caused by the failure of feedwater pumps in a secondary, non-nuclear cooling system. This, in turn, caused steam generators to no longer remove heat from the reactor. The turbine and reactor automatically shut down and a pressure relief valve automatically opened without the operators being aware that it was open. Cooling water escaped through the valve and the reactor overheated. Emergency feedwater valves were accidentally left closed after a test two days earlier and there was momentary confusion about what was happening. Even though the accident resulted from multiple mistakes, it was estimated that the escaped radiation had less of an effect on people than that of a chest X-ray.

On the other hand, the Chernobyl accident was catastrophic and affected an area extending from Ukraine to Poland, Germany and Scandinavia. Some radiation was detected as far away as the United Kingdom. It resulted in the evacuation of 360,000 people. The reactor used 1,870 short tons (1,680 tonnes) of graphite as its modulator, was water-cooled and had no containment vessel. Although some fuel did melt, there was no complete meltdown. It was believed to be caused by a massive power excursion and a subsequent steam explosion, loss of coolant and burning graphite.

There is considerable controversy over the use of radioactive fuels to generate power and widely varying opinions as to the dangers and effects of nuclear waste and the chance of catastrophic accidents. The controversy pits organizations for promoting nuclear power, such as the WNA, against several environmental groups, such as the Union of Concerned Citizens. A balance must be struck between our need for energy and the need to leave our land, water and air pollution free. In addition to pollution, there is the real danger in the proliferation of nuclear materials that could be used by terrorists to build so called "dirty bombs" or rogue nations to build nuclear weapons. By 2030, the EIA expects world energy consumption to increase by 50 percent to 695 quadrillion Btu. Electricity output is predicted to almost double and account for one-fourth of this increase. Presently, two-thirds of the world's electricity is generated using fossil fuels—one-half by coal. Fossil fuels, coal in particular, are primarily responsible for the excess carbon dioxide in the atmosphere that most scientists believe is the principal cause of the recent global warming. The people of the world and their governments must soon decide how this additional electrical energy is to be realized. It takes a long time to design, build and license nuclear power plants. In the United States, it typically takes 18 months just to complete the necessary Environmental Impact Study (EIS). Some countries, including some island nations, such as New Zealand, where a single nuclear accident could have devastating consequences, are even more cautious about constructing nuclear plants. Due to the Chernobyl accident, Eastern European nations are acutely aware of the dangers associated with nuclear power and are wary of constructing new nuclear plants.

The prospect of using fusion to generate electricity commercially is most probably at least 50 years and possibly over 100 years away. There are several problems that must be overcome before fusion will become a realistic alternative. It is not just a matter of controlling reactions at extreme temperatures and pressures. As discussed in an article entitled "Fusion's False Dawn" by Michael Moyer in the March, 2010, issue of *Scientific American*, there are two main avenues being investigated in the development of fusion reactors. One is that of containing a deuterium/tritium plasma in a magnetic field, called a *magnetic bottle*, and the other is to simultaneously fire several very high-energy lasers at deuterium/tritium pellets. For the magnetic bottle approach, the plasma must be held under extreme pressure by an extremely intense, evenly distributed magnetic field. As indicated in the article, keeping the magnetic field perfectly shaped is similar to squeezing a balloon while keeping it shaped the way you want. At the same time there must be a way of extracting the helium produced and injecting additional plasma. The plasma must be heated to 150 million degrees Celsius (270 million degrees Fahrenheit), 25,000 times the temperature of the sun's surface. A large plasma facility known by the acronym ITER is being constructed in Cadarache, France, and is expected to cost $14 billion.

For the laser approach, the lasers must be fired in short intervals at precisely the same times at precisely the proper locations and new pellets must be entered between the bursts. The "peppercorn-sized" pellets must be perfectly round and presently cost up to one million dollars each. While in operation, about 10 pellets per second will be needed to sustain the fusion. When the lasers fire the total energy needed by the lasers "outshines the nation's entire electricity consumption." The system built at the National Ignition Facility (NIF) uses 192 lasers and cost $4 billion.

With either approach the reactor must be enclosed in a shell capable of extracting most of the energy produced and using it to generate steam. This could be done by circulating molten sodium or another similar substance through pipes in the shell and an external heat exchanger in a manner similar to that used by a fission reactor. The entire enclosing material must be capable of continuously withstanding very high temperatures over long periods. A fusion facility must be able to run almost all of the time in order to justify its very high capital costs.

Also, although deuterium is plentiful, tritium is not and must be produced. Some tritium may be produced by fission reactors, but most must be gotten from the fusion facility itself. Unlike a fission chain reaction in which the neutrons produced by the fissures are used to cause other fissures, most neutrons produced by fusion escape (see Figs 5-3 and 5-4). It is lithium and these neutrons that are used to produce the required tritium. As shown in Fig. 5-20, when a lithium 7, ^{7}Li, atom is struck by a neutron of sufficiently high energy, the lithium atom breaks into a helium atom, a tritium atom and a neutron. Then those of these neutrons that strike lithium 6, ^{6}Li, atoms cause each ^{6}Li atom to break into a second helium atom and a second tritium atom. If done correctly, enough tritium could be collected to render a fusion plant tritium self-sufficient. To capture the neutrons escaping from the fusion reaction, this reaction must be surrounded by a blanket containing channels of molten lithium and there must be a means of removing the tritium and adding new lithium to the lithium flow.

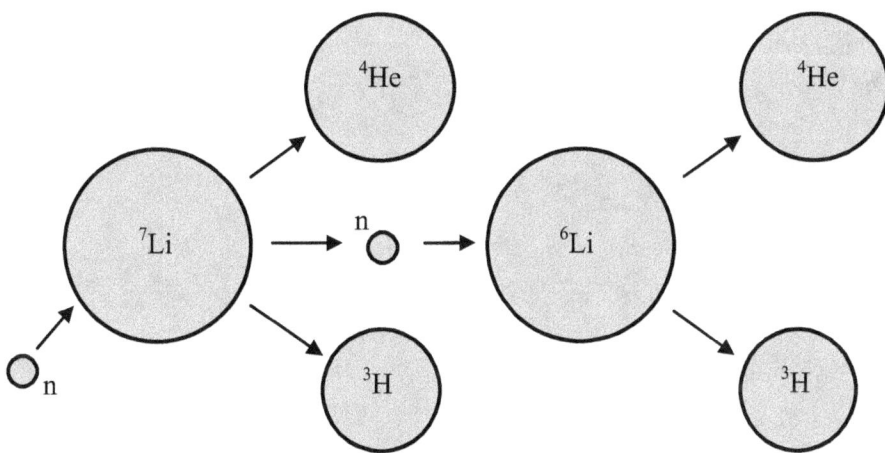

Fig. 5-20 Nuclear reaction for producing tritium.

Some, including the director of the NIF, Edward Moses, believe an intermediate solution is to build a hybrid reactor in which the neutrons in a laser-driven fusion reaction are used to increase the fission reaction in nuclear waste. Such a system would not only increase the amount of energy extracted from the uranium used in today's nuclear plants, which extract only about five percent of the energy, but would significantly reduce the amount of dangerously radioactive waste by increasing its decay into less radioactive isotopes. A system of this type is known as a laser inertial fusion engine (LIFE). Even though a LIFE system does overcome many of the problems associated with a purely fusion system, it still does not avoid the production of the very expensive pellets.

   Although the fusion reaction does not itself produce radiation, the escaping neutrons can trigger fissions and cause the surrounding vessels to become dangerously radioactive. Consequently, a fusion facility is not devoid of radioactivity and does produce nuclear waste. This waste, however, is not nearly as dangerous as spent fuel rods. All energy sources have drawbacks, including renewable sources (as is discussed in Chap. 6). Despite the fact that determining the optimal combination of sources to be used to satisfy our energy demands is not a realistic goal, it is imperative that we begin to change our past practices while we continually update our best guesses. If we wait until we believe we know the optimal combination, it will be too late. If the current trend of energy consumption is to be maintained, we face difficult decisions and time is growing short.

6

RENEWABLES

A renewable energy source is one that is virtually inexhaustible, such as heat from the Earth's core, or can be replenished within a relatively short time (one hundred years or so) as compared to fossil fuels that require millions of years to form. Included in these sources are biomass, geothermal, solar, hydro, wind and thermal differences in the ocean. Although they can be replenished, there is a limit to the rate at which they may be extracted. This physical limit may be very large, but the economic limit may be somewhat less. For example, taking advantage of wind power may be possible in many areas where the wind velocities may be too low to justify the cost of the wind turbines.

As our human ancestry progressed from hominids to Homo sapiens, it is impossible to determine exactly when humans first consciously began to utilize renewable sources of energy. The first such source that was used was undoubtedly the combustion of plant life to cook and provide warmth. The first evidence of using wind is the inscription of a sailboat on a vase that has been dated to 3500 BC, and windmills were known to exist as early as 500 A.D. and may have been used in China at least 500 years earlier. Waterwheels date back to the early Greeks and Romans. Both windmills and waterwheels were used to provide energy to grind grains and windmills were used to pump water. The first hydroelectric power plant opened on the Fox River in Wisconsin in 1882. Horizontal axis windmills appeared in Europe around 1300 AD and the first windmill for generating electricity was erected in 1887. Even though the ancient Greeks constructed buildings to capture heat and light from the sun, other direct uses of solar energy were very limited until the 20th century. It was not until 1953 that Gerald Pearson, Daryl Chapin and Calvin Fuller invented the first solar cell for directly converting solar radiation into electricity.

There is evidence that indigenous tribes used hot water from springs for cooking and bathing over 10,000 years ago. The first geothermal district heating system became operational in Idaho in 1892 and the first electric power plant that exploited geothermal heat was constructed by Prince Piero Ginori Conti in Italy in

1904. Energy obtained through the difference between the ocean surface and depth temperatures was known to be a possibility when electricity began to be used in the late 19th century, but was not given serious consideration until recently. Thermal differences are most likely to be used to generate electricity in coastal areas where only small systems are needed and there is a need to convert sea water into potable water, a possible byproduct of the generation process. Today, although renewable energy sources tend to require greater capital expenditures, they offer the best solution to our long term energy problems because they are continually replenished and cause the least amount of environmental damage.

In Fig. 1-6 it was seen that renewable sources accounted for only seven percent of the energy consumed in the United States in 2008, but it did not show how the seven percent was divided among these sources. Figure 6-1 gives this breakdown for the year 2009 in trillions of Btu and includes the energy needed to generate and distribute the electricity that was consumed. The total renewable energy used was 7,816 TBtu, with over half of it originating from biomass. Of this amount, 3,503 TBtu went toward generating electricity, but only 427 TBtu of the biomass energy was used for this purpose. Almost all wind power and hydro-power was for generating electricity.

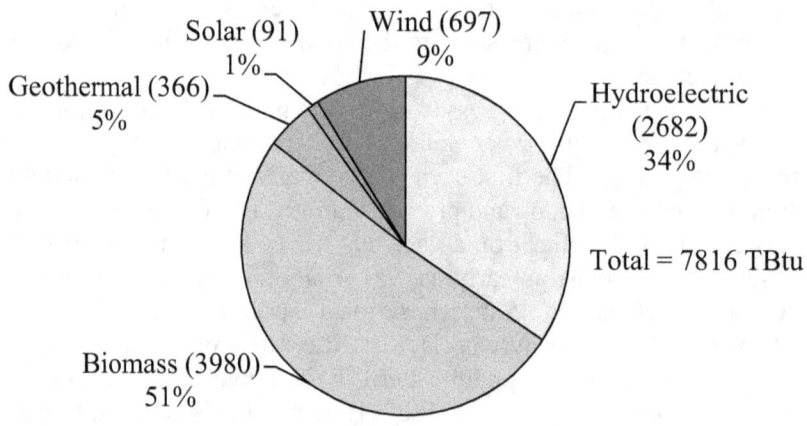

Fig. 6-1 2009 renewable energy consumption in the United States in TBtu [EIA].

Figure 6-2 illustrates the association between the renewable sources and the basic energy types and Figure 6-3 provides the relationships involved in the formation of biomass. Although these figures may seem a little daunting at first, they are central to the discussions in this chapter and it is worthwhile to examine them carefully. The basic energy types are represented by the large ovals and the arrows

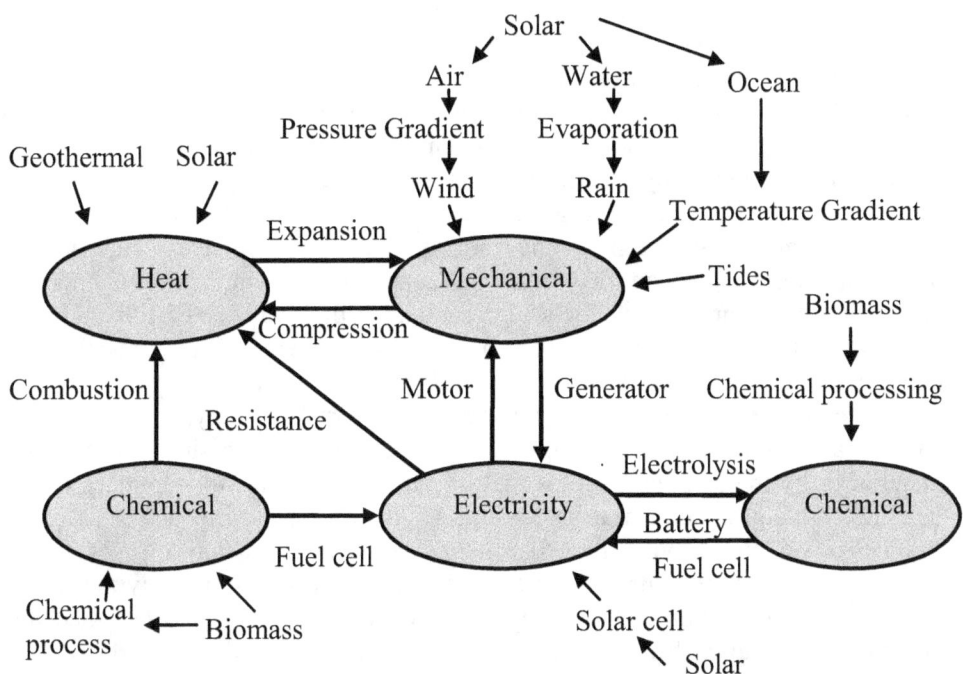

Fig. 6-2 Associations relating renewable sources with energy types.

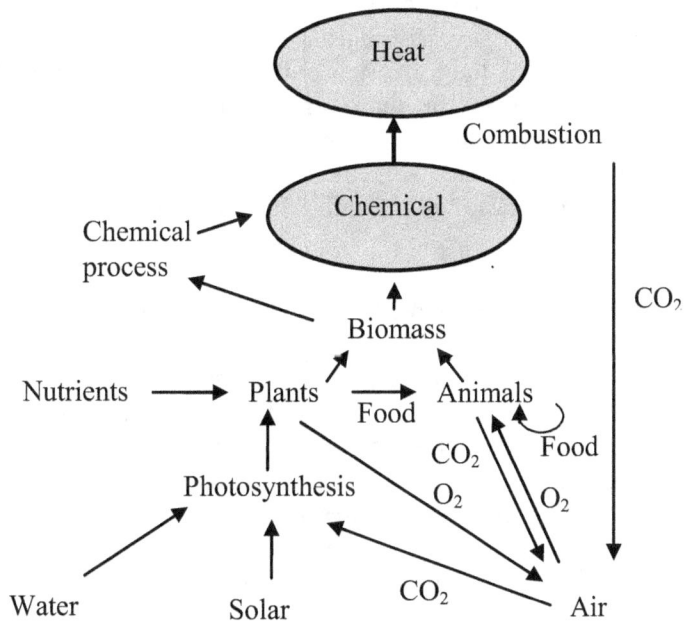

Fig. 6-3 Illustration of the formation of biomass.

between the ovals indicate the conversions between the energy types. The "Chemical" energy type is shown both to the right and left because there are two uses of this type. The oval on the left represents the fuels related to biomass. The oval on the right represents the temporary chemical energy storage to be converted into electricity, (i.e., storage in the chemicals in batteries and in hydrogen). Hydrogen is produced by electrolysis and converted into electricity by fuel cells. Batteries are used to store energy in chemical form by passing an electric current through them in one direction and later using them to produce a current in the other direction. Hydrogen may also be obtained from biomass by chemically processing the biomass. The fuels represented by the oval on the left include the biomass itself and those obtained from biomass by means of a chemical process, either natural or manmade.

Heat energy may be obtained from geothermal activity, from the sun, by passing an electric current through a resistance, by compressing a gas or other material or by the combustion of a fuel. It may be used directly for space or water heating, cooking, or industrial processing. Or it may be used to power machinery, turbines or vehicles through the expansion of gases such as steam or gases produced by the combustion of a fuel. The motion of wind or water, ultimately produced by the sun through evaporation or temperature differences, also provides mechanical energy. The radiation from the sun causes different areas to be heated differently, thus causing pressure differences that result in wind, and evaporation produces the rain that results in flowing water. Regardless of its source, some of this mechanical energy may power gas, wind, steam or hydro-turbines that, in turn, power electrical generators. Conversely, electricity may be used to drive electric motors.

As shown in Fig. 6-2, the formation of biomass begins with photosynthesis. Photosynthesis draws light from the sun, nutrients and water from the air and soil and carbon dioxide from the air and combines the carbon dioxide and water to produce plant cells. The primary ingredients of these cells are carbohydrates, $C_m(H_2O)_n$, principally cellulose and starch. The chemical equation for photosynthesis is

$$nCO_2 + nH_2O + \text{visible light} \rightarrow (CH_2O)_n + nO_2,$$

n being an integer. Although they have different chemical structures, both cellulose and starch are built from glucose molecules, $C_6H_{12}O_6$, that are bonded together to form $(C_6H_{12}O_6)_n$ molecules. Carbohydrates also include the sugars sucrose, fructose, lactose and maltose, which occur in various plants and food products. In addition to producing cellulose and starch, photosynthesis produces the oxygen, O_2, that exists in air. In addition, some biomass, particularly wood, contains lignin, a very complex molecule composed of carbon, hydrogen and oxygen that can be used as fuel or used to produce other fuels.

There is a symbiotic relationship between plants and animals in that O_2 is consumed and CO_2 is produced by animals through respiration, and CO_2 is consumed and O_2 is produced by plants through photosynthesis. Indeed, some scientists believe that the CO_2 in the primordial atmosphere made possible the existence of plants and plants produced the O_2 that made possible the existence of animals. Animals also survive by consuming the nutrients in plants and other animals. When the biomass from plants and animals becomes fuel (either directly or indirectly through a chemical process) and is burned, one of the products of the combustion is CO_2. If all CO_2 resulting from combustion and the respiration of animals is returned to plants through photosynthesis, then there is no net buildup of CO_2 in the atmosphere due to combustion and respiration, and the use of biomass as a fuel is often said to be *carbon neutral*. However, if the process of converting and transporting the biomass before it is used involves the use of fossil fuels, as is usually the case, then the consumption of biomass is not carbon neutral. When plant and animal life is relatively stable, as it had been for millions of years before the industrial revolution and the massive increase in human population, then the CO_2 in the atmosphere remained more or less in balance. But the industrial revolution and our heavy reliance on fossil fuels has unlocked much of the carbon formerly captured in the earth and caused a considerable increase in the CO_2 content of our atmosphere. By using biomass fuels and minimizing the amount of fossil fuels required for their usage, it is hoped that, along with the use of other renewable sources, the present CO_2 imbalance can be kept in check.

There are many definitions of *biomass* in use. The EIA definition is "Organic non-fossil material of biological origin constituting a renewable energy source," and the National Renewable Energy Laboratory (NREL) definition is simply "Organic material available on a renewable basis." These two definitions are essentially the same and comprise the definition assumed here.

Biomass is graded according to its moisture content. The *dry weight* of a material is its weight after it has been baked at 65 °C (149 °F) until its weight no longer changes. The moisture content of biomass varies drastically from sewage to dry grasses (e.g., wood milling waste has a moisture content of about 40 percent while construction waste consisting of wood that has already been processed has a moisture content of only 15 percent), but the average for biomass that is suitable for fuel is around 30 percent. The EIA's *Annual Energy Outlook 2002* estimates that 413 million dry short tons (372 tonnes), or 590 million wet short tons (531 tonnes), of biomass become available annually. But only a small portion of this amount is suitable for supplying useful energy at a feasible cost.

Much of the biomass must be transported and processed before it is used and this may require more energy than is usefully extracted from the biomass. Worse yet, this transportation and processing normally includes the consumption of fossil fuels which would tend to negate its main advantage of being carbon neutral. Most of the useful biomass is from organic wastes that are byproducts of a variety of processes such as the milling of timber, or from crops raised for the specific purpose of supplying energy.

Figure 6-4 gives the amounts and percentages of the biomass products consumed in the United States in 2008 in TBtu. The total amount is 3,414 TBtu. "Wood" accounts for 60 percent of the total and includes all wood waste and byproducts of wood processing as well as unprocessed wood. Wood provides almost two-thirds of the United States biomass energy. "Biodiesel" represents all materials used to produce biodiesel, primarily soy beans. Similarly, "Ethanol" represents the biomass for producing ethanol, which in the United States is mainly corn. "Landfill gas" is the methane that is naturally created in landfills as their contents decompose. "MSW" is municipal solid waste and is comprised of used wood products, paper, paper board, food, textiles, leather and yard trimmings. "Other" includes a variety of materials such as agricultural waste and byproducts, sludge waste and other biomass solids. In the past, the generation of hydrogen from biomass has been negligible but may become a factor in the future if fuel cells are used to power a significant number of vehicles.

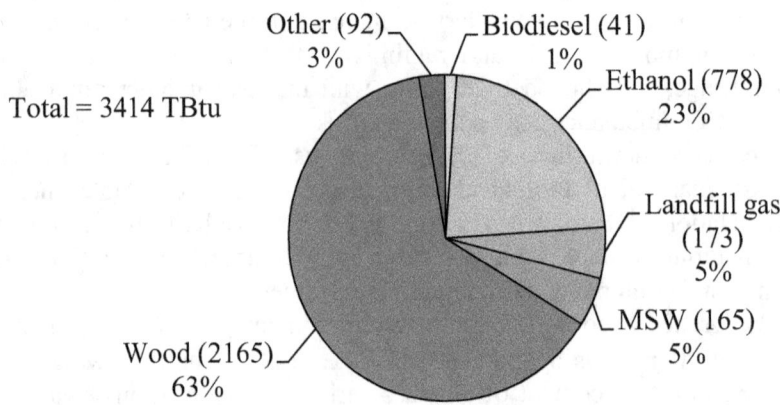

Fig. 6-4 2008 biomass energy consumed in the United States in TBtu [EIA].

Biomass primarily provides space, water and industrial heat, fuel for vehicles and fuel for generating electricity. How the consumption of biomass is divided among these three categories is shown in Fig. 6-5. Over half of our bio-energy goes toward providing heat directly to warm air or water or to provide heat for industrial processing. The application of biomass to supplying vehicle fuel more than doubled between 2003 and 2007, from 414 TBtu in 2003 to 923 TBtu in 2007. The use of biomass for direct heating and generating electricity remained almost constant during this period. The use of biomass for generating electricity almost always involves burning the biomass directly or gasifying the biomass within the electrical power plant before burning it. Over three-fourths of biomass is consumed with only a minimum amount of processing. To make the biomass suitable for burning, however, it must be dried and wood waste, paper and agricultural waste, such as straw and stalks, are normally pressed into pellets.

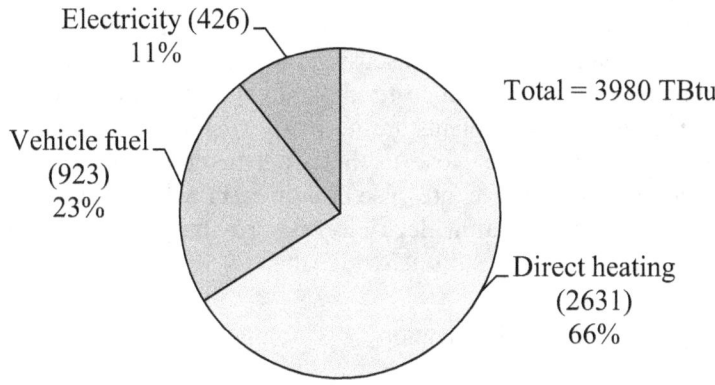

Fig. 6-5 2007 usage of biomass in the United States in TBtu [EIA].

There are two principal ways of employing biomass to produce electricity. One is to burn the biomass directly to heat a boiler and the other is to gasify the biomass and use it to drive a gas turbine. If biomass is burned directly, the plant is essentially the same as one that uses coal, except that the boiler and anti-pollution equipment must be redesigned. Biomass is often mixed with coal instead of being used by itself. A simplified diagram of a *co-fired* plant is the same as the nuclear plant depicted in Fig. 5-5, but with a boiler fed by a mixture of coal and biomass replacing the reactor. The biomass and coal may be mixed inside or outside the boiler. In most designs, the fuel is 1 to 8 percent biomass and 92 to 99 percent coal, but in some designs the percentage of biomass may be as much as 100 percent. One design allows 100 percent coal to be used during high demand periods and 100 percent biomass at other times. The capacities of existing co-fired plants range from 35 MW to 350 MW.

The main alternative to a co-fired power plant is a biogas integrated gasification combined-cycle (BIGCC) plant. In a BIGCC design, the biomass is gasified and the gas is then combusted to drive a gas turbine that operates under the same principle as a turbojet engine. The gasification portion of the design is shown in Fig. 6-6. Gasification is accomplished by thermal oxidation in an oxygen deprived atmosphere so that combustion is not complete. The combustion results in a combustible mixture of methane, carbon monoxide and hydrogen. In addition to the gasification unit, there are units for processing and drying the biomass and cooling and filtering the gas. The gasification process requires heat and the output gas must be cooled. The cooling unit acts as a heat exchanger and, instead of this heat energy being wasted, it is used along with the heat output by the gas turbine to create steam for driving a separate steam turbine. Both the gasification and filtering processes produce ash that must be disposed of as waste. Systems for which the heat output from a gas turbine is used to power a steam turbine are called *combined-cycle systems*. Combined-cycle designs are discussed more thoroughly in Chap. 7. BIGCC plants tend to be much smaller than co-fired plants, which could be as large as coal-fired plants. However, co-fired plants are normally smaller than those that use coal alone because they need to be close to their fuel in order to reduce the cost of transporting the biomass. There are also facilities in which the heat energy in biomass is used both for generating electricity and providing heat for other purposes. This is called cogeneration and is discussed in Chap. 7.

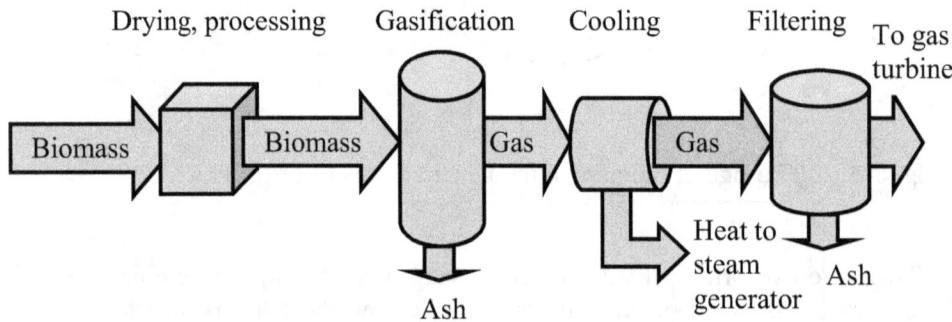

Fig. 6-6 Gasification equipment for a BIGCC electrical power plant.

The North American Electric Reliability Council (NERC), which studies the uses of biomass for generating electricity, divides biomass into four categories according to their origins: agricultural residues, energy crops, forestry residues and urban waste/mill residues. *Agricultural residues* are the leftovers from harvesting commodity crops and include chaff, straw and stalks of various kinds, and *energy crops* are the biomass grown with the intent of using it to produce energy. Energy crops are grown on currently cultivated cropland, idled cropland or pastureland. For generating electric power, these crops primarily include switchgrass, hybrid poplar and willow. *Forestry residues* are materials that remain after a forest has been harvested for timber and are composed of logging waste, dead wood and vegetation that is too small to be made into lumber. *Urban waste/mill residues* are waste biomass such as lumber mill residues and wastes that would otherwise be put into landfills. It includes a wide variety of organic materials.

The NERC rates the maximum availability of biomass by the amount in millions of Btu (MBtu) that could be affordably supplied at a given cost. NERC predicts that by 2020, when biomass electricity generation is expected to reach its maximum capacity, the biomass available at five dollars per MBtu will have an energy content of 7,100 TBtu. Some of the factors affecting the cost are the distance it must be transported, the inaccessibility of the location, the energy density per unit area, the volumetric energy density of the biomass and the percent of the biomass that can be removed from a site. The inaccessibility of the location is determined by the presence and types of roads and the roughness of the terrain. The energy density per unit area and volumetric energy density determine the ease and cost with which it can be collected and transported. Particularly important to the collection of waste forest vegetation, such as dead wood, and agricultural residues is the percent of the materials that can be removed. Some of the biomass, approximately 60 to 80 percent, must remain in order to maintain the soil's nutrients for future growth and to prevent erosion.

The predicted biomass available in the year 2020 in MBtu as a function of cost for the four NERC categories is given in Fig. 6-7. Because these estimates were made in 2001, the availability at the dollar value in the year 2000 was used. In 2020, inflation may require more money to be spent in order to attain the same levels of availability. Note that all four categories reach their maximums if the economics are such that there is a willingness by the electric power industry to spend the equivalent of six (year 2000) dollars per MBtu. Because of the considerable losses in processing the biomass, the EIA predicts that at five (year 2000) dollars per MBtu only about 98 billion kilowatt-hours (GkWh), or 334 TBtu, will be generated per year by 2020. This amount will be only 1.75 percent of the expected United States

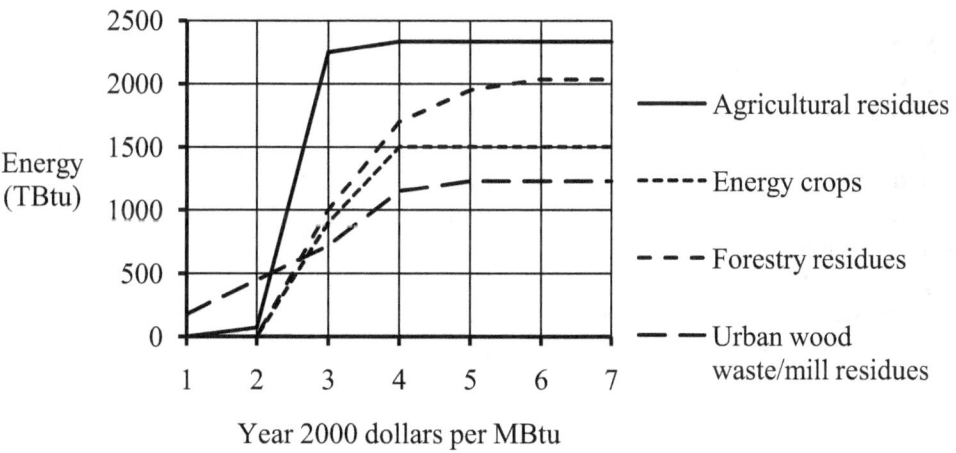

Fig. 6-7 2020 biomass energy available in the United States [EIA].

consumption of 5,476 GkWh (18,684 TBtu). For comparison, the total 2007 electricity consumption was 3748 GkWh (12,788 TBtu) and biomass contributed less than one percent of this amount. These predictions are highly speculative, but, by any measure, the percentage of electricity produced by biomass is likely to never exceed three percent. The cost of coal per MBtu is less than one (year 2000) dollar and this is not likely to change significantly in the foreseeable future.

The energy content by weight of several biomass materials is given in Table 6-1. Also given in the table are the energy contents of coal and petroleum, and, because plastics are in municipal waste, they are included even though they are not biomass. Several of the materials have at least half the energy of coal and all of them contain at least one-fourth the energy of coal. But their volumetric energy densities are substantially less than that of coal and this makes them more expensive to transport than coal. Adding to the expense of biomass fuels is the fact the agricultural and forestry residues are distributed over wide areas, but coal, by comparison, is concentrated in seams. When one considers that not all biomass that is harvested is suitable for use as fuel and energy is lost in the preparation and transportation of biomass, its overall efficiency in generating electricity is somewhat less than that of coal. The main advantages of biomass are that energy is extracted from matter that would otherwise be wasted and much of the carbon released when it is burned is recaptured by new plant growth (i.e., it may be considered carbon neutral). Municipalities like to burn as much waste as possible because it reduces the

Table 6-1 Average energy densities of biomass by weight.

Biomass type	Energy (MBtu/short ton)
Agricultural residues	8.2
Algae	18.0
Switchgrass	15.8
Poplar	14.9
Wood waste	10.0
Wood	16.4
Yard trimmings	6.0
Paper products	12.6
Food waste	5.2
Rubber, textiles, leather	18.0
Plastics	24.5
Coal	21.0
Petroleum	38.1

amount of waste they must put in landfills. Burning municipal waste reduces its volume by 80 to 90 percent. Landfill space has become increasingly difficult to find and, because no one wants to live near a landfill, municipalities often must truck the waste long distances. This increases the cost in both dollars and energy and the trucks normally burn fossil fuels that emit carbon dioxide and other pollutants. Each American produces about 1.5 cubic meters (2 cubic yards) of compacted waste per year. Also, landfill gas must be collected and disposed of (usually burned) even if it is not used because it is about 60 percent methane. Landfill gas has a noxious odor and methane is a major greenhouse gas.

But there are some disadvantages to burning waste as well. Regardless of whether it is solid waste or landfill gas that is burned to generate electricity, there is the danger of toxic pollutants being introduced into the air or water table. Municipal waste contains cleaning fluids, batteries, plastics and metals that produce toxins in both the gases emitted and ash residue. The ash is usually very fine and is much more likely to leach into ground water if it is simply put into a landfill. As a result, the most dangerous materials must be separated out and generating plants that burn municipal waste must include expensive anti-pollution systems and burn the waste at temperatures over 1,000°C (1,800°F). In addition, generating plants that use municipal waste for fuel produce more ash than coal-fired plants and the ash removal is more of a problem. For these reasons, it costs more to generate power from waste than from uranium or coal, even though coal generating plants must also include anti-pollution equipment. In the United States, it is the Environmental Protection Agency (EPA) that is charged with monitoring the gas and ash residues of generating plants that burn waste materials.

Humanitarians and environmentalists have several objections to using biomass as a fuel. If land is used for energy crops, then it is not being used for food production and this raises the cost of food for people and feed for animals. Environment groups point out that even though plants that are burned and then replaced by the same plants are carbon neutral, they are still a source of pollution. All plants contain sulfur and nitrogen oxides that are released when they are burned and the net result is that sulfur and nitrogen compounds are transferred from the soil to the air. Also, much of the municipal waste that is burned as biomass are plastics, which are neither renewable nor biomass. Agricultural residues and energy crops contain some herbicides and pesticides that contaminate the soil and water. How valid the objections of the environmentalists are is largely unknown. Many of these substances are burned whether the heat energy created is put to use or not and, if a generating plant uses the heat, then there is at least some gain. If the way in which biomass is produced, separated and burned is done properly, the resulting pollution could be acceptable. In any case, the percent of electricity generated from biomass will probably never exceed three percent.

The use of biomass to produce liquid or gaseous fuels for vehicles is another matter. Crops raised for this purpose could be almost carbon neutral, although they would introduce some pollutants into the atmosphere, and their vegetable oils would mostly be wasted. The primary fuels produced are ethanol, methanol, biodiesel, methane and hydrogen and the primary feedstocks are corn, switchgrass, sugar cane, sugar beets, seaweed, algae, soybeans, animal wastes, food wastes, vegetable oil and animal fat. Even kudzu, a nuisance plant that grows wild in the southeastern United States, is being researched as a possible source of biomass fuel. Currently, most attention is given to the production of ethanol. Ethanol is already an important fuel for automobiles in Brazil where some automobile engines are modified to run on fuel that is 100 percent ethanol. However, ethanol is just one type of alcohol that can also be used as fuel. *Alcohol* is any compound with a chemical formula of the form $C_nH_{2n+1}OH$. Alcohols may be obtained from hydrocarbons by adding an oxygen atom to form an OH group. The most common alcohols considered for use as fuels are methanol, CH_3OH, ethanol, C_2H_5OH, propanol, C_3H_7OH, and butanol, C_4H_9OH.

Conventional ethanol and cellulosic ethanol are the same chemically, but the different names indicate the feedstocks from which they are derived and the processes used to produce them. In both cases, the immediate goal is to extract sugars from plant matter. Conventional ethanol is primarily obtained from corn or sugar cane. There are four steps in the process of producing conventional ethanol: dry or wet milling, fermentation, distillation and dehydration. Dry milling involves liquefying the feedstock and heating it in water and enzymes to convert the liquefied starch to sugars. Wet milling separates the fiber, oils and protein from the starch. In either case the results are then fermented. The basic chemical equation for fermentation is

$$C_6H_{12}O_6 \rightarrow 2C_2H_6O + 2CO_2,$$

which converts a sugar molecule into two ethanol molecules and two carbon dioxide molecules. Distillation is used to eliminate 96 percent of the water and dehydration removes the remainder of the water.

Cellulosic biomass is from switchgrass, stalks, straws, wood and paper wastes, and contains hemicelluloses and lignin in addition to cellulose. In cellulosic biomass the sugars are in complex carbohydrates and either acid or enzymatic hydrolysis is used to extract the sugar needed for fermentation. *Hydrolysis* is the decomposition of a compound into other compounds by a reaction involving water and, therefore, requires significant amounts of water if done on a large scale. Heat is needed in this process, but can be provided by the lignin residues that make up the bulk of the dry mass produced by the process. Because ethanol is the same regardless of the feedstock from which it is obtained, the chemical equation for burning either conventional or cellulosic ethanol is

$$C_2H_6O + 3O_2 \rightarrow 2CO_2 + 3H_2O,$$

which indicates an ethanol molecule is converted into two carbon dioxide molecules and three water molecules.

Another method for producing ethanol from cellulosic biomass uses ammonia and is described by George W. Huber and Bruce E. Dale in the July, 2009, issue of *Scientific American*. The feedstock is first mixed with ammonia and heated under pressure to 100 °C (212 °F). This separates the cellulosic material from the lignin. When the pressure is released, the ammonia evaporates and is recycled. Enzymes are added to the residue to produce sugars that are allowed to ferment into a mixture of ethanol and water and distillation separates out the ethanol. The advantages of this process are that 90 percent of the cellulosic residue is turned into sugar and it is relatively inexpensive.

Other higher temperature processes produce syngas or a form of crude oil, called *biocrude*. Temperatures in the range 300 °C to 600 °C (636 °F to 1272 °F) produce biocrude that can be refined into gasoline, diesel or jet fuel. Processes that use temperatures above 700 °C (1484 °F) and are essentially the same as those for converting coal into syngas can also turn biomass into syngas. The syngas can then be refined into liquid fuel. A study by the United States Departments of Energy and Agriculture estimates that at least 1.3 billion dry tons of biomass could be grown on land that is not used for food crops and could be converted into more that 100 billion barrels of fuel each year. Fuels obtained from cellulosic biomass are sometimes referred to as *grassoline*. The primary disadvantage of all cellulosic feedstocks is that they are bulky and would tend to be grown in diffuse areas, thus causing considerable transportation costs to bring them to processing plants large enough to be economically viable. In addition, they are expensive to harvest.

Fuel that contains a percentage of a biofuel is labeled using the first letter of the biofuel followed by the percentage of the biofuel in the fuel, e.g., fuel that is 85 percent ethanol and 15 percent gasoline is referred to as E85 and fuel that is 20 percent biodiesel and 80 percent diesel is B20. The percentage of ethanol that is best depends on the engine characteristics, the ambient temperature and acceptable water content. The lower the temperature, the more difficult it is to start an engine with a high percentage of ethanol. The United States and Europe, except for Scandinavia, recommend no more than E85 and Sweden recommends no more than E75. Tropical countries can use E100 provided that the engines are designed for 100 percent ethanol. In very cold climates, engine heaters are recommended just as they are with gasoline engines in the northern states. High ethanol content can cause corrosion of ferrous parts and deteriorate some non-metal parts. Also, it loosens dirt, rust and slug from tanks that have been filled with gasoline or other liquids. The corrosiveness of ethanol makes it difficult to transport via pipelines and delivery by truck or railway increases the cost of delivery. Water in a fuel tends to cause engine stalling, but the higher the percentage of ethanol the less likely stalling will occur. Stalling is of no concern when the fuel is more than 70 percent ethanol.

Because the volumetric energy content of ethanol is approximately 80,000 Btu per gallon versus 125,000 Btu per gallon of gasoline, if an automobile running on E100 fuel has the same efficiency as a gasoline powered automobile, then its miles per gallon will only be 64 percent as much as the gasoline automobile.

However, the higher the percentage of ethanol, the higher an engine's combustion ratio can be and E100 automobiles can be designed to have miles per gallon ratings almost as high as gasoline automobiles. Ethanol has a high octane rating and engines have been designed with dual injection systems that allow anywhere from E0 to E100 to be used. E10, also referred as *gasohol*, can be used in ordinary gasoline engines. Because it is less polluting than gasoline alone, the inclusion of at least ten percent ethanol in some cities and states is mandated, especially during the winter. Brazil requires automobiles to use at least an E20 blend. E10 is sometimes used to replace the MTBE (methyl tertiary butyl ether) additive as an anti-pollutant because of complaints that storage leaks of the toxic MTBE contaminate ground water.

In 2007, the United States and Brazil produced 88 percent of the ethanol used as fuel. The United States produced 154 million barrels (6.47 billion gallons), or 89 million barrels of oil equivalent (boe), of ethanol and imported another 10.4 million barrels (6 Mboe) to meet its demand of 164 million barrels (95 Mboe). The world as a whole produced 312 million barrels (181 Mboe). The United States production was almost entirely from corn and Brazil's was almost entirely from sugar cane. In 2007, the United States consumption was more than three times the 50 million barrels (28 Mboe) consumed in 2001 and will most probably continue its steep climb in the years to come. An article by David Rotman entitled "The Price of Biofuels" in a 2009 special report in the *Technology Review* states that 93 million acres, 145,000 mi^2 (376,000 km^2), were dedicated to raising corn and approximately one fifth of the corn was used to produce ethanol. Thus, the area used to grow corn for producing ethanol is roughly that of half the state of Iowa. This article also indicates that "even proponents of corn ethanol say that its production levels cannot go much higher than around 15 billion gallons a year, which falls far short of Bush's goal." (President George W. Bush had proposed a goal of 35 billion gallons per year by 2017.)

There is controversy over just how much energy is gained from processing corn into ethanol. Some have claimed that it takes more energy to produce corn ethanol than is in the ethanol and some claim there is a 100 percent energy gain. However, most believe that the energy in ethanol is between 1.2 and 1.6 times the energy needed to raise the corn, transport it and process it. The variation is due to the differing assumptions made regarding what should be counted as input and output energies. Those that report less energy is output than is input often discount dual usage, e.g., some useful byproducts are ignored. Sugar cane produces about eight times as much energy as is required to grow, transport and process it. Sugar cane offers the additional advantage of having a residue, called *bagasse*, that can provide the heat needed to process the cane into ethanol, thus reducing the input energy. Unfortunately, the greatest energy gains are gotten by growing sugar cane in the tropics. There are only a few places in the United States where sugar cane will grow. The cellulose in cornstalks and bagasse could be used for producing ethanol but is more likely to be used to provide heat directly or through gasification.

Other candidates for producing ethanol are switchgrass, poplar and sweet sorghum. All three could be grown in the United States, but both switchgrass and poplar depend on the development of efficient cellulose technology in order to be processed into cellulosic ethanol economically. Table 6.2 gives the approximate yields per acre and the energy gains of the feed stocks being considered. An *energy gain* is the output energy divided by the total energy needed for the production of the output energy source. For corn ethanol, one Btu of energy is typically needed to produce ethanol containing between 1.2 and 1.6 Btu of energy. Figures such as these vary considerably with the information source. The yields per acre depend on where the crops are grown and the exact processing method and transport distances, e.g., sugar cane grows much better in Brazil than in the continental United States.

Table 6-2 Crop yields and input to output energy ratios for ethanol sources.

Ethanol source	Yield in gal/acre	Yield in MBtu/acre	Energy gain
Corn	330-430	26.4-34.4	1.2-1.6
Sugar cane	650-870	52.0-69.6	8-10
Cellulose*	400-1000	32.0-80.0	>2

*Experimental data only.

From a fundamental ethics perspective there has been a serious debate over the use of crop land to produce energy when so many people in the world do not have enough to eat. Also, there is a question as to whether tropical areas, Brazil in particular, should be encouraged to grow large amounts of sugar cane for fear that it would exacerbate the problem of deforestation. The argument is that rain forests are known to be the best absorbers of carbon dioxide and they should not be diminished in order to make room for sugar cane fields. Presently, the United States has a 54 cent per gallon tariff on imported ethanol, but this tariff is more to protect the American agricultural industry than the rain forests. The basic question is: Where is the land for producing ethanol crops to come from? Regardless of how much land is chosen to be used for ethanol crops, it is logical that the crop that produces the most net energy per acre should be grown on the available land. At the moment, that crop is sugar cane.

The other alcohols that can be used as fuel or fuel blends are methanol, butanol and propanol. Methanol can be produced from the renewable sources wood, wood waste, seaweed or organic garbage and has an energy content of 60,000 Btu/gal. Historically, wood has been methanol's primary feedstock and methanol is still sometimes referred to as wood alcohol. It has a higher octane rating than gasoline, but, as with ethanol, for equal sized tanks it has a shorter range than gasoline when used in an automobile. When produced from biomass it yields between 100 and 200 gallons per short ton depending on the feedstock. Table 6-3 provides the energy efficiencies with which methanol may be produced from the various biomass feedstocks. This table is from the website *www.forest.com/doc*, which obtained the data from the sources cited in the website. Methanol is poisonous and corrosive to metals, particularly aluminum. Engines can be designed to accommodate blends from M0 to M100 and in the 1990s up to 22,000 vehicles were running on methanol in the United States, but their number began to decline after 1997 as ethanol blends began to be favored. On the other hand, methanol is an important contender to power fuel cell vehicles because, unlike hydrogen, its natural state is liquid and this makes it relatively easy to store and transport. Methanol and ethanol share many of the same advantages and disadvantages. As liquids, both have much higher volumetric energy densities than compressed gaseous fuels.

Table 6-3 Energy efficiencies associated with
         producing methanol from biomass.

Feedstock	Percent efficiency
Pulp mill black liquor	65
Solid wastes	60
Forest residues	55
Wood residues	44-51
Liquid biofuels	65-75
Soybean-cake	45

Butanol or propanol blends could be used, but have not gotten as much attention as either methanol or ethanol. Butanol has an energy density of 91,000 Btu/gal and is primarily obtained from vegetable oils, animal fat and algae using the centia process, a three-step process that involves hydrolytic conversion (water separation), decarboxylation (splitting CO_2 from molecules that contain carboxyl groups) and reforming (changing a molecule's form). Propanol is considered as an additive or for low percentage blends only. All alcohols could be obtained from petroleum, but then they would not be considered renewable. In any case, alcohol blends with fossil fuel products are only partially renewable.

Most diesel fuel is from refined petroleum or synthesized from coal or natural gas and is simply referred to as diesel or petrodiesel. But, there is another form of diesel that is derived from vegetable or animal matter and is called *biodiesel*. Biodiesel is primarily obtained from soybeans, but may be gotten from other vegetable oils, algae, animal fat, cooking oil or animal waste. Other than soybean oil, the current most common feedstock is *yellow grease*, a greasy substance obtained from cooking oil or tallow from rendering plants. In order to use these oils in modern diesel engines they must be processed to reduce their viscosity. The production of biodiesel consists of converting the feedstock into fatty acids and then converting these acids into *alkyl ester* (the technical name for biodiesel), a chemical formed from an organic acid and an alcohol. The most common method is to react a vegetable oil or animal fat with methanol. Ethanol could be used, but methanol is less expensive. Roughly, ten parts of feedstock and one part of alcohol produce ten parts of biodiesel and one part of glycerol. The ratio of soybean oil to the amount of biodiesel produced is about 7.5 to 1 and for yellow grease it is about 8 to 1. Soybeans are approximately 20 percent oil, but the exact percentage depends on moisture content.

A 2005 report by the University of Wisconsin and made available through the EIA estimated that 85 percent of the variable cost of production of biodiesel is in the feedstock, five percent is in the cost of the methanol and catalyst and ten percent is in other costs. This report gave the total cost of petrodiesel to be $1.15 per gallon (year 2005 dollars), the cost of biodiesel from soybean oil to be $2.86 and the cost of biodiesel from yellow grease to be $1.74. Clearly, biodiesel will not be able to compete with petrodiesel on price alone within the foreseeable future unless there are subsidies.

A twenty year study of the use of algae as a biodiesel feedstock began in 1978 and was performed under the Aquatic Species Program and reported by NREL (report no. NREL/TP-580-24190). Although there are three major categories of algae (macroalgae, or seaweed, microalgae and emergents, plants that grow in bogs), the report concentrated on microalgae. The program's interest was mainly in land-based facilities that could be conveniently located, not in seaweed that is harvested from the ocean as an energy crop. In particular, it focused on diatoms, phytoplankton that grows in the ocean and brackish water, and green algae that can grow in fresh water. It found that microalgae produce up to 30 times more oil per square meter than other terrestrial oilseed crops and, although grown in water, it requires less water. Although the study concentrated on growing the algae in open ponds, other governments have investigated using closed aquatic systems. The growth rate of algae depends on enzyme activity which, in turn, depends on temperature. Also, the growth rate is affected by atmospheric contaminants including bacteria and unsuited strains of algae, factors that are better controlled in an enclosed system. Therefore, countries with cooler climates and a greater chance of contamination have a greater interest in closed systems.

A representative open pond system is illustrated in Fig. 6-8. In addition to sunlight and water, algae need nutrients, primarily nitrogen, phosphorous and potassium, and a large quantity of carbon dioxide. These ingredients are added as the paddlewheel circulates the aqueous solution around the tank. As the algae are grown, they are siphoned off and processed into biodiesel. Because as much as 90 percent of the carbon dioxide is absorbed by the algae, algae ponds can be a means of sequestering unwanted carbon dioxide (see Chap. 10).

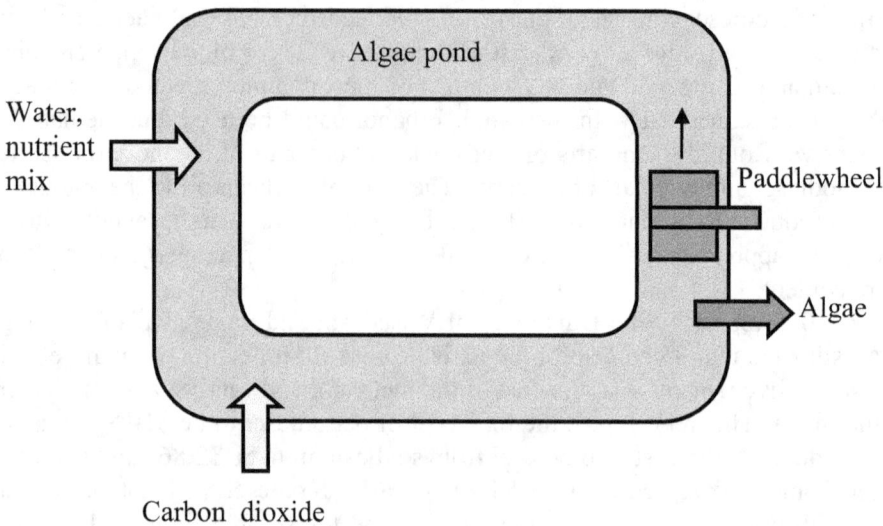

Fig. 6-8 An open pond for growing and extracting algae.

The natural oil produced by oilseed crops is in the form of triacylglycerols, commonly called *lipids*. For some species of algae up to 60 percent of the weight is due to lipids. The weight of lipids is approximately 0.9 kilograms per liter (7.5 lb/gal), their mass energy density is 37,600 Btu/kg and their volumetric energy density is 120,000 Btu/gal. In some of the Aquatic Species Program's experiments the growth rate was 50 grams of algae per square meter per day (the program's target growth rate), but the average for their experiments in Roswell, New Mexico, was closer to ten grams of algae per square meter per day. Even at 10 g/day and a lipid content of 50 percent, the energy production would be about 68,700 Btu per square meter per year. The study concluded that there is enough appropriately located land in the United States with a suitable climate to produce lipids containing between 2,000 and 7,000 TBtu per year, the energy equivalent of 345 to 1,200 million barrels of oil (345 to 1,200 Mboe). In recent years, the United States energy consumption has averaged about 100,000 TBtu. The study's conclusion implies that algae could supply up to seven percent of the United States energy needs in the form of biodiesel fuel and do so while temporarily absorbing unwanted carbon dioxide.

(However, the carbon dioxide would be released when the biodiesel is burned as fuel.)

Biodiesel may be blended with any percentage of petrodiesel. Biodiesel blends, even B100, do not require any engine or storage tank modifications and is generally cleaner burning than petrodiesel. Because it is a solvent, B100 does cause some deterioration of rubber and some other components in older vehicles. Biodiesel emits fewer particulates, hydrocarbons and carbon monoxide than petrodiesel and emits no sulfur, but does emit about the same amounts of carbon dioxide as diesel and a small increase in nitrogen oxides. As compared to petrodiesel, B20 has a two percent increase in nitrogen oxides, a ten percent decrease in particulates, an eleven percent decrease in carbon monoxide, a 66 percent decrease in hydrocarbons and a slight increase in efficiency. Sulfur is removed from petrodiesel during the refining process, but even low-level diesel blends reduce the need for this removal. Biodiesel odor is less noxious and the smoke is much less visible than that of petrodiesel.

Because few service stations provide biodiesel blends above a few percent, most biodiesel is used by government or private fleets that have their own storage facilities. Unless soybeans or other crops are grown for the sole purpose of providing biodiesel, the supply of biodiesel feedstocks are limited to the byproducts of soybean processing, cooking oils and animal matter. It is unlikely that high percentage blends will ever become common. If a significant amount of land is diverted to produce biodiesel, this land usage may face the same objections as those raised regarding other energy crops. However, algae can be grown in arid locations and in brackish water. No matter which feedstock is used, just how much biodiesel can be produced per acre of land depends on numerous factors and is quite controversial. Table 6.4 gives some typical ranges of values. If ten percent of the 1,325 million barrels of diesel supplied in 2009 in the United States had been replaced with biodiesel from soy beans and 50 gallons of biodiesel had been produced per acre, then 450,000 square kilometers (174,000 square miles–an area twice that of Kansas) would have been needed to grow the soybeans. If palm oil had been used, 34,600 square kilometers (13,400 square miles—roughly two percent of the area of Indonesia where much of the palm oil is produced) would have been needed.

Table 6-4 Biodiesel energy crop yields.

Feedstock	Gallon/acre
Algae	2,000-15,000
Palm oil	600-700
Rapeseed	200-300
Soybeans	40-60

As indicated in Chaps. 2 and 3, methane is the main constituent of natural gas, the fossil fuel that is extracted from either natural gas or oil wells or naturally emitted from the ground. But methane is also emitted from decaying organic matter. Landfill gas is from the organic matter in landfills and is collected by sinking pipes into the landfill. It contains 35 to 60 percent methane, 35 to 55 percent carbon dioxide, 1 to 10 percent water, 0 to 20 percent nitrogen and small amounts of other compounds. Figure 6-9 shows a typical landfill that is designed so that the gas can be collected and output to a natural gas pipeline system. Landfill gas may be burned directly for its heat content or undergo processing to remove its methane and the methane may then be fed into a natural gas pipeline system. It could also be used to produce methanol or biodiesel fuel. Even if it is burned directly for its heat content, it is normally processed to remove the carbon dioxide, nitrogen oxides and other contaminants such as mercury that may have been absorbed from the landfill. Its heat may be used to generate electricity or provide space heating or heat for industrial processing.

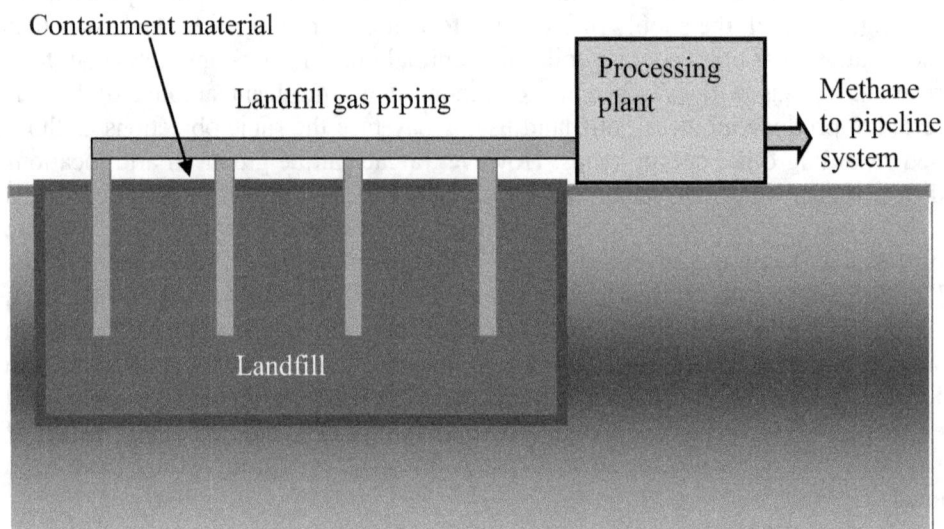

Fig. 6-9 Typical landfill gas extraction and processing to obtain methane.

The decay of organic matter to produce biogas is a three step process involving forms of bacteria. Proteins are broken down by fermentative bacteria that secret an enzyme, called protease, and, through hydrolysis, separate the proteins into amino acids. These amino acids are combined with oxygen to produce an acetic acid (the principal acid in vinegar), hydrogen, nitrogen and carbon dioxide, as indicated by the following chemical equation:

$$2C_3H_7NO_3 + O_2 \rightarrow 2HC_2H_3O_2 + 3H_2 + N_2 + 2CO_2.$$

With the help of another bacterium, called *methanogens*, the acetic acid, hydrogen, and carbon dioxide produce methane, water and carbon dioxide according to the equation

$$HC_2H_3O_2 + 4H_2 + CO_2 \rightarrow 2CH_4 + 2H_2O + CO_2.$$

This process not only creates landfill gas, but can also create methane from seaweed or crops grown for the purpose of producing methane. Methane obtained by such means is renewable. There is an incredible amount of the required bacteria that performs this overall process on decaying animals and vegetation naturally (even within the digestive tracts of living animals). The process is an important part of the methane and carbon cycles that determine the amount of methane and carbon dioxide in our atmosphere. Worldwide, the methanogens in nature emit between 530 and 790 million short tons of methane per year. To put this amount into perspective, this much methane has an energy content between 23 and 35 percent of the total energy consumption of the United States.

Hydrogen can be produced from fossil fuels, water or biomass in four principal ways: steam reforming, thermo-chemical processes, electrolysis and biological processes. The main interest in hydrogen is its use in powering fuel cells in vehicles. Although hydrogen is considered a renewable fuel, hydrogen produced by the steam reforming of coal or natural gas is not renewable because the fossil fuel feedstock is not renewable. Hydrogen from coal or natural gas is an alternative fuel, however, and deserves our brief attention. Hydrogen is a product of steam reacting with the methane in natural gas according to the equation

$$CH_4 + H_2O \rightarrow CO + 3H_2.$$

Additional hydrogen is obtained by the shift reaction,

$$CO + H_2O \rightarrow CO_2 + H_2.$$

Thermo-chemical processes for producing hydrogen are those in which heat is applied to a solution of water and a compound such as calcium bromine. Once the solution is made available, a thermo-chemical process can be quite efficient because only a relatively small amount of heat is required. But, to date, no commercial thermo-chemical production facilities have been built.

Electrolysis is any chemical reaction caused by the flow of an electrical current, (for a more complete discussion of electricity, see Appendix A). A device that performs electrolysis is called an *electrolyzer*. The basic principle for the electrolysis of water into hydrogen and oxygen is demonstrated by the electrolyzer depicted in Fig. 6-10. This electrolyzer is similar to a battery in that it consists of a container filled with an *electrolyte*, i.e., ionized matter capable of conducting an electrical current, and has a positive electrode, an *anode*, and a negative electrode, a *cathode*. However, it does not output an electrical current, but instead, is connected to an electrical source that causes a current to flow through the electrolyzer. The electrical source causes each pair of water molecules at the anode to be split into an oxygen molecule, four positive hydrogen ions and four electrons. The source then

$$2H_2O \rightarrow O_2 + 4H^+ + 4e^-.$$

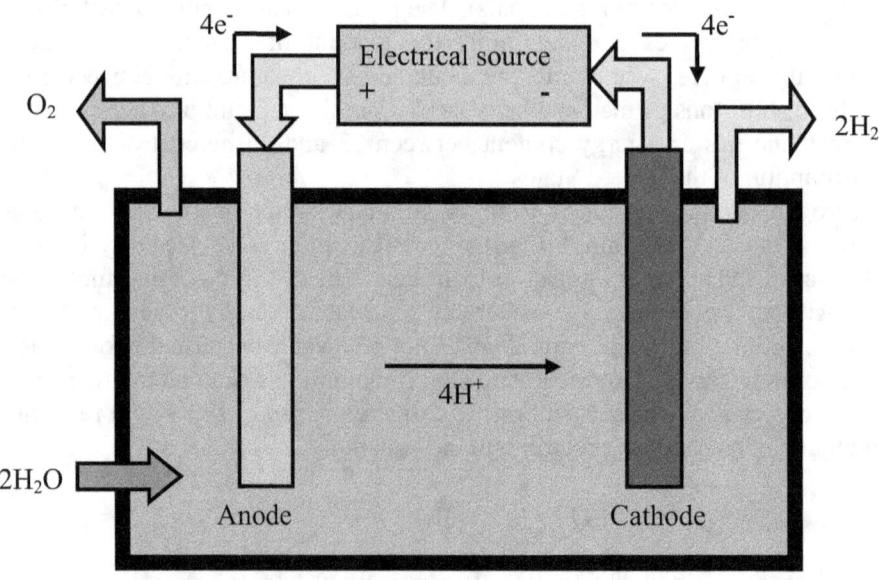

Fig. 6-10 A basic electrolyzer.

transfers the electrons from the anode to the cathode and the four hydrogen ions through the electrolyte to the cathode. For each four electrons and four hydrogen ions arriving at the cathode, two H_2 molecules are formed,

$$4H^+ + 4e^- \rightarrow 2H_2.$$

This cycle is repeated over and over as water is added. The net result is that the energy from the electrical source converts the water into hydrogen gas and oxygen gas. Although the figure implies pure water is used, the water is mixed with a compound such as sodium sulfate, Na_2SO_4, to provide conductivity. The principal types of commercially available hydrogen generators are *alkaline electrolyzers* and *proton exchange membranes* (*PEMs*). An alkaline electrolyzer is similar to the one shown in Fig. 6.10, but has a membrane separating the anode and cathode and uses potassium hydroxide, KOH, as its electrolyte. With a PEM, the electrolyte is a solid material.

Fig. 6-11 shows an overall alkaline electrolyzer system for generating hydrogen and preparing it for storage or transporting. First, the water must be highly purified and this purification requires heat. This heat could be provided in combination with the generation of electricity by using the output of a steam or gas turbine and, thereby, increase the overall efficiency by not wasting all of the heat expended by the generation process. After purification, the water is made into an electrolyte, which is then sent to the hydrogen generator. If the electricity is supplied

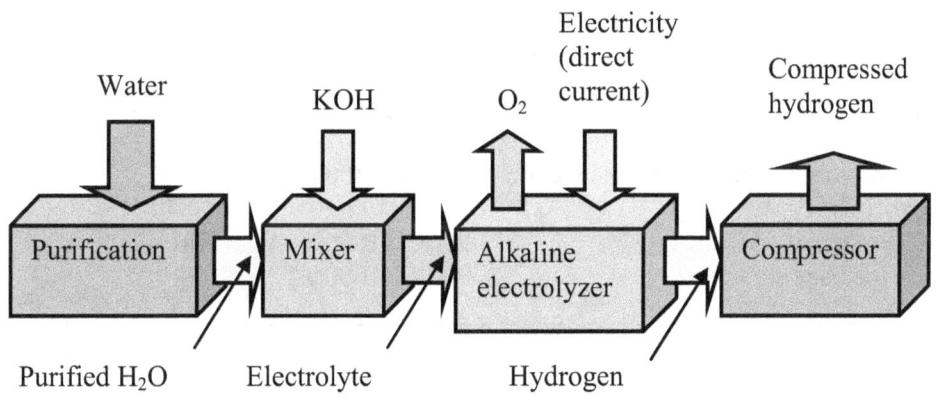

Fig. 6-11 Alkaline electrolyzer system.

by solar cells or directly from a wind turbine, then electrical conversion is not needed. But, if the electricity is from the grid then the alternating current, ac, must be changed to direct current, dc. The last stage is to compress the hydrogen for storing or transporting. Purification, mixing, electrical conversion and compression require energy and, together, they determine the overall efficiency of a hydrogen generation facility. The efficiency of the hydrogen generator alone may be as high as 95 percent, but the efficiency of an entire system ranges from 55 to 75 percent depending on the equipment needed and amount of compression. However, these percentages do not include the energy needed to generate and distribute the electricity.

Table 6-5 gives the operating temperatures and efficiencies, including average electrical losses, of a few methods for producing hydrogen. This table was obtained from a 2003 study by Argonne National Laboratories. If the electricity were generated from renewable sources, then the electrical losses would not reduce our natural resources or cause a significant amount of pollution. If the steam reforming method includes sequestration of the carbon dioxide, then its efficiency is reduced to 58 percent. A 2004 National Renewable Energy Laboratory (NREL) report (NREL/MP-560-36734) surveying the commercially available hydrogen generators indicates an output capacity range from 0.9 kilograms per day (or 0.112 MBtu/day) to 1046 kilograms per day (or 130 MBtu per day). The smaller generators would be enough to supply 1.6 automobiles and would be for a single household. The larger ones would supply 1909 automobiles and would be for a community.

Table 6-5 Temperatures and efficiencies for a few methods for producing hydrogen.

Method	Operating temperature	Efficiency
Calcium bromine thermo-chemical cycle	760°C (1400°F)	45-49
Electrolysis	90°C (194°F)	20-30
High-temperature electrolysis	900°C (1650°F)	40
Steam methane reforming	900°C (1650°F)	77

A problem common to both methane and hydrogen is the fact that both have very low boiling points, -162°C (-260°F) for methane and -259°C (-434°F) for hydrogen, and are liquids only when kept below these temperatures or at very high pressures. Therefore, both are stored as compressed gases, typically between 100 and 300 atmospheres. The greater the compression, the more energy can be stored in a given volume. The amount of compression is particularly important if the gas is to be used in vehicles. The compression and space used to store the gas ultimately determines the *vehicle's range*, i.e., the maximum distance a vehicle can travel before it must be refilled.

There are several biological processes capable of producing hydrogen. It was indicated above that the hydrolysis step in the decay of organic matter produces hydrogen and so any organic matter could be used as a source of hydrogen. In addition, if algae, which normally produces oxygen, is deprived of sulfur, then it will produce hydrogen instead. In any case, the research into the biological production of hydrogen is not yet mature enough to predict its commercial viability.

Figure 6-12 shows the 1985 through 2007 consumptions of ethanol and biodiesel in the United States. The other renewable fuels, methanol, butanol and hydrogen, were consumed in negligible amounts during this period and are not shown. Methanol and hydrogen are likely to be used to a significant degree only if fuel-cell automobiles become popular. For vehicular use, methanol has an advantage over hydrogen because it is not feasible to store hydrogen as a liquid in a vehicle and liquid methanol has a higher volumetric energy density than compressed hydrogen gas. Vehicular energy consumption is considered more thoroughly in Chap. 8.

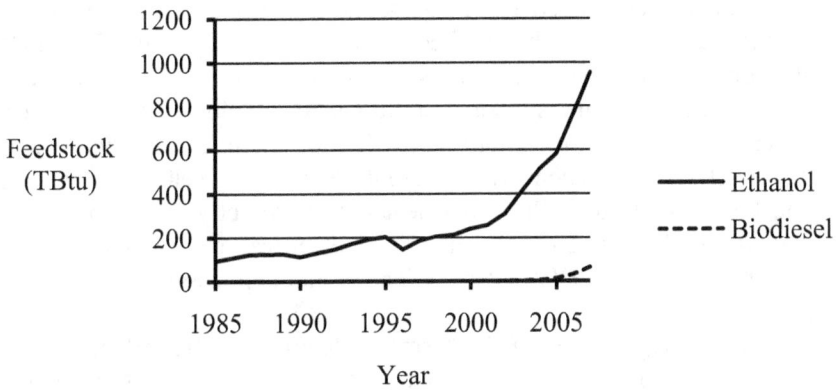

Fig. 6-12 United States consumption of biofuels in TBtu [EIA].

In 2007, the total energy consumption in the United States was 101,603 TBtu, the total renewable energy consumption was 6,831 TBtu, the total biomass consumption was 3,615 TBtu, the total biofuel consumption was 1,018 TBtu, the total ethanol consumption was 955 TBtu and the total biodiesel consumption was 64 TBtu. Therefore, only 6.7 percent of the total energy consumption was contributed by renewable energy sources and hydroelectric production accounted for 36 percent of the renewable share with much of the remainder being due to the direct use of wood for heating. Only one percent of the energy consumption was attributed to the biofuels ethanol and biodiesel, but the use of these biofuels is increasing rapidly as they are increasingly used in fuel blends. The principal impediment to using renewable sources other than hydroelectric power has been cost and the lack of infrastructure. However, as fossil fuels are depleted and the demand for cleaner energy increases, the renewable sources will become more and more competitive.

By and large, the application of renewable energy is to direct heating, powering vehicles and generating electricity. As seen from Fig. 6-2 and the discussions above, biomass is not only used to directly heat air, water and materials involved in industrial processing via combustion, it is also used to generate electricity and provide fuel for vehicles. Solar and geothermal energy are also utilized to provide direct heat and generate electricity, but not for powering vehicles. Wind and water power, although they have historically powered ships, pumps, grinding stones and other machinery, in today's world are used almost exclusively to generate electricity.

The sun emits such a tremendous amount of energy that 5.2 sextillion (trillion-trillion) Btu of its energy strikes the Earth's atmosphere each year. About 58 percent, or 3.01 sextillion Btu, of this energy makes it through the atmosphere to the Earth's surface, over 6,000 times as much energy as consumed by all of the world's citizens in 2007. Although this energy is free, capturing even a small fraction of it to serve useful purposes has proved challenging and expensive. This energy is not equally distributed across the Earth's surface and is only directly available to half of the Earth at a time. The amount of the sun's energy arriving at a particular location varies with time and the atmospheric conditions at the location. As a result, considerable energy storage, backup and redistribution facilities may be needed in addition to conversion equipment. There are two names given to solar energy depending on how it is used. If the sun's heat is used directly, it is referred to as *solar thermal energy*. If it is converted directly to electrical energy by photovoltaic cells, it is referred to as *solar PV energy*. Solar thermal applications are considered first.

The sun's energy arrives in the form of electromagnetic radiation and, as discussed in Chap. 7 and Appendix A, electromagnetic radiation is made up of waves of various frequencies with the radio and infrared frequencies being the ones most capable of heating air and water. In general, electromagnetic energy can be used to heat air or water by capturing the appropriate frequencies or converting radiation to the appropriate frequencies. In particular, the greenhouse principle can be used to capture the sun's radiation to produce useful heat. This is accomplished by allowing sunlight to pass through a transparent material, usually glass, into a container where it strikes an absorbing substance that heats up and reradiates the lower frequencies. The lower frequencies heat up the surrounding air or water in the container. The greenhouse principle is illustrated in Fig. 6-13.

In the discussions that follow, it is important to keep in mind that all matter that is above absolute zero (-273 °C or -459 °F) contains heat. An area that is described as *cool* or *cold* is simply an area that contains relatively less heat. In order to control the temperature in a location, it is necessary to convert another form of energy into heat or to move heat from one location to another. The principal means of converting chemical energy to heat is burning and of converting electricity directly into heat is passing an electric current through a resistance. The simplest form of moving heat is *convection*, which is achieved through the movement of a heated gas or liquid. An obvious example is the use of a fan to bring the relatively cool night air into a house while expelling the warmer air in the house. Convection is often used in conjunction with the conversion of another form of energy into heat (including the greenhouse conversion of electromagnetic energy from the sun's radiation).

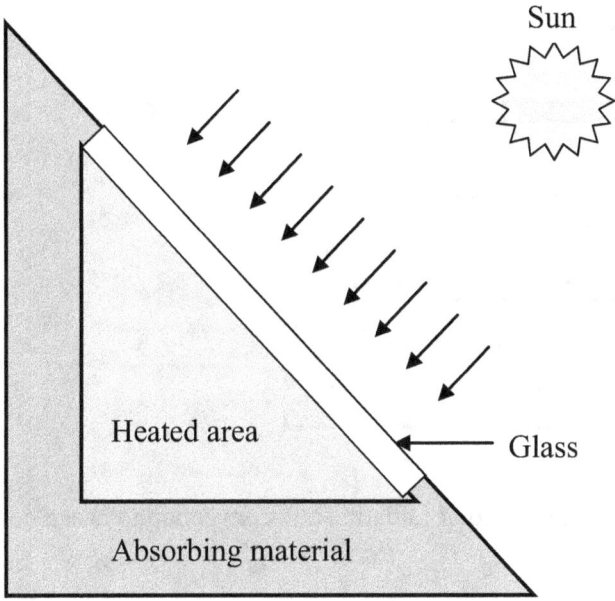

Fig. 6-13 Illustration of the greenhouse principle.

Figure 6-14 is an example of a house heated by the sun's radiation and convection within the house. The large window must face in the direction of the sun, which in the winter in the United States would be south. The radiation would warm the air and interior walls and floors of the house and convection would move the air so that the house would be warmed more or less uniformly. Because warm air is less dense than cool air, the air along the window would rise and, as it passes through the upper floor, it would cool. As it becomes less dense, it would descend along the back wall. The warm air rising along the window would pull the air along the back wall down under the lower floor and up toward the window and the cycle would be repeated. Fans may be used to aid the circulation. If fans were not used, the design in Fig. 6-14 would be described as *passive* and only the energy of the sun would be utilized. If fans are used to help distribute the warm air, then electricity would be consumed in addition to the sun's energy and it would be an *active* system.

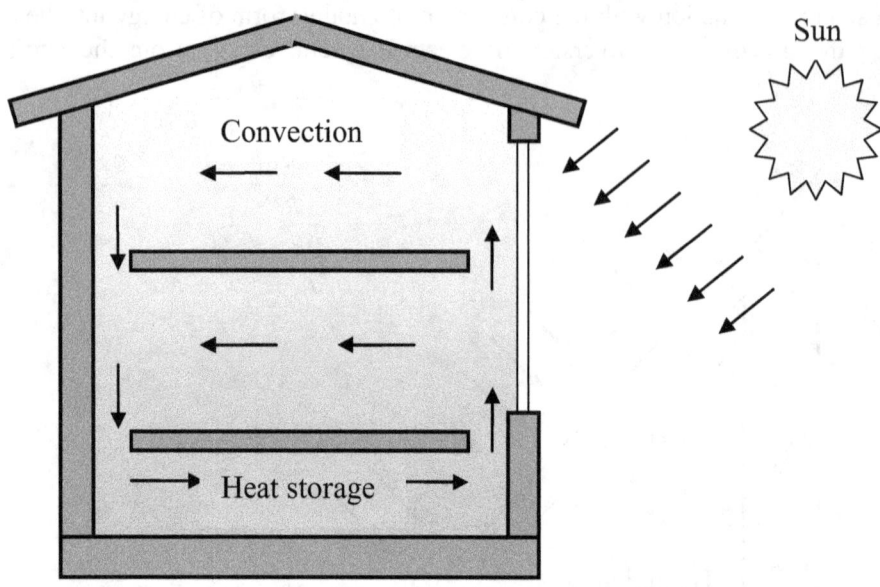

Fig. 6-14 Example of using solar radiation and convection to heat a house.

Normally, rocks would be placed below the lower floor and would be warmed by the air as it passes through them. Because solar heat is present only during the day and during clear weather, the rocks would preserve some of the heat for use during the night or when it is cloudy. Just as energy storage by batteries or other means is important in an electrical system, heat energy is frequently stored in rocks or tanks of water for later use.

A passive system for heating water is shown in Fig. 6-15. The sun's rays strike the collector's heat absorbent backing and are converted to heat which warms the water in the collector. Like air, the density of water decreases with temperature and causes the water to flow upwards into the hot water tank. But over time, the water in the tank cools and returns to the bottom of the collector where it is heated once again. Some of the hot water may be drawn off to supply hot water to a house or for other purposes and replaced by water from an external source.

There are two shortcomings of the design given in Fig. 6-15. One is that the sole storage is the hot water tank and there is no backup for keeping the water hot when there is no sunshine. This seriously limits the use of such a design, and such a system is mainly used to heat a swimming pool and the pool's pump makes it an active system. The other problem is that tap water contains corrosive minerals and is subject to freezing.

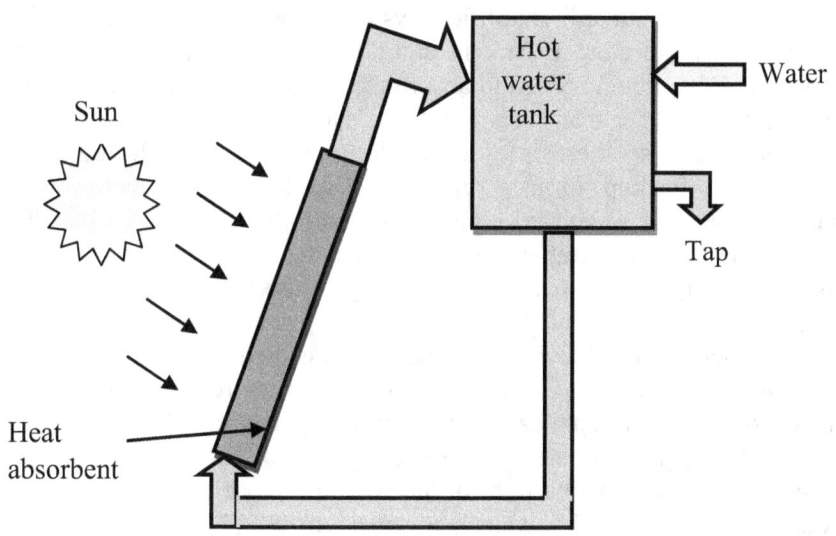

Fig. 6-15 Passive solar water heater design.

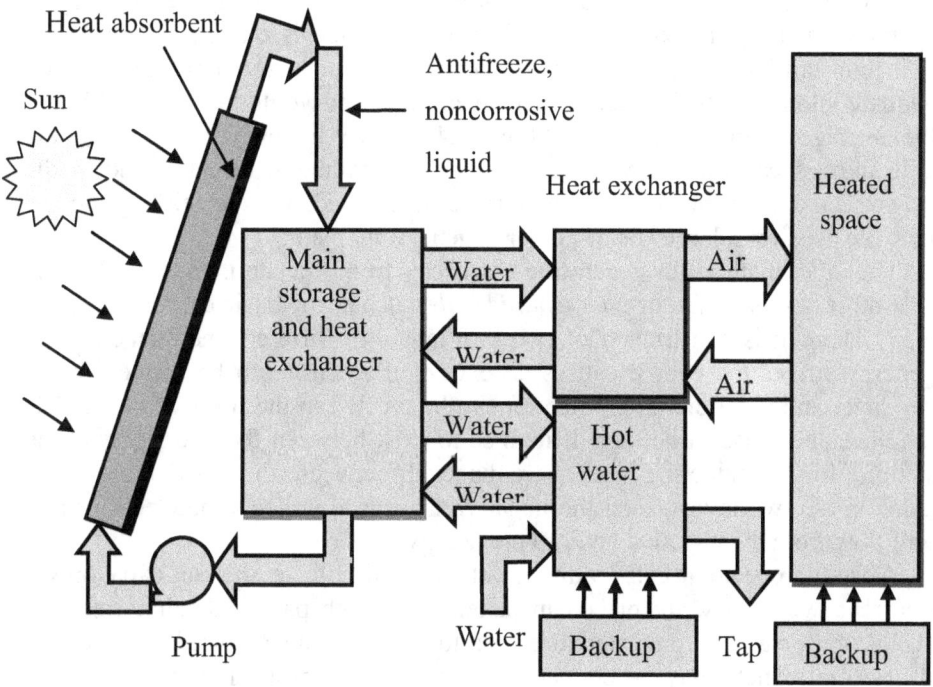

Fig. 6-16 Active solar system for heating both water and air.

A more realistic, but still simplified, system is the one illustrated in Fig. 6-16. Here there is a large storage tank in addition to the hot water tank and an antifreeze, noncorrosive liquid is used to absorb the sun's rays. The absorbing liquid is fed into a heat exchanger that heats the water. Not only does the heated liquid supply hot water, but also there is a second heat exchanger for heating air. In addition, there is a backup furnace and a backup heater for controlling the temperature of the air and water and a pump for controlling the water flow. Thus, this system provides both temperature regulated hot water and warm air.

There are two ways in which solar energy is used to generate electricity. One is to focus the sun's rays on a point or line where the concentrated heat is used to heat water or other liquid. The water or other liquid is then used to produce steam to drive a turbine. Essentially, the solar collector replaces the reactor in Fig. 5-5 and the system is a solar thermal application. Such power plants may be quite large. A plant located in the Mojave Desert is to have a capacity of 553 MW and cover 24 km^2 (9.27 mi^2). If, on average, one third of the system's capacity is output, then the facility would average about 7.7 watts per square meter (7.7 W/m^2) and 1,616 million kilowatt-hours (1,616 MkWh) of energy per year, enough to supply over a half million homes.

Another California project uses parabolic mirrors to drive Stirling engines that, in turn, drive 25 kW generators. (A *Stirling engine* is a reciprocating engine that operates on the principal shown in Fig. 6-19, which is discussed later.) The project is to install 20,000 such systems (giving it a peak capacity of 500 megawatts) on 18 square kilometers (4,500 acres). The average power would be about 9.3 W/m^2 and the average energy output would be 1,461 MkWh per year. Because the parabolic mirrors can track the sun better than the parabolic troughs used in the Mojave project, a better power per square meter is attained, and the Stirling engines eliminate the need for a large steam-powered generating plant.

The other method for generating electricity from the sun is to use solar PV. Solar PV utilizes *solar cells*, or *photovoltaic cells*, such as the basic cell illustrated in Fig. 6-17. The cell is constructed of two thin layers of different materials. As the sun's energy strikes the cell, it causes an excess of electrons to be created in the negative layer and a deficiency of electrons to be created in the positive layer. This charge imbalance causes a voltage difference to exist between the two layers at the pn-junction. If a conducting path is connected between the conducting grids embedded in the two layers, then the negative layer grid collects the electrons and they will flow from the negative layer to the positive layer.

Although one solar cell produces only a small voltage and can carry only a small current, when they are placed in panels with each panel containing several solar cells, they can be connected together to produce the required voltage and current. Several panels may be connected together to form a large system that is capable of outputting a large amount of energy. The output of a solar PV system is direct current, dc, electricity and, except for small dedicated systems, must be connected to equipment that converts the output to alternating current, ac, electricity.

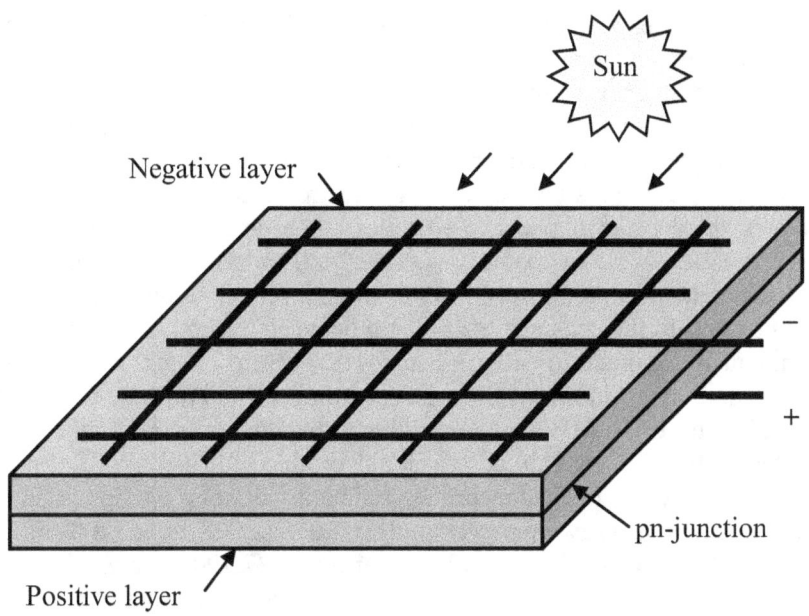

Sun

Negative layer

−

+

pn-junction

Positive layer

Fig. 6-17 Basic solar cell design.

The single layer cell shown in Fig. 6-17 is designed to collect only one range of frequencies and has an energy collection efficiency (i.e., output electrical energy divided by the incident solar energy) up to 20 percent. There are multi-layer cells that collect the energies in multiple ranges of frequencies that can reap over 40 percent of the incident energy, but they are very expensive to make and are primarily used to supply energy to satellites. It is estimated that four-layer cells could be developed that would be as much as 50 percent efficient. The theoretical limit is over 85 percent. There are also very large, thin cells, called *thin film cells*, that are produced in very thin sheets backed by supporting material such as glass. Thin film cells tend to have lower efficiencies, but are relatively inexpensive to construct and could be bonded to window panes. This would make it possible for many buildings to provide much of their electrical needs. Thin film cells currently have efficiencies between 8 and 12 percent, but if they could capture lower frequencies than their thinness currently allows, then they could be made more efficient. Research in Australia by Kylie Catchpole involves studying the use of silver nanoparticles to overcome the thinness problem. Research is also being conducted on organic solar cells that would be inefficient but be very inexpensive to produce. The best such cells currently have efficiencies between five and six percent.

The amount of energy arriving from the sun on a square meter that is perpendicular to the sun's rays is about 1.4 kilowatts. After taking into account that only half of the Earth is exposed to the sun at a time, the Earth is not a flat plane perpendicular to the sun, on average the Earth's atmosphere absorbs or reflects about 42 percent of the incident energy, and the land in North America is at an angle to the sun, it is estimated that from 125 to 375 watts per square meter of the sun's power is available at the surface in North America. This amount clearly depends on the location with the southern areas that have clear skies receiving the most energy. In the United States there is an abundance of areas, such as the Northern Arizona desert, that average 200 W/m^2 of usable solar energy. If an array of solar cells that are 16 percent efficient were at such a location, it would produce an average of 32 W/m^2. But the contacts to the cells and the spacing between the cells and panels may reduce this amount by as much as 40 percent to 19.2 W/m^2. The conversion and distribution system may reduce the electrical output by another 12 percent to approximately 16 W/m^2 or 16 MW/km^2. Since this is an average over all time, a one square kilometer system having this design could produce 0.140 billion kilowatt-hours (0.478 TBtu) per year. To demonstrate the potential of solar PV, note that the total delivered electrical energy in the United States in 2008 was 3,879 billion kilowatt-hours (13,275 TBtu). Theoretically, an array of 27,700 square kilometers (10,700 square miles), 9.1 percent of the area of Arizona, could supply all of the electrical demand in the United States.

However, because neither the supply nor demand is the same all the time, there would also have to be facilities capable of matching the supply to the load. Such huge storage facilities are unrealistic (as is discussed in Chap. 7). The most optimistic estimates indicate no more than 70 percent of our electricity could be supplied by solar energy and some claim no more than 20 percent should be supplied by solar means. Then there is the additional problem of transporting the electrical energy from sunny areas to the populated areas where it is most needed. In the near future it is anticipated that most solar electricity will be generated by residences and businesses, not by large centralized facilities. These residences and businesses would receive backup power from our existing electrical grid or onsite generating facilities. They would not require large storage facilities and may not have any energy storage.

An example of a large commercial solar panel array is the 0.364 km^2 (90 acre) facility constructed by General Electric in Portugal for $75 million. An article by Katherine Bourzac entitled "Good Day Sunshine" in a 2009 special report in the *Technology Review* published by The Massachusetts Institute of Technology indicates the array has an 11 MW capacity and produces about 21,340,000 kWh per year. This implies that the facility cost is about seven dollars per watt, has a capacity of 30 MW per square kilometer (78 MW per square mile) and averages 22 percent of its maximum capacity. It would take a similar facility to cover an area of 66,200 km^2 (25,600 mi^2) to produce the 2008 electricity demand of the United States. Such a facility would cost $13.7 trillion, which was approximately equal to the United States total debt in 2010. If this amount were borrowed using a 50-year loan at five percent per annum compounded monthly and the electricity consumption rate were

3,800 GkWh per year, then the cost of the loan would be roughly 20 cents per kWh. This is three to four times the 2010 grid cost of electricity and does not include maintenance and operation costs (which would be small compared to the cost of the loan), but there never would be a fuel cost regardless of the lifetime of the array. However, if a carbon penalty is added to the cost of generating power using fossil fuels, then such a solar array would be much more competitive (see Fig. 6-26 below and in Chap. 9). Also, the cost would be much less if solar panels were to be purchased in such massive quantities. Because the panels are built high enough above the ground to allow sheep to graze underneath, the facility does, in fact, permit dual usage of much of the land. If solar panels were to replace a significant portion of United States generating capacity, many of the panels undoubtedly would be located on rooftops.

Like solar energy, geothermal energy may be used for direct heating or for generating electricity. Heat from the earth has provided warmth for homes and baths for centuries. Most noteworthy were some of the Roman bathhouses. Even when it is used to generate electricity, the heat from geothermal activity is applied directly to producing steam. In some locations, such as Iceland and New Zealand, geothermal areas are used both directly and for generating electricity. At present, only locations where geothermal activity is near the surface is it being significantly exploited, but in the future, deep wells may be drilled to extract the heat from as much as ten kilometers (six miles) below the surface. Such a system includes an injection well for pumping water into a hot rock formation where it circulates through the formation and is then removed through production wells placed near the injection well. At the surface, the water passes through a heat exchanger that creates steam to drive a turbine. The water is then recirculated through the injection well. Manmade geothermal designs of this nature are called *hot dry rock*, or *enhanced geothermal systems* (*EGSs*). Australia has drilled a four-kilometer deep pilot well in its Cooper Basin and has had positive results in widening the cracks in the rocks, a requirement for producing enough heat to operate a commercially viable EGS. A report released by MIT in 2006 indicates that the United States potential EGS capacity could be much as 100,000 MW. Whether a natural or manmade geothermal source is used, the electrical generating plant is similar to the one in Fig. 5-5 with an underground piping system replacing the reactor. A discussion of geothermal generating plant design is given in Chap. 7.

The advantages of geothermal generating systems are that the heat energy itself costs nothing and the output capacity is constant regardless of the time of day or year or the weather conditions. The disadvantages of geothermal systems, particularly EGSs, are the corrosiveness of the circulating water, the fact that the temperature of the output tends to be lower than the required amount, the large capital expense of drilling deep wells and the fact that, in time, they lose their heating value as the rock formation cools. This cooling may take decades and depends on the rock formation and the rates at which the heat is extracted and replenished. The rate of replenishment may be very low and some argue that the energy from EGSs is not renewable. Others contend the potential energy from EGSs is so vast that it may be considered inexhaustible. The corrosiveness of the water increases the maintenance costs and requires that a percentage of the water be continually replaced. The replaced water is polluted from its contact with the rocks and should be sequestered or treated before it is released.

However, it is not just the high temperatures of geothermal active areas or hot areas deep underground that are of value, the warmth of the ground in general can be useful. The ground absorbs and acts as storage for heat energy. Except for active geothermal areas where the heat comes from magma near the surface of the Earth, much of the heat in the Earth's upper crust is due to the sun's radiation. At between 2 meters (6.6 ft) and 15 m (49 ft) below the surface, the ground is kept at the approximate average temperature of the air just above the surface, normally from 10 to 16 °C (50 to 60 °F). This near constant temperature may be used to provide, or at least supplement, a space heating and cooling system. The basic design that uses the temperature of the ground is shown in Fig. 6-18. In this design, a liquid, perhaps water, is pumped from a large horizontal array of underground pipes through the floor of the area to be heated or cooled. In place of the ground, a pond is sometimes used to store and extract solar heat.

Heat releasing area

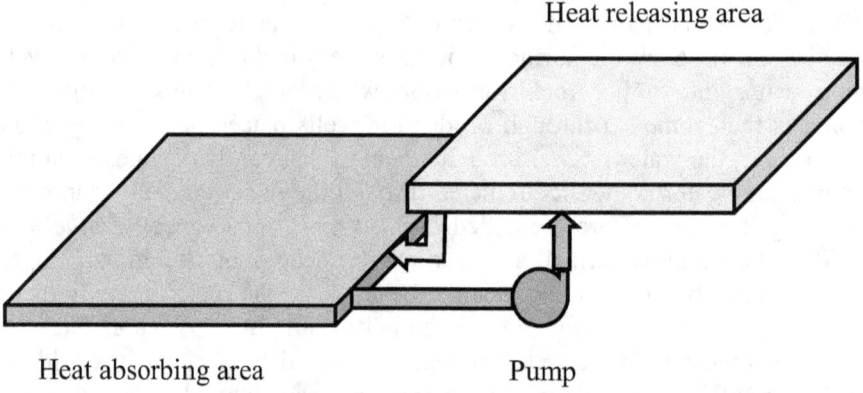

Heat absorbing area   Pump

Fig. 6-18 Basic ground heating system.

If the available area for underground piping is limited, as it is in many parts of Europe, vertical boreholes are employed to tap shallow geothermal resources up to 400 m (1312 ft) below the surface. Typical boreholes are between 15 m (49 ft) and 200 m (656 ft) deep and the deeper the borehole the more geothermal heat flux dominates that of the sun's radiation. Where convenient, boreholes may use groundwater directly or use a closed circulation pipe and the ground as a heat exchanger. The liquid in the circulation pipe may be water or a less corrosive liquid.

As an example, in one development in Europe, for each house five kilowatts of heat was provided by drilling four 30 m (98 ft) boreholes.

When designing any space heating or cooling system, an important consideration is the number of heating and cooling degree days of the location. The number of heating degree days (HDDs) in degrees Fahrenheit is computed by subtracting from 65 °F the average of the high and low temperatures for each day for which this difference is less than 65 °F, and then adding these differences. For example, if the averages for five days are 72, 63, 59, 65 and 76, then the number of HDDs during this period is 0+2+6+0+0=8. The number of cooling degree days (CDDs) is computed similarly by subtracting 65 °F from the average of the high and low temperatures for each day. For the previous example, the number of CDDs is 7+0+0+0+11=18. In Key West, Florida, typical numbers of CDDs and HDDs for an entire year are 4,820 and 68, respectively, while in Bismarck, North Dakota, these typical numbers are 499 and 8,932. Ground source heating and cooling alone tends to be most effective where the number of HDDs exceeds that of CDDs.

The thermal cycle depicted in Fig. 6-19 can be used to transport heat in any climate. The cycle is based on the fundamental thermodynamic laws of Boyle, Charles and Gay-Lussac (see Appendix A) and has five components, a compressor, a condenser, an expansion valve, an evaporator and a gas, or refrigerant. Both the

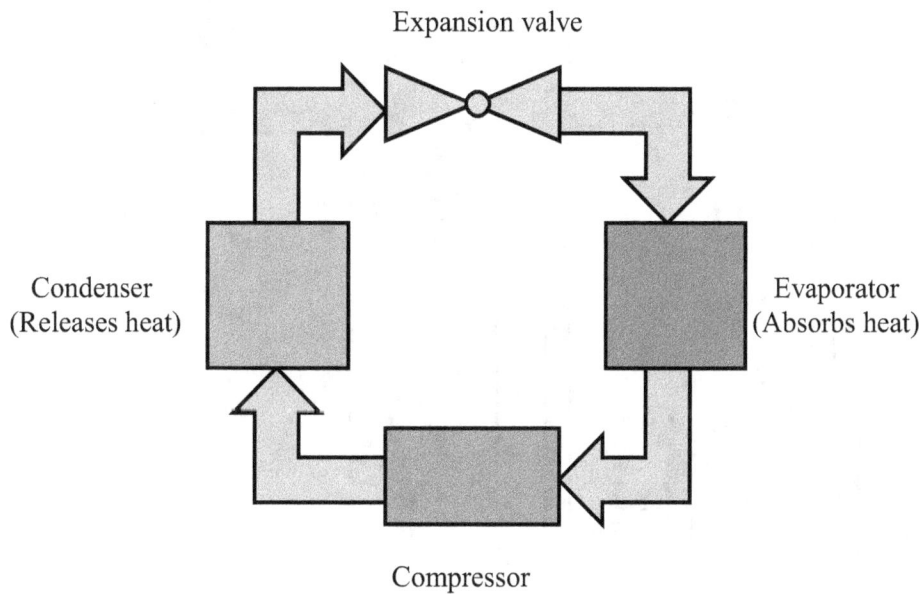

Fig. 6-19 The thermal cycle for moving heat from one location to another.

evaporator and condenser are heat exchangers. The pressure of the refrigerant entering the expansion valve decreases rapidly as it exits the valve and this causes the refrigerant to be cooled. The evaporator causes the refrigerant to be warmed as it absorbs the heat of the surrounding medium (e.g., air or water). The compressor then increases the pressure of the refrigerant, but also increases its temperature further. The refrigerant is then cooled by the condenser, thereby releasing the heat in the refrigerant in the vicinity of the condenser. But most of the pressure is retained so that the pressure entering the expansion valve is maintained. The net result is that the mechanical energy applied to the compressor enables the heat near the evaporator to be released by the condenser. The cycle continues as long as the compressor is running. Freezers, refrigerators, and air conditioners are based on this cycle.

Figures 6-20 and 6-21 illustrate heat pumps, systems that, in effect, use the cycle depicted in Fig. 6-19 to pump heat from one place to another. The system in Fig. 6-20 uses the outside air to vent the heat absorbed by the evaporator, which is in the area to be cooled. If heating is desired, then the cycle can be reversed so that the

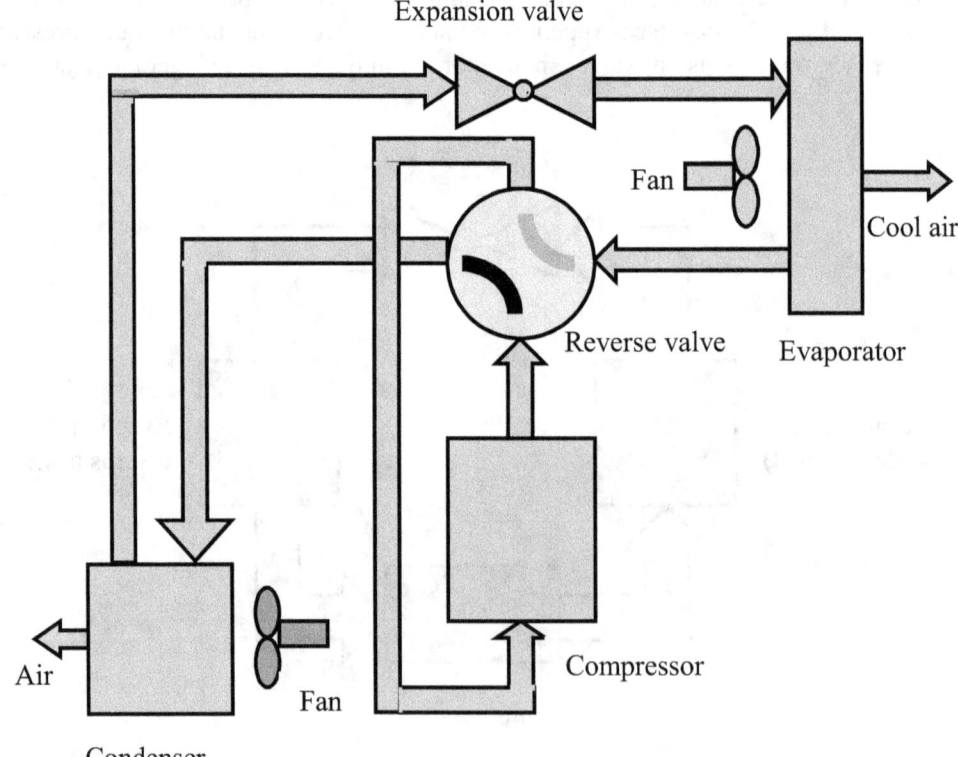

Fig. 6-20 A heat pump system that uses air for both heating and cooling the refrigerant.

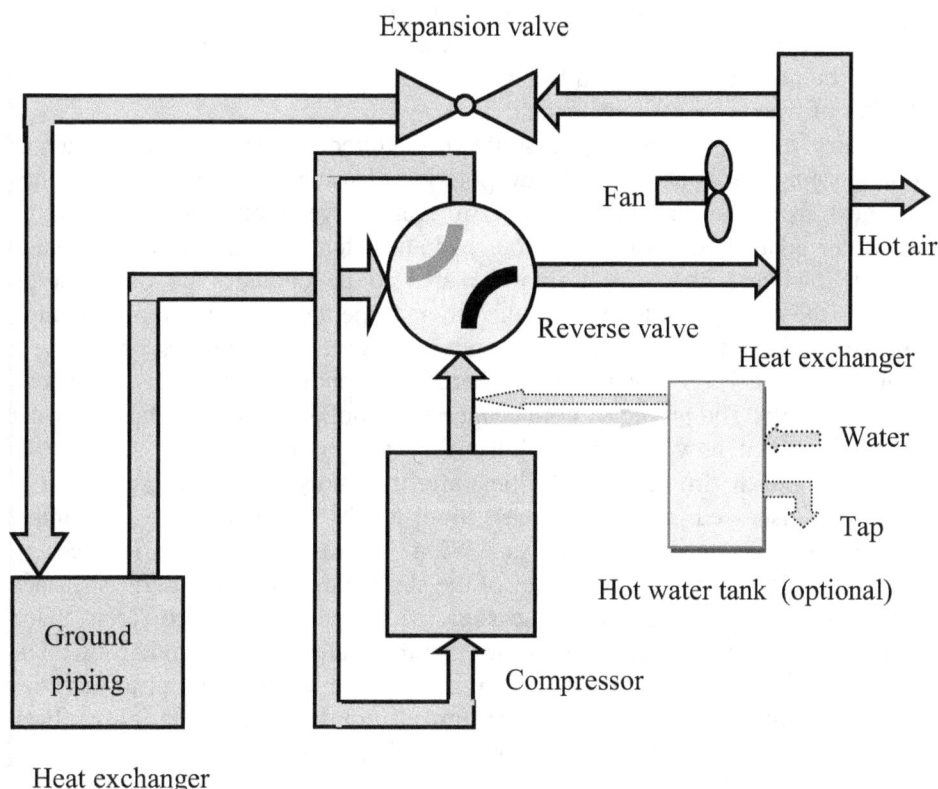

Fig. 6-21 A heat pump that uses the ground as one of the heat exchangers.

evaporator becomes the condenser and the condenser becomes the evaporator. When the flow is reversed (as it is in Fig. 6-21) the heat from the condenser warms the inside air and heat is absorbed by the outside air. The system in Fig. 6-21 is different in that the ground instead of the air surrounds one of the heat exchangers and an optional hot water tank has been included. The ground piping may be either horizontal or vertical and the amount of piping may depend on the relative numbers of HDDs and CDDs for the given location.

Anywhere the sun creates a more or less constant difference in temperature there is opportunity for exploiting this difference. Another example of taking advantage of such a difference is the use of the difference in ocean or lake temperatures between the surface and deeper water. The surface water is used to boil a substance with a low boiling point and the resulting gas is then passed through a turbine and down to where it is cooled by the colder water far below the surface. After the gas is condensed it once again is returned to be heated by the warmer water. Although the pressure difference across the turbine is much less than that of a steam turbine, it is sufficient to turn an electrical generator. Systems of this type are in the experimental stage and could be constructed in southern locations that are near large bodies of water (e.g., tropical islands).

In dry climates such as the southwest United States, there is a way of providing cooling that is based on the evaporation of water. When water evaporates it takes heat from the surrounding air. An evaporative cooling system is quite simple. Water is pumped through a porous pad and a fan sucks air from the outside through the pad and channels it into the area to be cooled. The only energy consumed is the energy needed to turn the fan and operate a small pump. Although evaporative cooling uses less electricity than refrigerative cooling based on the cycle shown in Fig. 6-19, it only works well where the humidity is low and the air can absorb water easily. The problems with evaporative cooling are that it requires water and is quite limited in how much it can lower the temperature.

Wind and hydroelectric facilities are the remaining renewable energy facilities to be discussed. As with biomass, both wind and water energy are almost entirely, but indirectly, due to the sun. Wind is caused by the differences in temperature resulting from the variations of the sun's radiation at the Earth's surface with the time of day and changing seasons. Most hydroelectric power is made possible by the sun's rays evaporating water that then falls to the ground as rain. The net effect is that the water is lifted to a higher altitude and hydroelectric facilities convert the potential energy of the water into electricity as the water flows back toward the ocean. An exception to this means of hydroelectric power is the use of tidal energy. Although the sun has some effect on the tides, most of the tidal energy is not caused by the sun, but is due to the moon and the Earth's rotation.

The days when wind and water were primarily used to sail ships, grind grains, power machinery or pump water are gone. Today, wind and water are used almost exclusively to generate electricity. Instead of a turbine that gets its power from exploding gases or steam, wind turbines and hydro-turbines get their energy directly from the air or water pushing against their blades. Hydroelectric facilities may get their energy from large dams that store water in lakes, the movement of tidal water or the water in rivers. The main problem with tidal dams is that there are very few places where they can be installed and be economically viable. There is some interest is using ocean or river currents to power underwater windmill-like energy collectors, but again there are a limited number of places where they could be profitably constructed.

The largest hydroelectric facility is the Three Gorges Dam in China, which has a generating capacity of 18,200 MW. Of the industrialized countries, only Canada has plans for additional large hydroelectric projects. On the other hand, Southeast Asia, including India and China, has substantial plans for additional projects. Also, some South American countries, including oil rich Venezuela, intend to increase their hydroelectric capabilities. There are a few locations in the United States where additional hydroelectric dams could be built (including the lower part of the Grand Canyon), but environmentalists oppose creating large lakes in most of these places. Hydro-turbine design is discussed in Chap. 7.

Wind turbines are windmills connected to electrical generators. Some are small and used for a specific, localized service, such as a single isolated residence. But others are for the commercial generation of electricity and are very large, some reaching 130 m (430 ft) in height and having output capacities up to six million watts (6 MW). Wind turbines may occur individually or distributed in small groups, or centralized in *wind farms* made up of a large group of turbines. The spacing and locations of the turbines are very critical, even in an area known to have good wind characteristics, and are carefully determined after a number of wind statistics have been gathered. Even in a good area, the average energy output of a wind farm is normally not more than 35 percent of its maximum capacity because of the variations in the wind. Wind farms vary in size with several of the larger ones having total capacities of about 80 MW. However, the Horse Hollow Wind Energy Center in Texas, which covers 190 km^2 (73 mi^2), has a capacity of 735.5 MW, or 3.87 W/m^2. The best locations in the United States are in the Great Plains, in the Great Lakes, off the New England coast and on high, barren ridges, such as those in California and the Northwest. In general, the southeastern states are not well suited to wind-powered electricity generation.

Although other countries are increasingly planning on wind power, the United States and Australia have the advantage of having vast open areas with good wind conditions. Some nations, particularly Denmark, are locating wind farms in the shallow areas of the sea. Figure 6-22 provides the total installed wind turbine capacity at the beginning of 2007 for some of the nations that are making a concerted effort to increase their wind-powered generation capabilities. Germany, despite its limited space and fewer optimal locations for installing wind turbines, is a leader in the efforts to develop wind power. Europe as a whole had a capacity of 48,000 MW versus 13,191 MW for the United States and Canada, although the United States had 4,500 MW of capacity under construction in 2007.

The New York State Energy Research and Development Authority (NYSERDA) divided wind energy users into three classifications. The residential scale users tend to have wind turbines with capacities of 400 W to 50 kW, the industrial scale users have wind turbines with capacities of 50 kW to 250 kW and the utility scale users have turbines with capacities greater than 900kW. The residential users are mainly those in remote locations and the industrial users primarily employ wind turbines as backup or to reduce their reliance on the electrical grid during peak loads. The utility users are those that sell wind generated power to the grid. Residential and industrial users have relatively small systems, often with only one turbine, and their contribution to the total wind energy output is small compared to the utility users. It is the utility group that constructs large wind farms and it is this group on which we will concentrate our discussion.

Wind farms are generally located on ridges, very large flat areas or in an ocean or large lake. To be profitable they should be in an area where wind averages

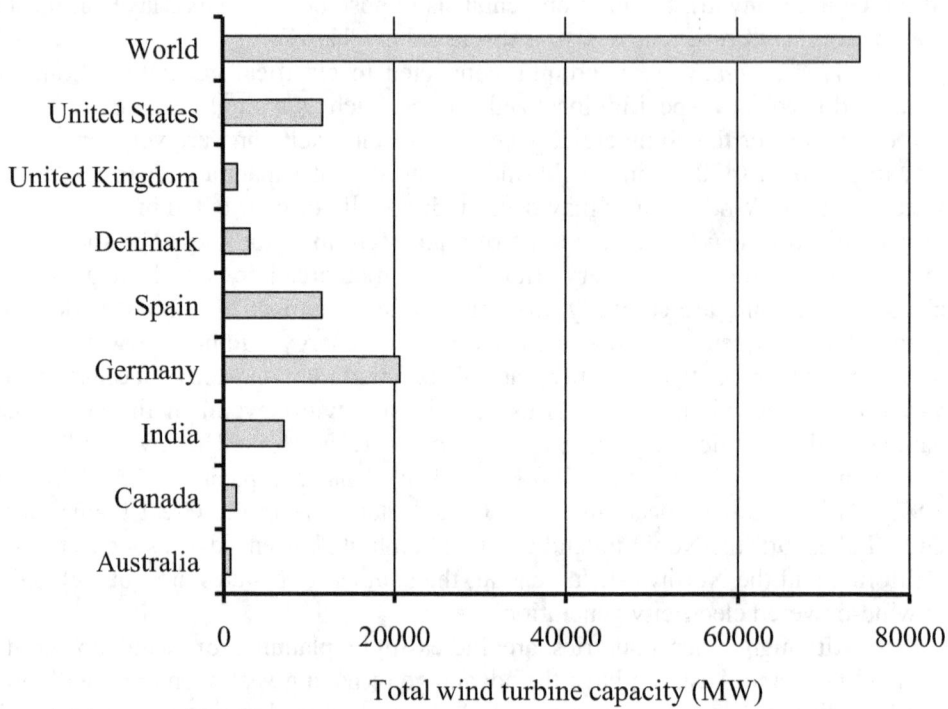

Fig. 6-22 Installed wind turbine capacities at the beginning of 2007.

at least four meters per second (9 miles an hour). A large wind farm may be connected to the grid through its own high voltage, high capacity transmission line, but a small farm should be located close to an existing grid line (within 16 kilometers, 10 miles) so as to avoid an expensive grid connection. The Europeans have placed many of their wind farms in the shallow areas of the Baltic and North Seas. Although the Rocky Mountains offer some of the best wind conditions in the United States, they also offer the greatest difficulties because of rough terrain and opposition from environmental groups. Locating wind farms at sea tends to be more expensive. Germany is building a 60 MW wind farm off its northern coast in the North Sea that will cost $4.20 per watt. Under the assumptions made for the General Electric solar panel array example given above, the loan costs would be less than nine cents per kWh, making such a wind farm reasonably competitive with our current means of generating electricity if a carbon penalty is included. But maintenance and operations costs, which tend to be more than those for large solar arrays, are expected to add one cent per kWh (see Fig. 6-26 below).

New wind turbines, particularly those to be located in the sea, may have blades 63 m (207 ft) long and stand as high as 190 m (620 ft). A wind farm should have a capacity of at least 30 MW to be economical and most existing farms are in the 20 to 300 MW range. Those living near land-based wind farms are concerned about both the visual and audio impact of farms, and people with seafront property are worried about their ocean views.

The two basic types of wind turbine are the vertical axis and horizontal axis turbines. Almost all large wind turbines are of the horizontal axis type for which the shaft is horizontal and the windmill, or rotor, is perpendicular to the shaft. The turbine assembly is mounted on a tower and the ratio of the maximum height of the rotor tip to the tower height is normally between 1.3 and 1.5. The wind speed increases as the distance from the ground increases making it advantageous to construct tall towers. But in order to withstand the force of the maximum expected wind velocity, the cost of tower materials tends to be proportional to the cube of the height. Large wind turbines must have massive bases and the pitch of the blades is adjustable so that there is some control over the force on the rotor. To reduce the torque on the shaft and force on the tower the blades may be fully furled under high wind conditions.

Because of the large amount of turbulence created by a wind turbine, turbines must be spaced far apart. Figure 6-23 shows an array of turbines made up of rows and gives the recommended spacing in terms of rotor diameter. The spacing is set according to whether the prevailing winds tend to be unidirectional or vary significantly. If they vary significantly, then the turbine is designed to swivel so that it is always facing the wind. On water or flat land where a more or less rectangular array may be used, the ratio of area to capacity is usually in the range from 0.08 to 0.24 km^2/MW (20 to 60 acres/MW). By taking the reciprocal, this ratio implies a range of 4.2 to 12.5 W/m^2.

For a single wind turbine, the ratio of the area circumscribed by the rotor to the turbine's capacity in square meters per watt tends to be in the range from 320 to 470 W/m^2. If the rotor diameter is 80 m and this ratio is 400, then the capacity would be about 2 MW. If the wind is multidirectional, the spacing within the rows is 6 diameters and the spacing between rows is 7 diameters, then the average area for each turbine would be 0.27 km^2 (66.6 acres) and the capacity-to-area ratio would be 7.44 W/m^2. Assuming 0.135 km^2 (33.3 acres) per megawatt, a 1,000 MW wind farm would require 135 km^2 (52 mi^2). Assuming an average utilization of 30 percent of capacity, this farm would produce approximately 2.63 GkWh/yr, or 0.07 percent of the 2008 United States electrical energy consumption. Therefore, an array covering 199,000 km^2 (76,900 mi^2) would be required for the electrical output to equal the United States total 2008 electrical energy consumption. This is equivalent to an area slightly larger than that of North Dakota. If the array were in unidirectional winds and three diameters were used within the rows and 10 diameters between rows, then the capacity-to-area ratio would be 10.4 W/m^2 and the area to satisfy the 2008 demand would be 142,000 km^2 (55,000 mi^2).

The dominant renewable energy sources in the long run will almost certainly be those that are used to generate electricity. The demand for electricity is likely to

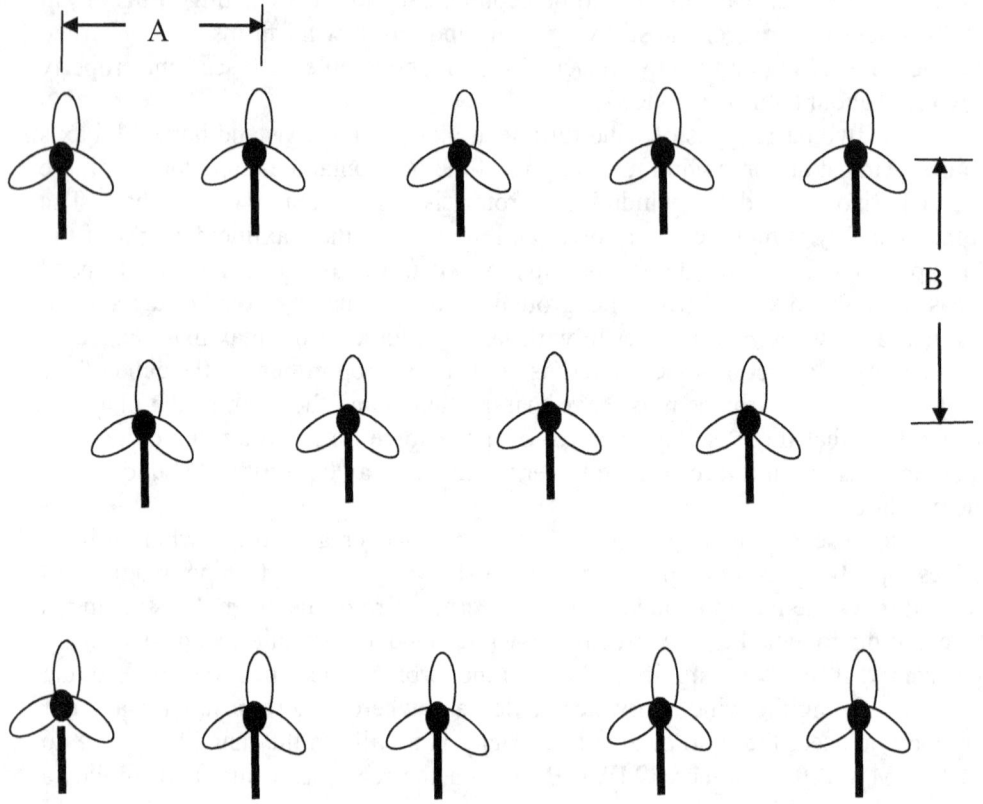

Unidirectional wind: A = 3 to 4 rotor diameters, B = 10 rotor diameters
Multidirectional wind: A = 5 to 7 rotor diameters, B = 7 to 8 rotor diameters

Fig. 6-23 Typical turbine spacing.

increase dramatically, especially if electric modes of transportation become the norm. Although biomass will be increasingly employed for generating electricity, except for waste biomass it requires land to be set aside for producing energy crops. Because forest areas are needed to absorb carbon dioxide and produce lumber and other forest products and land is also used to raise food, there is a limit to the amount of land that should be dedicated to producing energy crops. Other than biomass, the choices for providing the energy to generate electricity are geothermal, hydroelectric, solar and wind power. All renewables offer great potential for replacing the fossil fuels for generating electricity, but all are more expensive than coal and lack the infrastructure to provide electricity throughout the United States. A technology begins to become competitive with coal when it can deliver electrical power below 15 cents (year 2008 dollars) per kilowatt-hour. Hydroelectric power, in particular, is limited and already has exploited the most advantageous locations.

Most of the emphasis is currently placed on solar PV and wind power. According to the California examples given above, solar thermal requires more land per megawatt of capacity than either solar PV or wind turbines. Even though most of the best locations for these sources are not near large population centers and the availability of both vary considerably with time, between them they could supplement each other and supply over half of our electricity needs. However, the proper infrastructure would need to be constructed.

The above examples suggest that solar PV energy has the advantage in that its average power could be as much as 16 W/m^2 as opposed to a maximum of 12 or 13 W/m^2 for wind energy. In addition, research may improve solar efficiency within the next decade. But the examples given for wind energy are deceptive. Solar PV plants may require that 100 percent of the land be dedicated to generation facilities. Although between 0.08 and 0.24 km^2/MW may be required for wind farms, only about five percent of the area is actually taken out of service by the towers, other needed structures and access roads. If the turbines are on farm or ranch land, 95 percent of the land could still be used. In addition, wind farms offer the advantage of supplementing the farmers' or ranchers' income, which is often marginal. If only the land taken up by the facilities is considered, then the capacity-to-area ratio is increased by a factor of approximately 20. Neither solar PV nor wind facilities require water or pollute the atmosphere, land or groundwater once they are constructed. The United States total mid-year capacities of solar PV and wind from the summer of 2002 to the summer of 2006 are shown in Fig. 6-24.

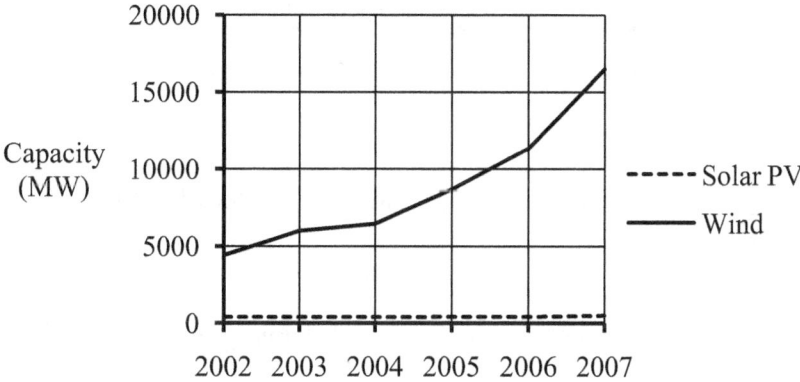

Fig. 6-24 United States mid-year capacities in MW [EIA].

Figure 6-25 shows the history of the United States consumption of the renewables from the summer of 1985 to the summer of 2009. Note that, due to the variability of rainfall, even the production of hydroelectric power is not constant. Hydroelectric and wood are the most common sources of renewable energy, but the use of biofuels is increasing rapidly and wind energy use began to increase in 2005. Wood is the only renewable source with a declining consumption. The use of solar energy is expected to rise significantly as research makes it more efficient and more price competitive. Biofuels will continue their steep rise in the near term due to their ability to substitute for fossil fuels in vehicles, but will reach their limits in the more distant future. The use of geothermal energy will undoubtedly increase, although it is difficult to predict its costs relative to other energy sources.

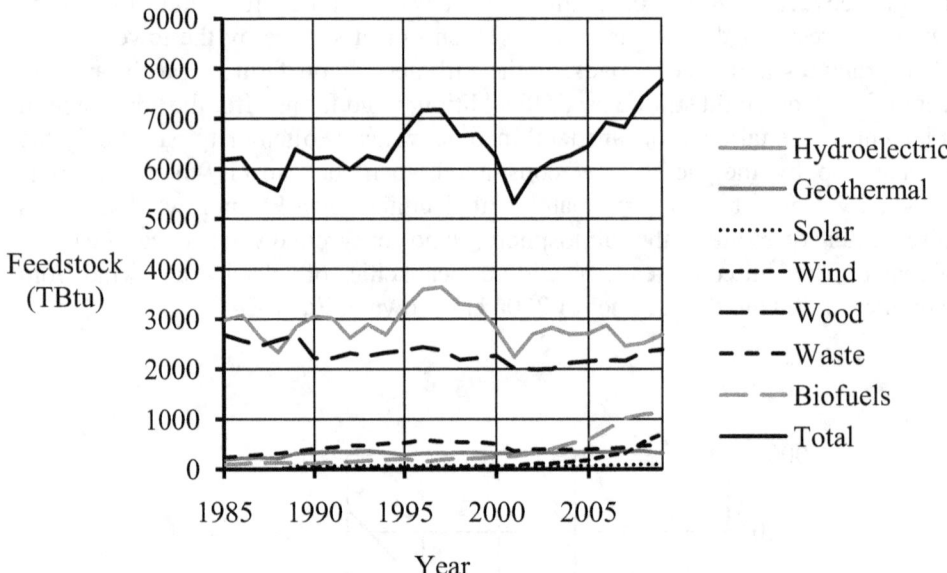

Fig. 6-25 Energy consumption in the United States by feedstock in TBtu
[EIA].

Figure 6-26 summarizes the results of a 2003 European study of costs for generating electricity presented by the World Nuclear Association (WNA). It assumes a five percent interest rate, a five percent discount rate, a 40-year plant life, a €20/tonne penalty for carbon dioxide emissions, and a 91 percent load factor (8,000 hours of use per year) for fuels other than wind and a 28 percent load factor

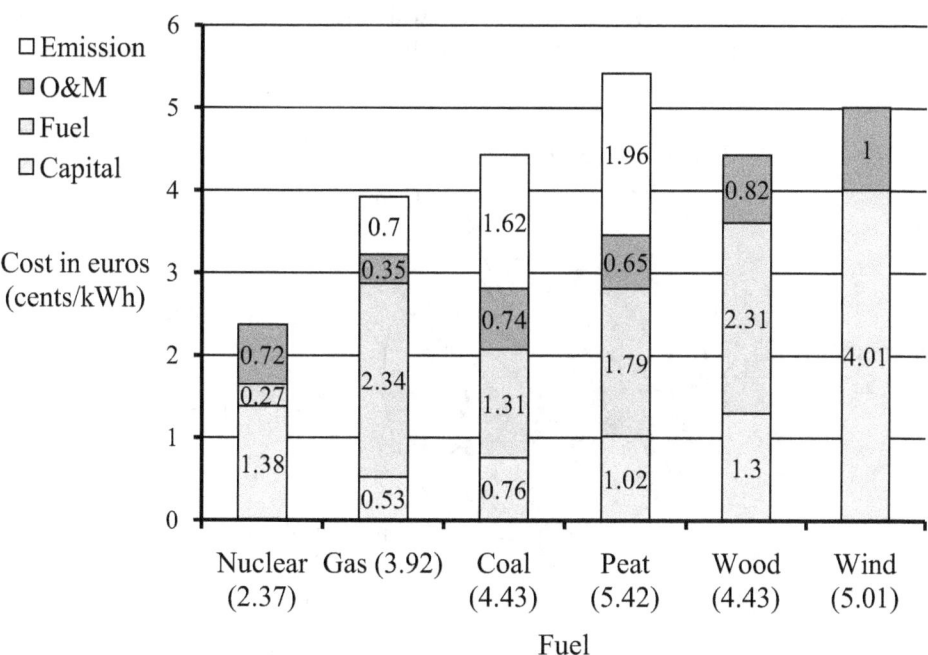

Fig. 6-26 2003 European costs for generating electricity in euro cents/kWh [WNA].

(2,200 hours of use per year) for wind. Solar was not included because of the uncertainties in its costs which tend to decrease with time, but are still comparatively expensive. As the volume of wind turbines has gone up, the price has come down so that, if a penalty is given for carbon dioxide emissions, the price of wind energy becomes quite competitive. According to this study, nuclear generation of electricity is the least expensive even if there were no carbon dioxide penalty, but there was no monetary penalty assigned to the geopolitical costs associated with nuclear energy. Other than wood, biomass sources were not included in the study. As usual, the carbon dioxide created by burning wood was considered to be recycled and no emissions penalty was assigned to wood burning.

The study also included the costs associated with nuclear, coal and gas generating plants in the United States. Assuming an 85 percent load factor, a five percent discount rate and a 40-year plant life, the costs in dollar-cents per kilowatt-hour was estimated to be 3.01 for nuclear, 2.71 for coal and 4.67 for gas. Because the United States had no penalties for carbon dioxide emissions, these figures do not include any penalties.

The 2009 division of renewable energy usage among the sectors and electricity generation in the United States is given in Fig. 6-27. Electricity generation accounted for over half of the renewable energy usage due to the fact that it included hydroelectric power consumption, the source of 34 percent of the renew-

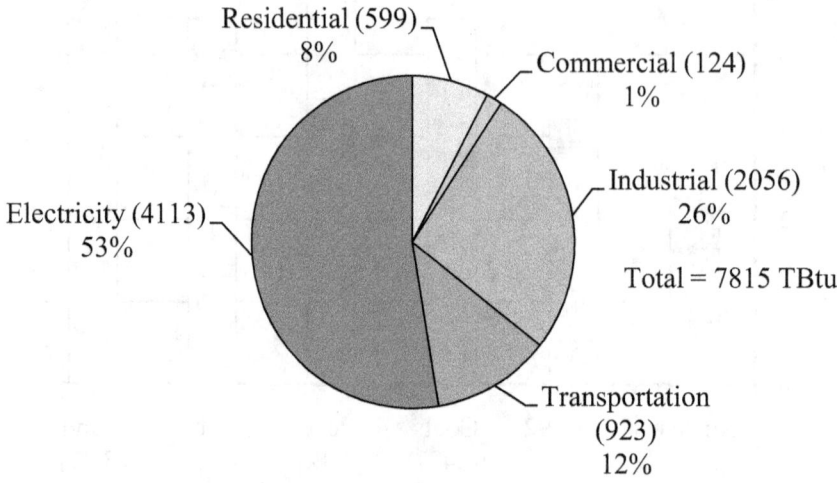

Fig. 6-27 2009 United States electricity and sector renewable energy usage in TBtu [EIA].

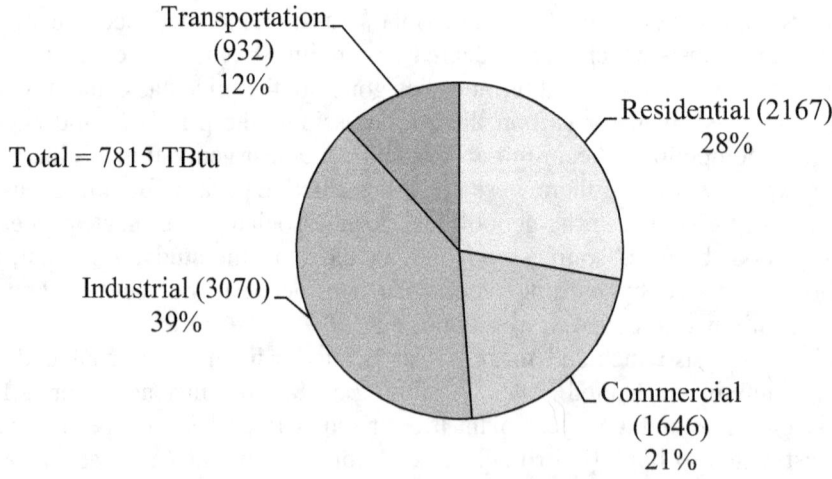

Fig.6-28 2009 United States sector usage of renewable energy in TBtu [EIA].

able energy used. Industry and residences obtained most of their renewable energy from wood and transportation got its renewable energy from biofuels, primarily ethanol. Figure 6-28 was obtained by assigning the electrical energy to the four basic sectors according to the total amount of electrical energy generated by renewable sources and the percentages of electricity consumed by the various sectors. Upon dividing electricity among the sectors, transportation's percentage of renewable energy increased very little because it used little electricity. It is seen that the bulk of the commercial sector's consumption of renewables was through electricity.

In the near term the most probable scenario is that wind and nuclear power will become a much bigger factor with respect to electricity generation and ethanol will replace an increasing amount of petroleum products in the transportation sector. A carbon dioxide penalty will be mandated in the United States that will cause a shift from coal to wind and natural gas (which produces only 56 percent as much carbon dioxide as coal when it is burned) as the means of generating electricity. In the longer term, solar energy will be a more significant source for both heat and electricity, and most new coal-fired plants will sequester their carbon dioxide or utilize algae ponds to produce biodiesel. Electricity consumption is likely to increase dramatically as electric and hybrid-electric vehicles become common. This increase will provide the impetus for an even greater expansion of wind and solar generating capacity.

7

ELECTROMAGNETISM

Most of the energy we use today can be traced to the electromagnetic energy we have received from the sun since the birth of our Earth over four and a half billion years ago. About eight percent is derived from the uranium deposited in the Earth at the time it was formed and somewhat less than one percent is due to the motion and heat derived from the radioactive elements in the Earth. The remainder, including the energy from fossil fuels, plants, animals, wind and water, owes its existence to the electromagnetic radiation of the sun.

Although all of the electromagnetism that we receive from the sun is in the form of radiation, there is a second form of electromagnetism that plays a much greater role in our ability to manage and use it, electricity. Humans have been aware of the presence of electric and magnetic effects for centuries, but these phenomena were not well understood until James Maxwell developed his equations describing electric and magnetic fields in the 1860s. *Electromagnetism* is a combination of these fields, their effects and the effects of their changes. *Electromagnetic radiation* is the propagation of these fields through space and matter and can be viewed as waves that comprise a continuum of frequencies. The lower frequencies make up the *infrared spectrum*, which can be sensed as heat, and a higher band of frequencies is the *light frequencies*, the ones we can see. Above the light frequencies are the ultraviolet frequencies, X-rays and so on (see Appendix A). In accordance with Maxwell's equations, a changing magnetic field causes an electric field, which, in turn, causes charged particles, normally electrons, to move. A net flow of charged particles is called a *current*, and currents and their effects is what is referred to in this book as *electricity*. In return and also in accordance with Maxwell's equations, electricity (moving charged particles) causes a magnetic field and electromagnetic radiation. It is our understanding this interplay between radiation and electricity that has enabled us to harness electromagnetic energy and use it so for many diverse purposes.

Our control of electromagnetism has been made possible by managing electric and magnetic fields to generate electricity, and then using electricity to supply energy to a large variety of devices and equipment. About 40 percent of the total energy we consume is used to generate electricity. Because so many purposes for which electricity is employed could be met by other means, why go to the trouble of generating so much electricity, especially considering that two-thirds of the energy may be lost in the generation and distribution of the electricity? First, it must be realized that not all electrical applications can be satisfied by other types of energy and there must be an electrical system to meet these needs.

The most important reasons for expanding an electrical system to include uses that could be satisfied in other ways are convenience and economics. It is much more convenient to use a light bulb than a kerosene lamp or electric motors instead of gasoline engines to power home appliances. Once electricity is provided to your home or business, why not take advantage of its controllability, quietness and nonpolluting attributes to do as many things as possible. For an American born after 1930 it is hard to imagine a world without a ubiquitous electrical infrastructure. It is so easy to convert electricity to other forms of energy that it may be employed in a large variety of situations. Although most of the ways in which we generate electricity create considerable pollution, there are ways that are nonpolluting, and once generated electricity is essentially pollution free. In cities where there are underground transmission lines, even the visual obtrusiveness of years past is avoided. The economic advantages are gained through the economics of scale made possible by building large centralized generating plants and an efficient distribution system.

The difference in potential for moving charged particles between two points in an electric field is called the *voltage difference* between those points and is measured in *volts* (*V*). Voltage is analogous to pressure in air or water. Pressure difference is a measure of the potential for moving the air or water and voltage is the potential for moving electrically charged particles, i.e., creating a current. In a closed water system, a pump provides the pressure difference needed to circulate the water. In an electrical circuit, a generator, solar cell or battery provides the voltage that causes the current to flow. It is the difference in voltage, which is commonly referred to as simply *voltage*. Current is measured in *amperes* (*A*) and the greater the voltage, the greater the resulting current. As indicated in Chap. 1, *power* is the generation or consumption of energy per unit of time. Electrical power is most often measured in watts (W) or kilowatts (kW), with a watt being one joule per second (J/s). A kilowatt is a thousand watts, a thousand joules per second (kJ/s), and a kilowatt-hour (kWh) is the amount of energy expended or generated over the course of one hour by one kilowatt. Also one kilowatt-hour (kWh) is equivalent to 3,412 British thermal units (Btu), or the amount of energy expended by burning 94 mL (3.2 oz.) of petroleum. But due to losses, the equivalent of approximately 280 mL (9.5 oz.) of petroleum is required by a typical generating facility to generate one kWh.

Figure 7-1 is a simplified diagram showing an electrical generating plant and the distribution equipment needed to transmit its electricity to places where it is utilized. Although the figure shows only one power plant and a single line of distribution, an actual system may consist of a large complex *power grid*, or simply *grid*. The grid may extend thousands of miles and contain numerous plants and an interlocking array of transmission lines, substations for changing voltage levels and other supporting equipment.

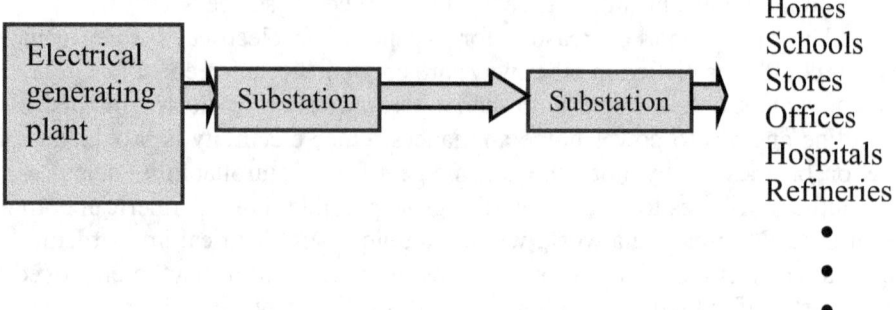

Fig. 7-1 Generating and distributing electricity to users.

There are two kinds of electricity. For *direct current* (*dc*), there is a constant net flow of electrons in one direction, but for an *alternating current* (*ac*), a voltage that reverses direction causes the electrons alternately go in one direction and then the other in the form of a wave. Batteries supply dc electricity only, but there are both ac and dc generators. In the United States, the direction changes 60 times per second, i.e., the current has a *frequency* of 60 *Hertz* (*Hz*), but in Europe and many other parts of the world the frequency is 50 Hz.

The advantage of ac electricity is that the voltage can be raised or lowered by equipment called *transformers* and, by transmitting electricity at a much higher voltage than is output by generators or is needed by the users, distribution losses can be reduced substantially (see Appendix A). Usually, a generator's output is a few thousand volts and residential and commercial customers use 110 and 220 volt electricity, but electricity is distributed at much higher voltages. Typically, long distance transmissions use 765 kilovolt (kV) lines. Therefore, located next to each generating plant is a *substation* containing transformers with high-voltage outputs and there are substations scattered throughout the distribution system for raising and lowering the voltage as required. Substations also include circuit breaker equipment for protecting the electrical system from overloads and disconnecting parts of the system. Also, *power factor* equipment is needed to keep the voltage and current aligned and some applications need equipment called *inverters* to convert dc electricity to ac electricity and equipment called *converters* to convert ac electricity to dc electricity. (See Appendix A for a more detailed discussion of ac and dc electricity.) All of this additional equipment represents added losses in the distribution system and in the electricity's final consumption.

Figure 7-2 shows the major components of an electrical power plant that uses steam to generate electricity. The fuel may be anything that produces the heat necessary to provide the steam. It may be anything that burns, usually a fossil fuel, but as seen in Chaps. 5 and 6 it is sometimes biomass (wood, solid waste, etc.) or uranium that produces heat as it disintegrates. If the fuel is coal, pulverizing equipment is needed to break down the coal before it enters the boiler. If it is uranium, it is not fed in continuously, but is put in the reactor's core until it has expended its useful energy.

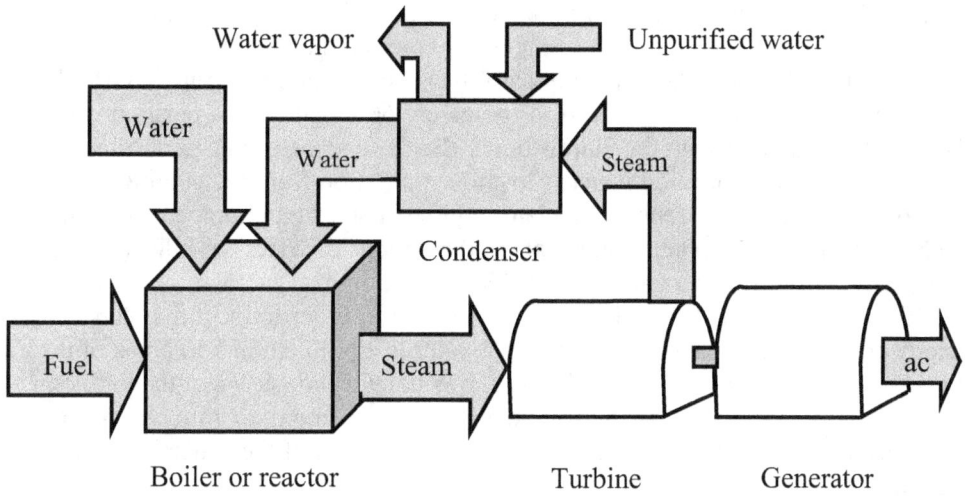

Fig. 7-2 Steam-powered electrical power plant's major components.

The steam is superheated to about 540 °C (1000 °F) and the high-pressure steam is used to drive a turbine. The steam drives the turbine in the same way an exploding gas drives a gas turbine. In fact, simple gas turbines that avoid the need for boilers and water may be used in combined-cycle systems (discussed below) or for applications that do not require a great amount of electricity. The turbine spins the generator and the interactions between the electric and magnetic fields within the generator cause a voltage difference at the generator's output terminals. If a *load* (i.e., something that provides a conductive path) is connected to these terminals, then a current will flow, thus supplying electricity to the load. The *load* may be an entire electrical distribution system or any equipment or system to which electricity is supplied.

In order for a turbine to provide the torque (i.e., turning force) required to turn the generator, the input pressure to the turbine must be much greater than the exhaust pressure. To accomplish this pressure difference the turbine's output steam must be condensed back into water. The output steam is fed into a condenser and the water output from the condenser is returned to the boiler where it is turned back into steam. To avoid corrosion, this steam/water stream must be purified. The *condenser* is a heat exchanger that transfers the steam's heat to a separate unpurified water stream in the cooling tower. The most visible part of the condenser is the cooling tower that accompanies the power plant. The unpurified water is cooled by the atmosphere and often produces a visible cloud of vapor. Because some of both the purified and unpurified water is lost, additional water must be continuously input to both streams. Although Fig. 7-2 depicts a simplified system with a single-stage turbine, most turbines have two or three stages and after the first high-pressure stage some of the steam is diverted to preheat the water entering the boiler.

A 500 MW power plant operating at full capacity requires about six GBtu/hr of input fuel (e.g., 250 tons of coal). Approximately 6000 gal/min of purified water needs to be circulated through the plant's boiler. Because some of this purified water is lost to leakage and unpurified water is lost to evaporation from the cooling tower, a steam-powered electrical generating plant requires a substantial amount of water. A 500 MW steam turbine plant requires about 0.5 gallons of water per kilowatt-hour of energy generated. If a person uses 200 kWh per month, then each month about 100 gallons of water is lost to leakage or evaporation in order to meet that person's electricity consumption. Supplying this much water is a problem in arid areas of the world such as the southwestern United States. It is noted in Chap. 9 that the problem is worsened if equipment is used to remove the carbon dioxide from the plant's exhaust because this equipment typically requires 18 percent of the generated energy to operate this equipment.

Although simple gas turbines tend to be inefficient, there are combined-cycle turbines that employ a heat exchanger to use the heat of the exhaust gases from the gas turbine to heat the water used by a steam turbine. This use of the heat in the exhaust gases increases the efficiency considerably. A diagram of the gas turbine portion of a combined-cycle system is given in Fig. 7-3. The steam turbine may drive a separate generator, as implied by this diagram, or a single shaft may be used and both turbines would then drive the same generator. Because exhaust gases provide some of the heat instead of circulated water, a 500 MW plant without carbon dioxide capturing equipment that uses combined-cycle turbines would require only about 0.3 gallons of water per kilowatt-hour of generated power.

Alternatively, if the plant is located in an active geothermal area, the steam from the earth may be supplied to a heat exchanger as shown in Fig. 7-4. There may or may not be a boiler depending on the temperature of the steam obtained from the geothermal well. As with other steam systems, the outputs of the turbine and heat exchanger must pass through condensers and the water output from the heat exchanger is normally pumped back into ground.

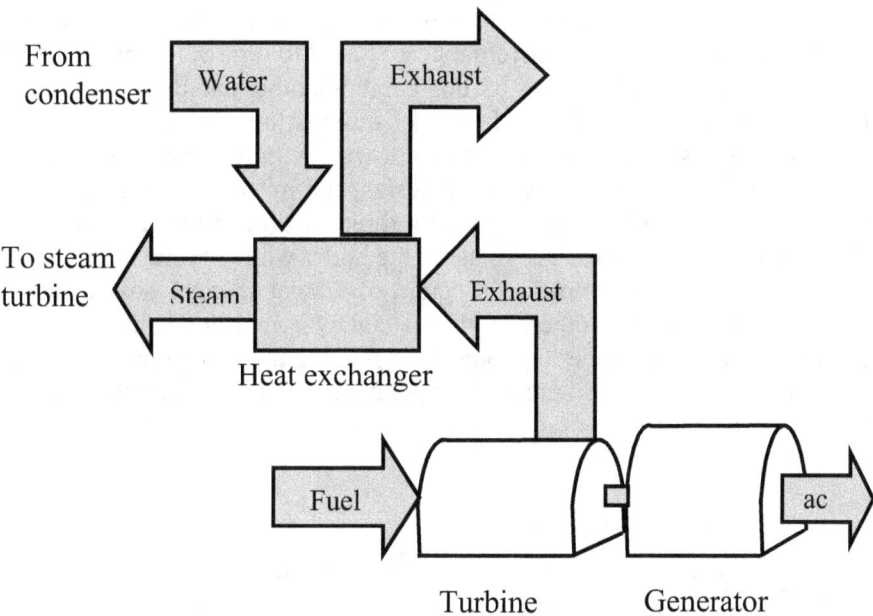

Fig. 7-3 Gas turbine portion of a combined-cycle system.

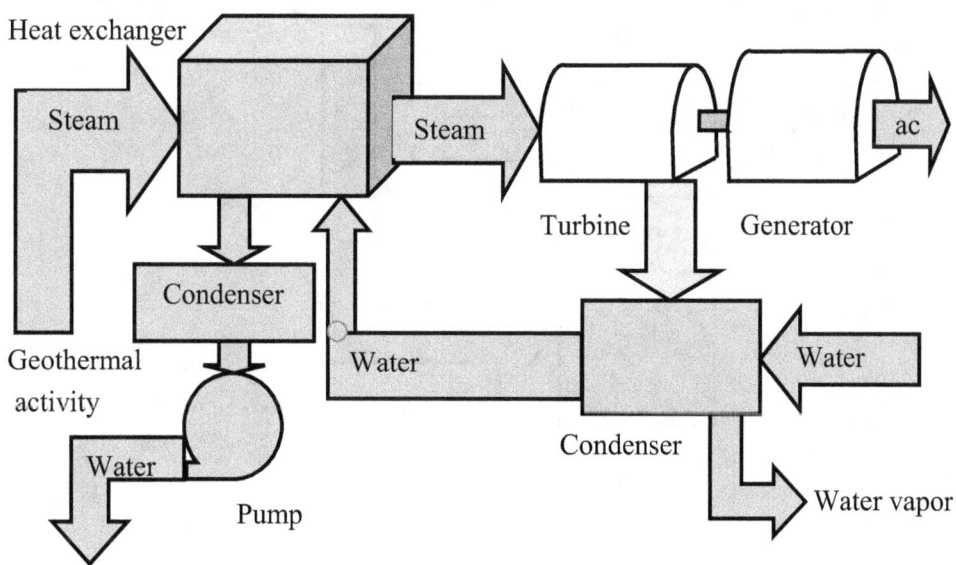

Fig. 7-4 A typical geothermal power generating plant.

There are two fundamental means of collecting energy from the sun directly. One method uses a large array of mirrors to direct the energy to a focal point where the concentrated energy heats a special fluid, called a *heat transfer fluid* (*HTF*), which is sent to a *heat exchanger* where the heat is used to turn water into steam. In some arrays the mirrors are like halves of long parabolic tubes and reflect the energy toward pipes that contain the HTF, and in other designs the energy is collected at one central point that is located at the top of a tower. A third method is to use an array of parabolic dishes. Regardless of the design, the mirrors move so that they are, within the limits of their design, optimally directed toward the sun and, along with the heat exchanger, either replace or assist the boiler while the sun is shining.

A representative but simplified diagram of a solar thermal power plant is shown in Fig. 7-5. To this design there may be added a thermal storage tank, an HTF heater, a steam preheater and other equipment for making the system more efficient. Also not shown is the control equipment for tracking the sun. Another possibility is

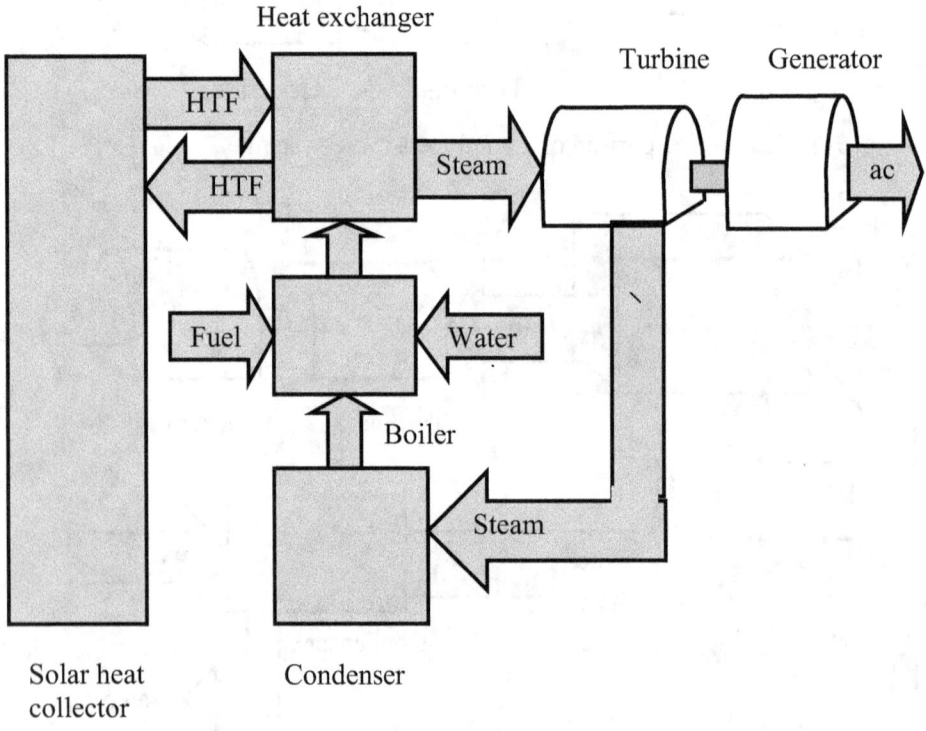

Fig.7-5 Simplified solar thermal power plant.

to heat water directly from the sun's radiation. But an HTF with a high specific heat capacity is used because it can hold more heat at a lower pressure. The solar array would need to be designed to withstand much greater pressures if the water were heated directly. Such power plants may be quite large. A plant to be located in the Mojave Desert is to have a capacity of 553 MW and cover 24 km^2 (6000 acres).

As introduced in Chap. 6, the other means of converting solar radiation into electricity is to construct panels containing large photovoltaic sheets or arrays of solar cells and connect them together in such a way that they produce a suitable voltage and current. To date, most solar cells are for generating the electricity requirements for a single home or business. A diagram of such a system is given in Fig. 7-6. Because solar cells produce dc electricity, a dc to ac inverter must be included to obtain the ac electricity required by the appliances and other equipment in the home or business and to match the electricity on the power grid. The system may be connected to the power grid, as shown in Fig. 7-6, or have a bank of batteries for storage that could provide electricity when the sun is not shining. If it is connected to the power grid, then a meter is included that allows the home or business to receive credits when the solar cell system is producing excess power and

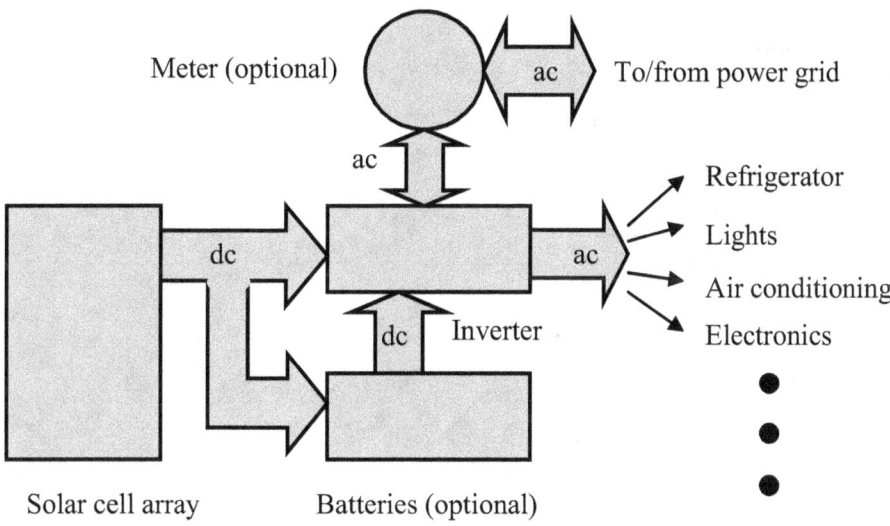

Fig. 7-6 Diagram of a solar cell system for a home or business.

debits when power must be drawn from the power grid. How the credits and debits are translated into money depends on current state and national laws. While steam powered plants tend to be large, centralized systems that take advantage of the economy of scale, solar cell systems are relatively small and distributed, but reduce the transmission costs. Large solar cell systems for the sole purpose of supplying power to the grid will become common when they become economically competitive.

The steam turbine in Fig. 7-2 may be replaced by a hydro-turbine or wind turbine, in which case the generator is driven directly by water or wind and there is no need for a boiler or heat exchanger. There are two basic systems that use water to drive the turbine. The more prevalent one is found in dams that are associated with large lakes and is shown in Fig. 7-7. The other option is to put the turbines in dams

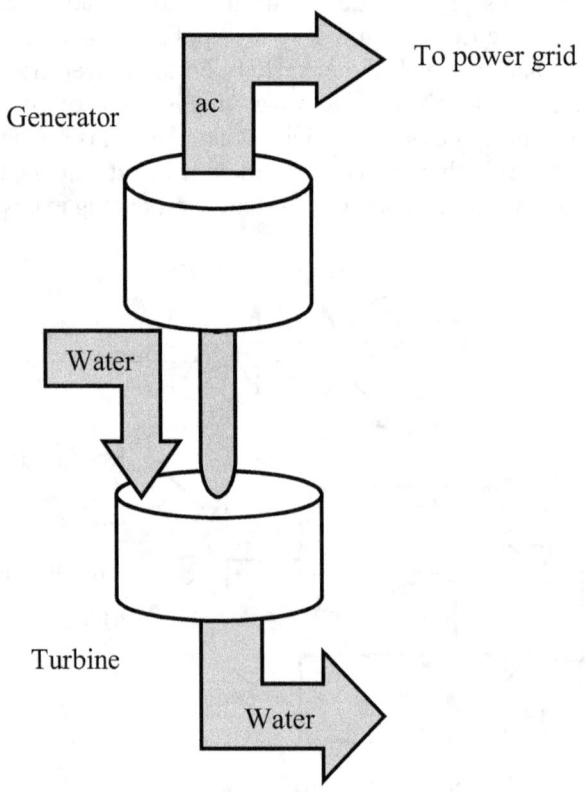

Fig. 7-7 Hydroelectric dam generating design.

located at ocean inlets or narrow bays where they could take advantage of the tides. But in this case, they are constructed and controlled more like wind turbines because the water pressure varies considerably and is much lower than that below a lake. The main problem with tidal dams is that there are very few places where they can be installed and be economically viable. There is some interest is using ocean or river currents to power underwater windmill-like energy collectors, but again there are a limited number of places where they could be profitably constructed.

The largest hydroelectric facility is the Three Gorges Dam in China, which has a generating capacity of 18,200 MW. The largest hydroelectric system in the United States is at the Grand Coulee Dam on the Columbia River and has a capacity of 6,800 MW. The total energy generated by hydroelectric facilities in the United States in 2008 was 737 GkWh, or about 19 percent of the 3,879 GkWh of electrical energy that was generated that year. Of the industrialized countries, only Canada has plans for additional large hydroelectric projects. On the other hand, Southeast Asia, including India and China, has substantial plans for additional projects. Also, some South American countries, including oil rich Venezuela, intend to increase their hydroelectric capabilities.

The energy of the wind is emerging as the most likely renewable replacement for fossil fuel in the near future. A typical system based on a wind turbine is shown in Fig. 7-8. Because the rotation of the windmill is only 5 to 20 revolutions per minute (rpm) and the wind velocity varies so much, a gear box is

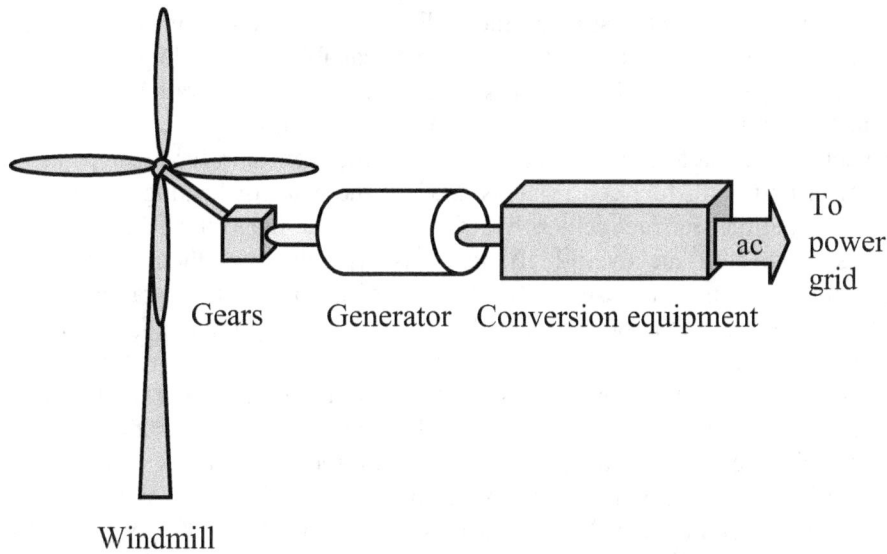

Fig. 7-8 Representative wind turbine design.

needed to increase the rotational speed and the first stage is a variable speed generator. To match the frequency and voltage of the power grid, the output of the generator is connected to conversion equipment. The design of this equipment varies from system to system and may be broken into stages with some of the stages being located in the tower and some of them at a central wind farm location. Not shown in the figure are the controls for maintaining the optimal direction of the windmill and braking. Although the efficiency of a wind turbine tends to increase with the number of blades, this increase is offset by the structural cost of the blades and support tower. It may be better to spend the money on more windmills or bigger blades. Small windmills may have two or three blades, but larger windmills almost all use three blades. Increasing the blade count to four is not justified by the small amount of increased efficiency. As indicated in Chap. 6, a single wind turbine may have a maximum capacity of as much as 6 MW, but more typically has a capacity between 1.5 and 4 MW.

In addition to the wind turbines, a wind farm must have an electrical collection system, access roads, a maintenance and operations building and a substation where the farm's output is connected to the grid. The generator voltage is normally between 550 and 690 volts and the collection voltage is typically between 25 and 35 kilovolts. The substation would boost the collection voltage to that of the grid.

Another means of generating electricity that is currently used very little, but holds promise in some situations, is the fuel cell. Fuel cells are generally associated with providing energy for electric cars, but systems capable of providing megawatts of power have been developed. As with steam turbine-based generation, fuel cells produce heat as well as electricity and are normally less than 70 percent efficient if only the electricity is put to useful work. However, some homes in Japan supplement their grid electricity with fuel cell systems and use the heat for heating water. The problem is that the fuel for fuel cells is hydrogen and when the system for obtaining hydrogen is considered the overall efficiency is typically less than 50 percent. However, larger systems that use molten carbonate or solid-oxide ceramic as the electrolyte may be more than 60 percent efficient. If the heat produced is also used, then the efficiency may be as much as 85 percent. The most convenient means of supplying hydrogen to a home or business is to obtain it from the natural gas that is already distributed to the home or business, although it could be produced from liquid methanol. The production of hydrogen from either natural gas or methanol results in the byproduct carbon dioxide. An example fuel cell system is the one in the 4 Times Square Building in New York City. It includes two 200 kW fuel cells for heating water and generating much of the building's electricity. The principal downside to fuel cells is cost. Fuel cells are discussed in greater detail when transportation is considered in Chap. 8.

Figure 7-9 gives the percentages of the various sources used to generate the world's electrical energy in 2007. Also, given in parentheses are the amounts of electricity generated by these sources in giga-kilowatt-hours (GkWh). However, these amounts do not include the losses incurred in generating and distributing the electricity; they are only the amounts consumed by end-users. The worldwide total amount of consumed electricity in 2007 was 16,990 GkWh, which is the energy equivalent of a little over ten billion barrels of oil or 2.57 billion tonnes (2.83 billion short tons) of coal. It is seen from the figure that fossil fuels are used to generate about two-thirds of the world's electricity.

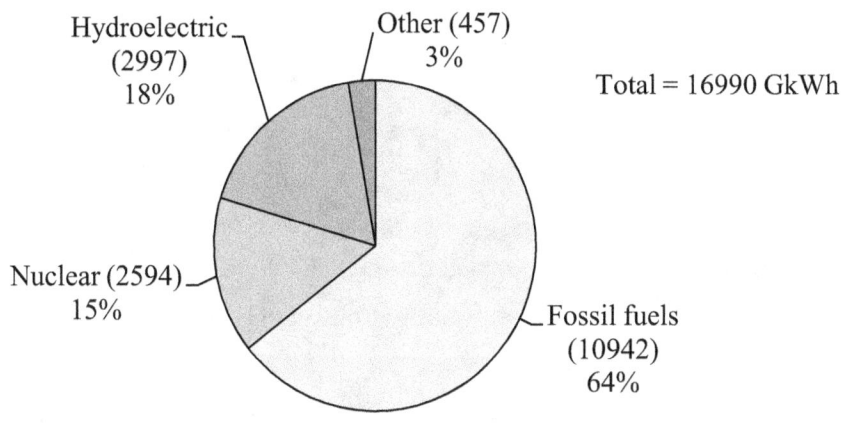

Fig. 7-9 2007 world consumption of electricity by energy source [EIA].

The 2009 amounts and percentages of the sources used to generate electricity within the United States are given in Fig. 7-10. As in the United States, about half of the world's electricity is generated by coal. The choice of energy source is largely dependent on its availability. The United States and China have vast coal reserves, while Iceland gets virtually all of its electricity from either hydroelectric or geothermal power plants, and 98.5 percent of Norway's electricity is generated from hydropower. France, because of its shortage of fossil fuels and limited suitable hydroelectric sites, depends on nuclear power for almost 80 percent of its electricity. Due to the polluting nature of coal, both China and the United States are in the process of expanding their nuclear power capabilities substantially within the next decade. Denmark gets much of its electrical power from wind turbines and Germany, which also has limited fossil fuel resources, is currently investing heavily in its solar power capability. As in Fig. 7-9, the amounts in Fig. 7-10 do not include generation and distribution losses

The efficiency of a power station is dependent on the energy source it utilizes and the sizes and ages of its generators. Newer generating plants tend to have larger generators that are more efficient. The bar graph in Fig. 7-11 gives the United States 2008 average sizes of generators according to the energy sources used. To be economically viable, nuclear facilities are usually built on a grand scale with

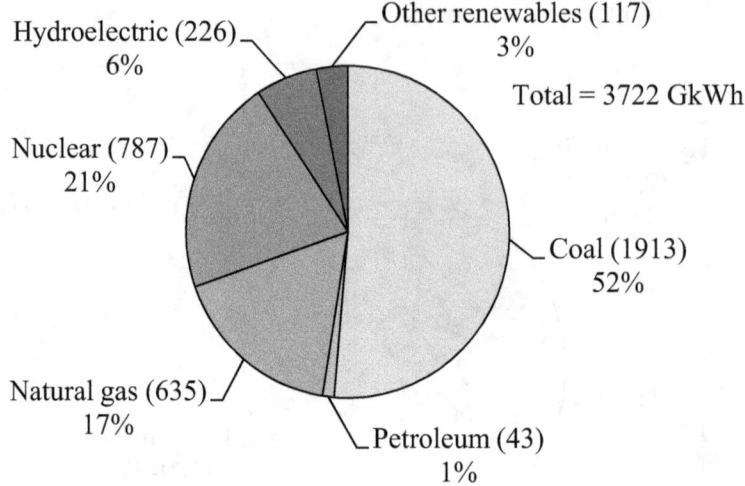

Fig. 7-10 2009 United States consumption of electricity by energy source [EIA].

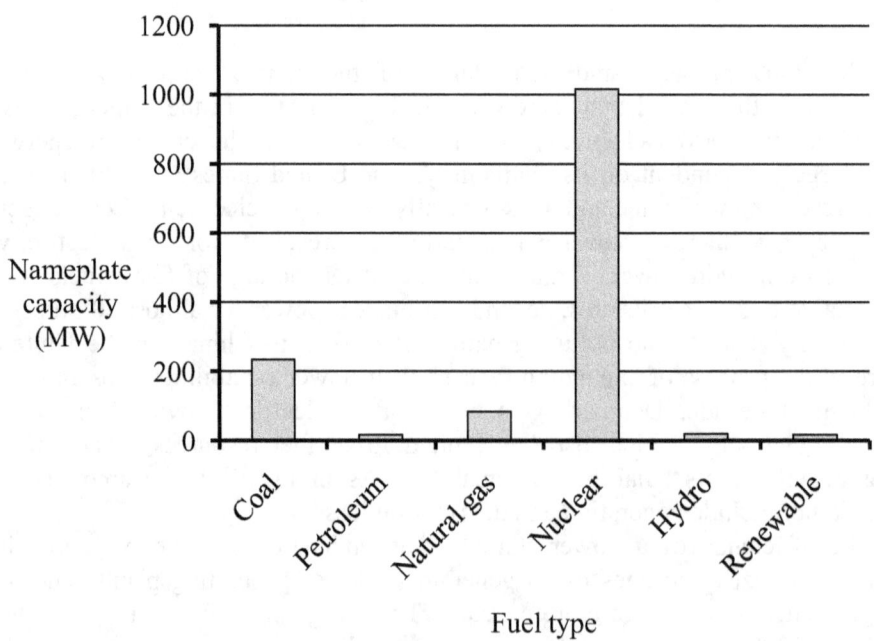

Fig. 7-11 2008 average generator capacity by energy source [EIA].

large generators. The generators in coal-fired plants also tend to be fairly large. Plants that are fired by petroleum are generally smaller and their generators are correspondingly smaller. Hydro-turbines have less capacity, but modern hydroelectric plants may have generators with capacities as large as 800MW. The capacity of a wind turbine is limited by the size of its blades and, as mentioned above, wind farm wind turbines tend to have capacities between 1.5 and 4 MW.

A typical boiler has an efficiency between 85 and 90 percent and the generator is even more efficient. Most of the energy loss in a plant that uses steam turbines is due to converting the steam back into water. But if the heat energy in the steam could be extracted for a useful purpose while it is being cooled, then the efficiency of the plant could be improved substantially. Use of this heat energy for another purpose while generating electricity is called *cogeneration*, or *combined heat and power* (*CHP*). CHP is most often used to heat buildings or to assist industrial processes that require heat. It is also used for in situ heating of tar sands or tertiary extraction of petroleum from depleted oil wells and, in the future, may be used to extract petroleum products from oil shale deposits. For an in situ operation, the fuel for firing the boiler is taken directly from the site itself. Being able to claim more of the output energy as useful can boost the efficiency of a plant to between 40 and 80 percent, but is typically in the neighborhood of 60 percent. The reason for the wide variation is the varying requirements of the heat applications. For many applications the steam cannot be allowed to condense and, thus, not as much of the heat can be recovered. Also, for some applications the turbine's output pressure and temperature must be carefully controlled and this may reduce the efficiency of the system. If the excess heat produced by a gas turbine is used to produce steam for a separate steam turbine (i.e., the turbines form a combined cycle system) then the efficiency of the overall system is typically raised to over 50 percent. The average efficiency of generating plants in the United States in 2009 was approximately 35 percent.

While using standardized methods for comparing efficiencies of hydro-turbines or wind turbines of different designs makes sense, comparing these efficiencies with those of steam turbines is somewhat dubious. The reason for this is that the computation of the input energy is so different. For a steam turbine the efficiency is computed from the usable chemical energy of the fuel or nuclear energy of the uranium, but for a hydro-turbine the input energy may be estimated by using the pressure difference across the turbine and the water flow through the turbine. Hydro-turbines have efficiencies between 80 and 90 percent. For a wind turbine, using pressure difference and air flow must be highly standardized just to make comparisons between wind turbine designs.

If the generation of electricity were spread evenly over time, the required generating capacity for the world would be 1,938,000 MW. But there is a limit to the economically feasible distance that electricity can be transmitted and, for a given area, the demand for electricity is much more at some times than at others. Less electricity is required at night than during the day and in the winter than in summer if air conditioning is needed. Steel mills and other large industrial consumers can cause surges that can be satisfied only by drawing power from several generating plants. The generating capacity for an area must be large enough to supply the peak demand for that area. As a consequence, the generating capacity for an area must be considerably larger than it would be if all generators could be used at their maximum capacities all the time. The problem is partially alleviated by constructing huge power grids extending over hundreds of thousands of square miles and providing some interconnections between the grids for emergency purposes. One such grid covers the northeastern United States, some of the Midwest and part of Canada. This grid failed in August, 2003, and the entire area was blacked out for the better part of a day before power was fully restored. The economic loss during this period was enormous.

The worldwide 2007 generating capacity was 4,147,000 MW, two and one-eighth times the average power demand indicated above. The United States 2008 capacity was 1,022,000 MW, which is similarly two and one-third times its average demand. Even though this excess capacity is costly in terms of capital expenditure, it is necessary to provide a stable system and prevent blackouts such as the one that occurred in 2003. There is a tradeoff. The larger the grid the less the total generating capacity needs to be, but the grid becomes more complex and difficult to control and this tends to increase the need for a greater cushion in the total capacity of the grid.

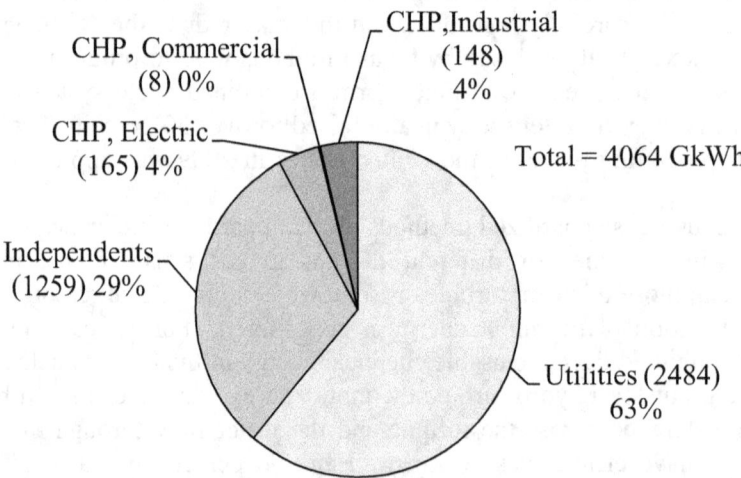

Fig. 7-12 2006 power generated by various entities in the United States (EIA).

The EIA divides those who generate electricity into five categories as indicated in Fig. 7-12. Three of the categories are for those who employ CHP. Those who use CHP and whose primary businesses are for generating electricity are in the category "CHP, Electric". "CHP, Commercial" and "CHP, Industrial", indicate the amounts generated by commercial and industrial establishments. There are some residences that generate electricity as well, but at present their combined contributions are insignificant. The categories "Independents" and "Utilities" include all generating plants that do not use CHP. The total electricity produced by all generating plants in 2006 was 4,064 GkWh and the total for plants that use CHP was 321 GkWh, or eight percent.

In addition to the transmission line losses in a grid, there are losses in the transformers and other equipment that the electricity goes through before it is consumed. The losses in an electrical distribution system are those due to the resistance of the transmission lines and equipment, electromagnetic radiation, and leakage current. The losses depend on the distances and amounts of equipment traversed as well as the transmission voltages. The resistive losses produce heat and are the reason high-voltage transmission lines are used to reduce the current. Large pieces of equipment, such as transformers, need fins and fans to disperse the heat into the atmosphere. Leakage currents exist because there are no perfect insulators, not even the air. High-voltage transmission lines are surrounded by intense electric fields that cause some leakage currents through the air. These leakage currents, accompanied by radiation, sometimes appear as corona. High voltage transmission systems are three-phase, three-line systems—the three phases being equally displaced voltage waveforms. Under extreme conditions, corona may result in arcs between the lines, the lines and the ground or the lines and metal transmission towers. Radiation increases with frequency and is relatively low at 50 Hz (Europe) or 60 Hz (United States). Total distribution losses are between 5 and 12 percent. The average 2008 distribution losses in the United States were approximately 6.6 percent.

If there is a 7 percent energy loss in producing and processing the fuel, a 64 percent loss in the generating plant and a 7 percent loss in the distribution system, then

$$0.93 \times 0.36 \times 0.93 = 0.31,$$

or 31 percent, is the overall efficiency from the extraction of the fuel to the delivery of energy to the consumer. Clearly, an improvement in any one of these efficiencies would improve the overall efficiency. In particular, note the importance of the generation losses. If CHP raises the plant efficiency to 60 percent, then the overall efficiency would be 52 percent. The consumer's equipment would further reduce the actual energy that produces useful work. For example, it may be that less than five percent of the energy supplied to a light bulb is converted to visible light. A light bulb efficiency of five percent would result in an overall efficiency of 1.56 percent. Clearly, end-use equipment also has a major impact on the overall efficiency measured from fuel source production to useful work.

One means of reducing the losses in a distribution system is to use superconducting transmission lines. A material is *superconductive* if it has zero resistance to current flow. In ordinary conductors the free electrons bounce off the atoms that make up the conductor and create heat, but in superconductors they do not. Superconductivity was first discovered in 1911 by Heike Onnes while experimenting with mercury at cryogenic (near absolute zero) temperatures. Since then several other materials have been found to be superconductive at extremely low temperatures, and in 1986 cuprates were found to be superconductive at -181 °C (-294 °F). The importance of this discovery is that the material can be kept superconductive by liquid nitrogen, which condenses at -196 °C (-321 °F). By refrigerating a transmission line with liquid nitrogen, electrons can be transported with no resistive heat loss. As of 2008 superconducting transmission was experimental, but New York City has plans to connect some of its large substations using superconducting lines so that blackouts may be avoided by transporting energy between them quickly.

One theoretical design indicates a 1,600 km (1,000 mi) line with a 5,000 MW capacity could be built that would transport energy at about the same cost as that of a natural gas pipeline. This design has refrigeration stations every 16 km (10 mi) and vacuum pump stations every 1.6 km (1 mi). Including the refrigeration channels, structural support and conducting material, the line would be 67 cm (26.4 in) in diameter. It would transport energy at a rate equivalent to moving about 71,000 barrels of petroleum 1,600 km (1,000 mi) per day. Such a line would be able to connect the densely populated eastern United States to areas more amenable to electricity generation by the wind and direct sunlight. This design, however, is theoretical and to claim superconductivity could compete with the mature technology of a gas pipeline is highly speculative. Also, the superconducting cable would be surrounded by an intense magnetic field while it is transporting 5,000 MW and no one knows the precise dangers this might pose to humans and other life.

Figure 7-13 gives the amount of electrical energy taken from the power grid and consumed in the United States in 2009 by the four EIA sectors. The total amount consumed was 3,722 GkWh, but this total does not include the 88 GkWh that is produced and consumed on site and is not output to the power grid. Also, it does not include the electricity produced by the generators in vehicles. These figures include only the electricity supplied by the power grid and not the electricity lost by the grid. The total amount used by the transportation sector was less than half a percent and was almost entirely due to the electricity consumed by light rail. This will change in the future as we begin to rely more and more on plug-in hybrid and purely electric vehicles.

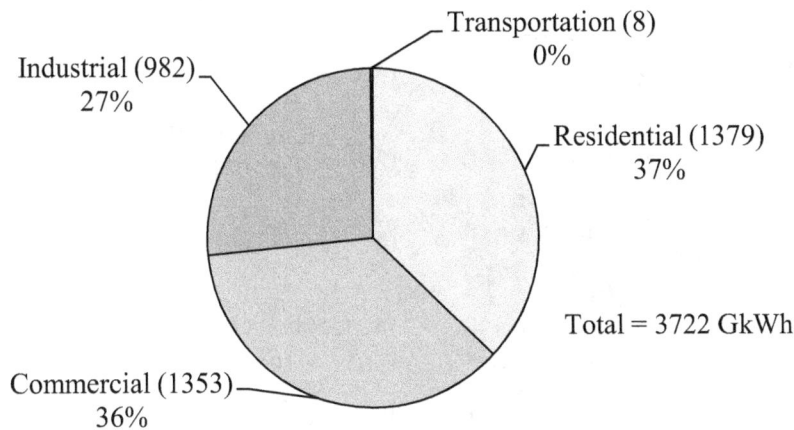

Fig. 7-13 2009 sector electricity usage in the United States in GkWh [EIA].

The residential sector was the largest user of electricity, but was closely followed by the commercial sector. The results of an EIA sponsored survey of residential electricity usage in the United States conducted in 2001 are shown in Fig. 7-14. This figure does not include the 62 GkWh of the wiring losses within the residences. "Refrigeration" includes both refrigerators and freezers and "Cooling" includes both refrigerated and evaporative cooling (the type of cooling often used in the dry southwest). The "Kitchen" category is comprised of all kitchen equipment not accounted for by refrigeration. "Electronics" represents all equipment that is primarily electronic in nature, such as televisions, computers, printers, stereos and so on. Refrigeration and cooling are responsible for one third of the residential electricity consumed and heating the air and water account for almost an additional one-fourth. In 2001, the average home consumed about 29 kWh of electrical energy per day at an average rate of 1.21 kW. Because more power is needed at some times than others, a typical home should have access to at least 3 kW of power.

The results of a similar EIA survey conducted in 1999 for the commercial sector is given in Fig.7-15. Again the internal wiring losses are not included. The "Cooking" category has replaced the residential "Kitchen" category. "Ventilation" was not considered significant in the residential sector. Refrigeration and cooling again consume almost one third of the electricity, but the electricity used for heating is much less. This does not imply that heating uses less energy in commercial buildings, it means that electricity is not the source of the heat. At 41 percent, lighting and electronics play a more important role in the commercial sector than in the residential sector.

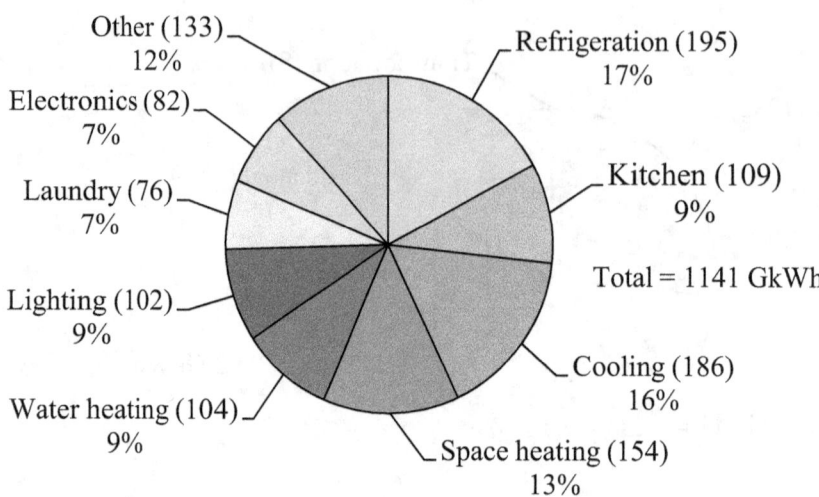

Fig. 7-14 2001 residential electricity usage in GkWh [EIA].

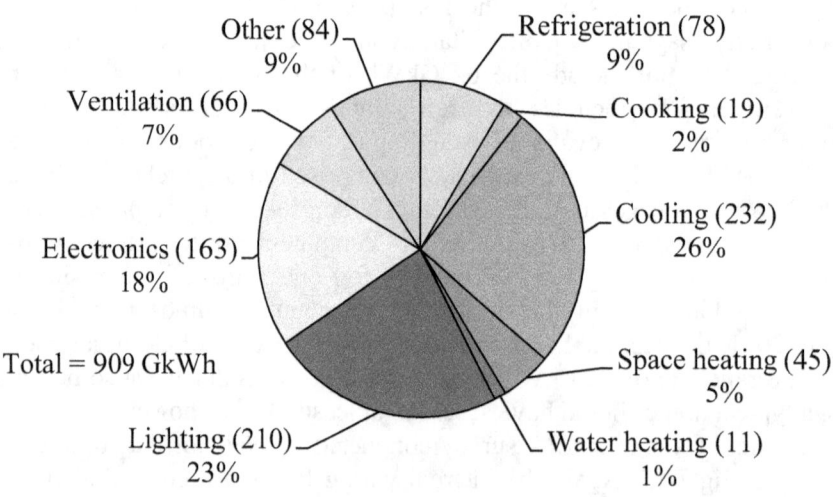

Fig. 7-15 1999 Commercial electricity usage in GkWh [EIA].

Figure 7-16 shows the 2003 electricity amounts and percentages consumed by the various EIA-defined types of commercial establishments. Some of the EIA types have been combined in this figure. (Note that the term "commercial" is somewhat misleading because religious, education and government buildings have been included.) "Mercantile" encompasses all types of retail stores, except those that mainly sell food and are indicated by the "Food sales" category. "Lodging" represents commercial enterprises, such as hotels, and does not include homes, which are in the residential sector. The "Office", "Mercantile" and "Education" categories account for slightly over half of the electricity used.

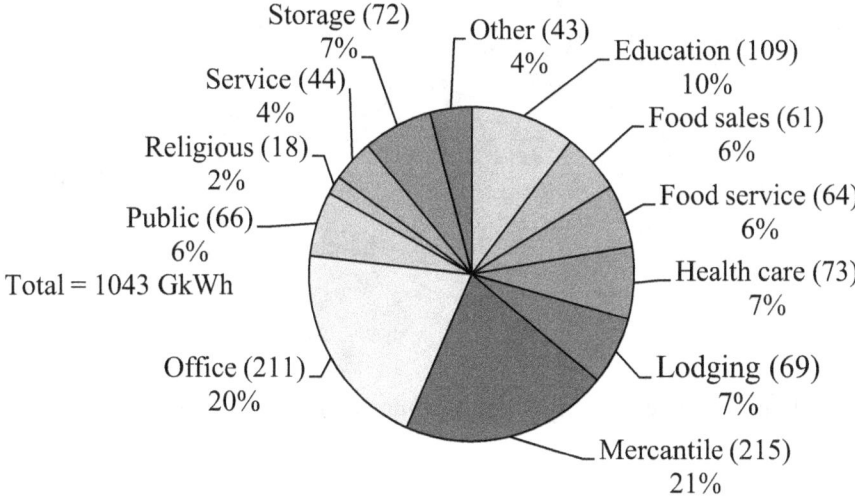

Fig. 7-16 2003 commercial electricity usage by type of consumer in GkWh [EIA].

A 2003 survey of the industrial sector produced the amounts and percentages given in Fig. 7-17. Once again EIA categories are used, but some of the categories given in the figure are combined EIA categories. "Food/drink" includes tobacco, "Wood" indicates the lumbering, kiln and wood milling industries, "Fossil fuels" encompasses the extraction and refining of coal, petroleum and natural gas and "Machinery" includes metal fabrication as well as the manufacturing of machinery. "Transportation" indicates the manufacture of all types of vehicles—automobiles, trucks, buses, trains, boats and airplanes. The EIA categories represent industries involved in processing or manufacturing. Over half of the total is for the extraction and processing of fossil fuels, chemicals, metals and nonmetals such as plastics. All four of these components require a disproportionate amount of heat and this provides an opportunity for CHP generation of electricity.

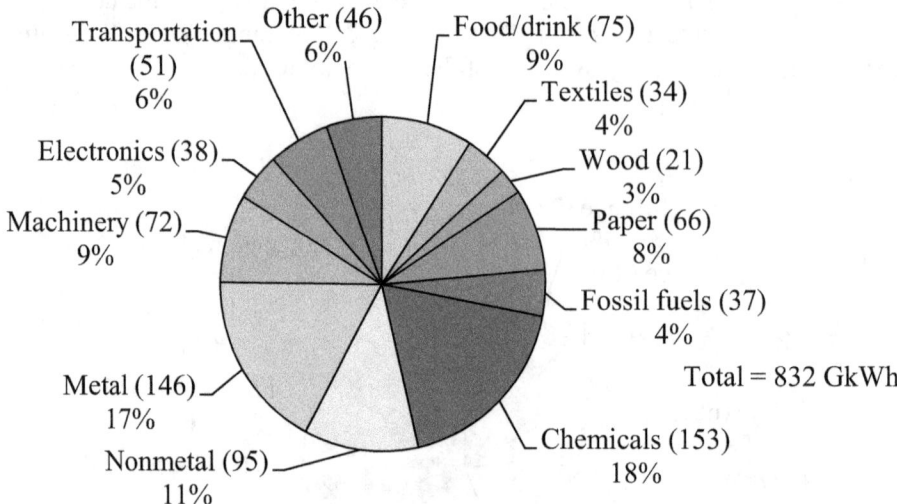

Fig. 7-17 2003 industrial electricity usage by consumer in GkWh [EIA].

There are numerous possibilities for saving energy in our power system and many of these possibilities involve making more efficient end-user products. The products receiving the most publicity are those for producing light. There are three basic means for producing light, incandescent bulbs, fluorescent bulbs and light-emitting diodes (LEDs). Incandescent light bulbs are constructed of a bulb containing a filament surrounded by an inert gas such as argon. When a current is passed through the filament it radiates electromagnetic energy. Some of this energy is in the band of light frequencies, but, unfortunately, most of it is in the band of infrared frequencies and incandescent lighting is notoriously wasteful of energy because of the heat they produce. Essentially all of the energy input to any lighting fixture is dispersed as either heat or light and for incandescent bulbs only two to three percent is radiated as light.

Fluorescent bulbs consist of tubes internally coated with phosphorous and filled with a vaporous form of mercury. When a voltage is applied to the ends of the tube, electrons are released from the vapor leaving positive ions behind. The electrons form a current that flows out of the tube and are replaced by electrons entering from the other end of the tube. The entering electrons combine with the positive ions and the recombination process gives off electromagnetic frequencies in the ultraviolet band. These frequencies are absorbed by the phosphorous and reradiated as light frequencies. Although fluorescent bulbs require associated circuitry that uses energy, they are 7 to 9 percent efficient. Their improved efficiency is due to producing a higher percentage of their energy as visible light. The primary disadvantage of fluorescent bulbs is that they contain mercury, which is extremely poisonous, and their disposal presents problems (see Chap. 9).

The construction of an LED is similar to the solar cell shown in Fig. 6-17, but the materials used are different. As with fluorescent bulbs, when a voltage is applied to an LED ionization occurs and there is a flow of electrons out of and into the LED. The incoming electrons recombine with the positive ions left behind by the outgoing electrons and electromagnetic energy is radiated. The LED design is simpler than that of a fluorescent bulb and an LED produces a much greater percentage of light frequencies than an incandescent bulb. LED lighting is up to 13 percent efficient.

In addition to the quantity of light indicated by efficiency, the quality as perceived by the human eye must be considered. More important than efficiency is luminous efficacy. *Luminous efficacy* is luminous flux divided by input power. According to an August, 2009, article by Richard Stevenson in the *IEEE Spectrum*, the luminous efficacy for incandescent bulbs is about 16 lumens per watt, for fluorescent bulbs it is over 100 lumens per watt, and for LEDs it is as much as 250 lumens per watt. As research continues, affordable, high-efficiency and high quality LED lights are expected to become available. Some LEDs are already in use in traffic lights and car indicator lights. The basic problem with LEDs is known as droop. *Droop* is the property of blue light LEDs (the type of LEDS needed to produce high quality white light) that results in them losing efficacy as their power input increases. Therefore, either they are good for only low power applications or it takes several LEDs to produce an adequate amount of light.

Between one fifth and one sixth of the electricity consumed is for lighting. However, much of this lighting is in the commercial and industrial sectors which already use fluorescent lights, so the overall benefit of converting to fluorescent and LED lighting is limited. If the electricity needed for lighting were reduced to half the 2008 amount by using more efficient bulbs and being less wasteful, then the fossil fuels for producing that electricity would be reduced by about three and a half percent. Since 70 percent of the electricity generated by fossil fuels is generated by coal and only 4 percent by petroleum, this reduction would mainly affect coal consumption and have very little effect on our overall petroleum usage. On the other hand, burning coal is the most polluting fossil fuel and emits more carbon dioxide than the other two fossil fuels. Any technology that reduces the amount of coal consumed is certainly beneficial.

Another area in which significant improvements are being made is in electronics and semiconductor manufacturers and internet equipment providers are continually trying to reduce the energy requirements of their products. There have been drastic changes since the days when computers of modest capabilities required their own cooling systems and most electronic equipment needed at least a fan to dissipate the heat. With respect to electronics the 2001 data given in Fig. 7-14 for the residential sector and the 1999 data in Fig.7-15 for the commercial sector are outdated. The spiraling usage of information and communications devices has caused an exponential growth in the energy consumption by electronics. A prime example is the usage of the internet. Internet servers consume an enormous amount of energy and the personal computers connected to the internet consume even more. One problem is that most electronics devices, particularly those connected to the internet, use energy even when they are supposedly turned off. To have an internet presence, at least some of a device's circuitry must remain operational. Considerable effort is being expended to reduce the energy wasted by internet-related equipment while they are in sleep mode. A single personal computer in sleep mode consumes very little energy, but worldwide there are billions of personal computers. It is difficult for the improved efficiencies in electronic equipment to keep up with the expansion of their usage.

Although employing CHP does not reduce the demand for electricity, it does reduce the demand for energy. CHP can greatly relieve the use of fossil fuels in the commercial and industrial sectors. However, CHP generation must be in the vicinity of where the heat is needed and is applicable only in industrial plants, cities and towns and sites such as those where heat is needed to extract oil from tertiary wells or tar sands. In large, crowded areas such as Manhattan there are companies that sell heat as well as electricity and Iceland uses both CHP and its geothermal activity to provide much of its heating requirements. Expansion of CHP generation offers a notable avenue to reducing our use of fossil fuels.

One of the greatest impediments to tapping wind and solar energy for generating electricity is the problem of storing the energy when conditions are good or demand is low for times when conditions are bad or demand is high. Wind is constantly changing and the sun is up only during the daytime and the length of time there is daylight varies with the seasons and weather. Also, means are needed to level out the short term electricity flow such as a sudden power outage in part of the grid or an event that requires a surge in demand, e.g., during startup of a steel mill or an aluminum producing plant. Table 7-1 summarizes the characteristics of some of the energy storage technologies suitable for power grid applications. It gives the energy mass densities, energy volumetric densities and efficiencies of these technologies.

All values given in the table are typical values and there is considerable disagreement on these values in the literature. In addition, precisely what has been included in their calculation is not always clear. For example, for batteries the mass and volume of the electrolyte only may be included and the weight of the enclosure may have been ignored. The values for a particular facility depend on its size and design and may vary substantially from one facility to another. The table is meant to provide a crude estimate only. The first three entries store energy in mechanical

Table 7-1 Characteristics of electricity storage technologies.

Technology	Energy mass density kWh/kg	Energy volumetric density kWh/l	Efficiency percent
Pumped water[*]	0.000278	0.000278	75-80
CAES[**]	0.034	0.64	60-70
Flywheel	0.12	0.21	85-95
Lead acid	0.025	0.04	70-75
NiCd	0.039	0.061	78-83
NiMH	0.055	0.22	64-68
NaS	0.76	0.34	87-92
Li-ion	0.14	0.35	93-97
Li-ion nanotube	1.75	3.5	93-97
Vanadium flow	0.025	0.03	78-85
ZnBr flow	0.075	0.08	65-75
Fuel cell system	—	—	<50
Super capacitor	0.003	0.02	95-98
Ultra-capacitor	0.29	0.06	95-98
SMES	—	—	>95

[*]Based on the potential energy of water raised by 100 m.
[**]Based on air compressed to 1500 psi.
Note: Compressed Air Energy Storage (CAES), Nickel-cadmium (NiCd), Nichol-Metal-Hydride (NiMH), Sodium-sulfur (NaS), Lithium (Li), Zinc-bromide (ZnBr), Superconducting Magnetic Energy Storage (SMES).

form, the next nine in chemical form and the last three are the only ones that store energy in electromagnetic form. In any case, the energy must be converted back into electrical form of the required voltage and frequency before it is used.

The capital cost per kWh of constructing a storage installation is high and varies so much with the installation and time that no averages are listed here. Pumped water storage costs the least despite the fact that pumped water facilities sometimes require the construction of a large dam. Compressed air systems cost a few times more than pumped water systems and battery storage, which is currently used for short term storage only, is many times that of pumped storage.

Besides the actual storage mechanisms, storage systems must contain supporting electronics and other equipment for making the outputs compatible with the power grid. The efficiencies indicated are roundtrip efficiencies that give the percent of energy put into the storage mechanism to store and retrieve the energy that is returned as electrical energy. They do not include the inefficiencies due to the supporting electronics and other equipment. This equipment may further degrade the efficiencies up to 15 percent. Clearly, the storage and later retrieval of electricity would cause the average losses of a system to increase. If a solar thermal system had to store and retrieve 30 percent of its electricity in a storage facility that is 80 percent efficient, then the overall efficiency of the system would be effectively reduced by

$$1 - (0.7 \times 1 + 0.3 \times 0.8) = 0.06,$$

or six percent.

In addition to energy density, efficiency and capital cost, other characteristics that are important are maintenance costs, reliability, recharge time and output capability. Output capability is the amount of power the system can safely provide and for power grid applications is normally given in megawatts. For a grid the demand for power is more during the day than at night and to avoid excess generating capacity there is a need to store energy during the night for use during the day. Systems that are to balance power grid loads for extended periods of time must be very large and have high output capabilities. They are almost always pumped water or compressed air energy storage (CAES) systems. Flywheels are for smoothing short duration electrical fluctuations and are most prominently used with wind turbines to compensate for wind variations. Smaller systems for storing electricity for a single home or business may consist of banks of batteries.

Pumped water facilities raise the potential energy of water by simply pumping the water up into a lake when demand is low and then letting it flow back down through a hydro-turbine to produce electrical energy when it is needed. The motor/pump may be reversible and also serve as the hydro-turbine/generator. Because the energy stored per liter is small, the amount of water raised must be large. But in the right location, the storage reservoir can be large enough to provide a huge amount of energy. The supporting equipment is minimal so that the overall efficiency is close to that given in the table. Worldwide, pumped water systems provide 90,000 MW of backup power and in the United States they have a combined capacity of over 19,000 MW.

CAES systems pressurize air and regain the energy when it is released through a turbine. Large systems often use underground caverns or abandoned mines to hold the pressurized air, but smaller systems for leveling out fluctuations may use tanks. CAES systems that use tanks typically compress the air to pressures that are three times those that use caverns or mines. There are, of course, a limited number of locations where natural caverns and mines are suitable for storing compressed air. The largest CAES system is being constructed in a limestone mine in Ohio. It will be 670 m (2,200 ft) underground, store air at between 55 and 110 atmospheres and have a peak capacity of 2,700 MW. Although large CAES systems can hold considerable energy, they are more costly, less efficient and contain less energy than the more widely used pumped water systems. Much of the energy lost is through the heat that is generated while the air is being compressed, a problem that does not exist in a pumped water system.

     Flywheels store energy in the rotating mass of the wheel and come in a large variety of sizes. The amount of energy a flywheel can store depends on its maximum rotational speed, its mass and how the mass is distributed relative to its center. The idea is to place as much mass as possible as far from the center of the wheel as possible because this maximizes the wheel's ability to store energy. As energy is taken from the wheel, it slows down, so transmission equipment is needed to compensate for the wheel's varying rotational speed. However, because the connection to the flywheel is basically mechanical, there is little heat loss and flywheel systems tend to be quite efficient. There are flywheels that store as much as 40,000 kWh.

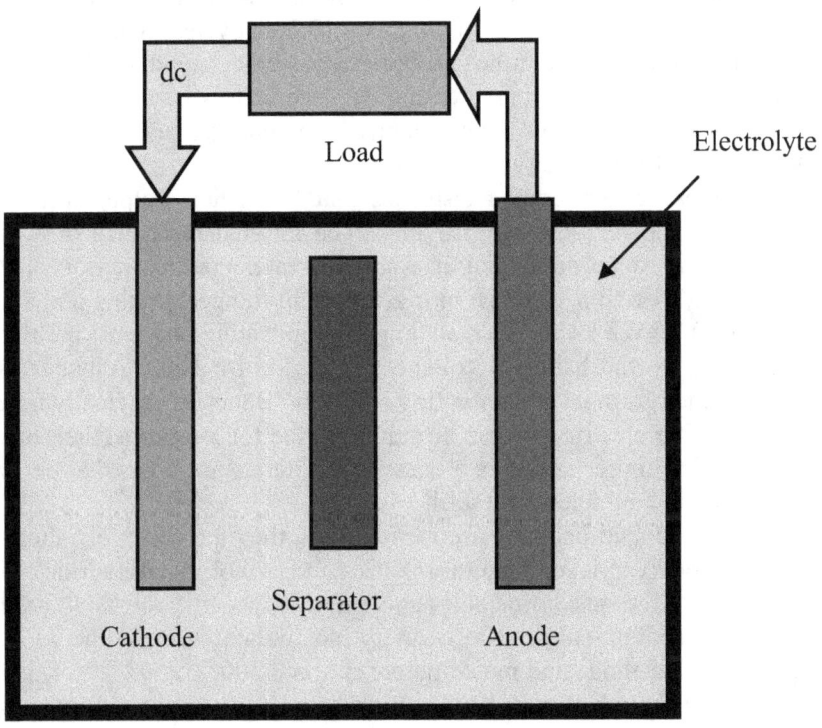

Fig. 7-18 Typical conventional battery construction.

Batteries store their energy in chemical form. As shown in Fig. 7-18, a single-cell battery consists of two electrodes, an anode and a cathode, a separator and an electrolyte. The chemical differences between the electrodes and the electrolyte cause a voltage to exist between the anode and cathode. When a load is connected between the electrodes, dc electricity flows through the load. The voltage depends on the electrolyte and electrode materials used, but higher voltages may be attained by adding alternating electrodes and separators to form multi-cell batteries. The separators are to prevent shorting between the electrodes.

As current flows out of a battery, the chemicals in the electrolyte and surfaces of the electrodes change and the voltage decreases until an adequate external current can be no longer maintained. However, an important property of batteries is that their voltages do not drop significantly until they are almost discharged. Some batteries can be recharged by applying a reverse voltage that forces a current to flow through the battery in the opposite direction. This current causes the chemical characteristics to be restored, but each time the battery is discharged and recharged it is damaged and the number of discharge/recharge cycles is limited. The lifetime of a battery is measured in the number of these cycles it can withstand. The life of a lead-acid battery is in the hundreds of cycles and the life of a lithium-ion (Li-ion) battery may be more than 10,000 cycles. The lifetime of a battery determines the average time between replacements, which is the major maintenance cost of a battery storage system. Replacement is particularly important in applications for which batteries are constantly being charged and discharged, e.g., automobiles powered by electricity.

All batteries have internal resistances that limit their output current and, thus, the amount of power they can provide. The internal resistance of a battery produces heat that must be dissipated at a suitable rate. For this reason, batteries have an energy (kWh or ampere-hour) rating, a current (ampere) rating and a power (kW) rating. To get a kWh rating from an ampere-hour rating one must multiply the ampere-hour rating by the battery's voltage and divide by 1000. A battery's kW rating can be obtained from its ampere rating similarly. Batteries especially need fast recharge times. Before electric cars can be suitably used for long trips, their batteries must be capable of being recharged in a matter of minutes, not hours. No one wants to wait more than a few minutes for a refill.

Although lead-acid batteries are inexpensive, they are relatively short-lived and are heavy and bulky. Nickel-cadmium (NiCd) and nichol-metal-hydride (NiMH) batteries have better lifetimes, efficiencies and energy to weight ratios, but are much more expensive. A sodium-sulfur (NaS) battery has molten sulfur at the anode and molten sodium at the cathode and must operate at about 300 °C (572 °F). This limits its application even though it may have a high energy to weight ratio and power rating. A large NaS system is relatively inexpensive and may have a rating as much as 360 kWh. Lithium-ion (Li-ion) storage systems are expensive, are presently very limited in size and overheat if charged improperly, but offer excellent energy to weight and energy to volume ratios. Within the last few years considerable research has been done toward the development of batteries with nanotube electrodes. The capacity of a battery depends on the surface area of the electrodes and nanotubes increase this area thousands of times. One such battery is the Li-ion nanotube battery. Although Li-ion nanotube batteries have outstanding physical characteristics, they are currently extremely expensive and are not presently a viable alternative, but their cost will come down as they move from experimentation to mass production.

A much different battery design that has some similarities to that of a fuel cell (discussed in Chap. 9) is the flow design depicted in Fig. 7-19. In this design two different electrolytes are stored in separate tanks and pumped through two sides

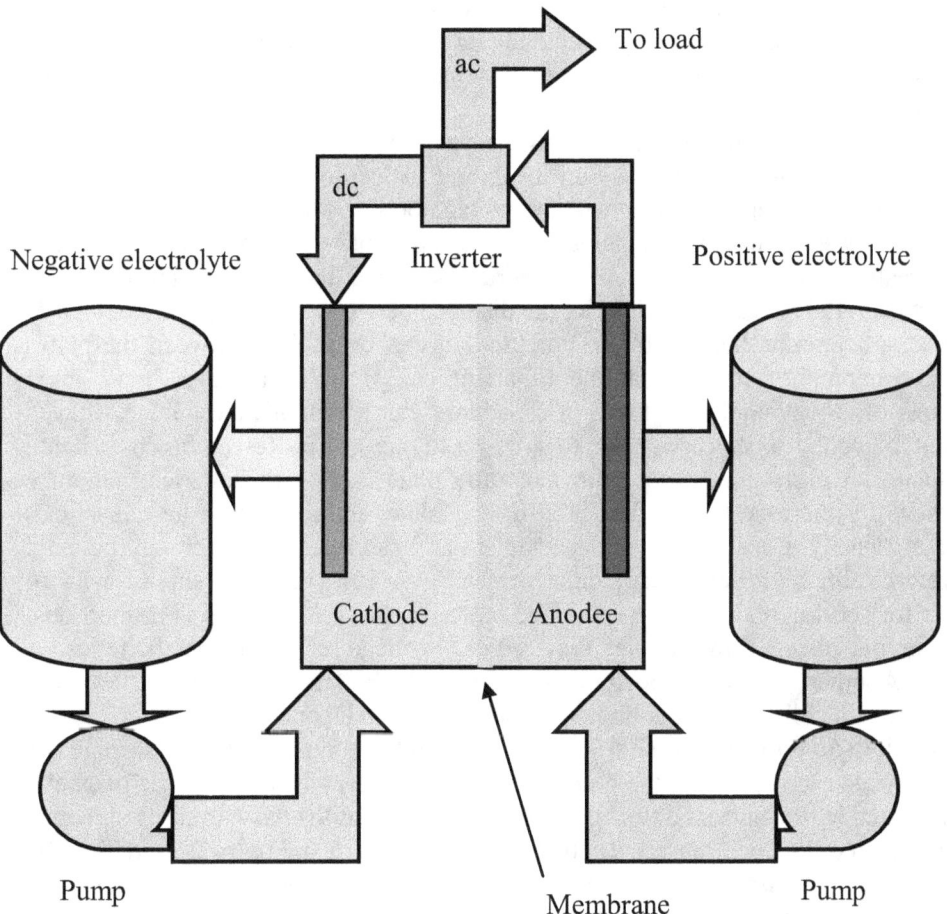

Fig. 7-19 Basic flow battery design.

of a container that are separated by a membrane. The difference in the electrolytes causes electrons to flow through the membrane and collect on the cathode. A connection between the anode and cathode then allows dc electricity to flow from the anode to the cathode. As the current flows, the electrolytes change and the battery is discharged. The battery may be recharged on site when there is excess power available or the electrolytes may be replaced. Some flow battery designs hold up to 120 MWh and can supply 12 MW of power. But the energy densities of the electrolytes are relatively small and large flow batteries require a lot of space. Flow cells may use vanadium or ZnBr.

Fuel cells themselves can be used as a basis for storing energy that is later used to produce electricity. One such system is a Molten Carbonate Fuel Cell (MCFC) system. Although an MCFC system can store a reasonably large amount of energy, it has a low efficiency when the production of hydrogen is taken into account. There are other fuel cell based systems discussed in Chap. 8, but all have low efficiencies.

Capacitors store their energy in an electric field that is created by applying a voltage across plates separated by an insulating dielectric. The exact configuration varies, but the principal is the same. The voltage between the plates cause electrons to be transferred to one of the plates while the other plate is left electron deficient, thus creating the electric field. The energy is stored in this electric field. The properties of the plates and dielectric, the distance between the plates and the area of the plates determine the maximum voltage that can be applied between the plates and the amount of energy that the capacitor can store. In the past, only small amounts of energy could be stored in capacitors, but recent advances have allowed capacitors, called *super capacitors*, to store up to one MWh. Research is now being conducted on *ultra-capacitors* with nanotube plates. As with batteries, nanotube plates vastly increase the surface area of the plates and, thereby, the capacitor's storage capacity. Capacitors are very efficient and can release their energy quickly. Therefore, although their energy capacities may be low, they can supply a lot of power for short periods. However, unlike batteries, the voltage drop is proportional to the output current flow and they require better voltage regulation than batteries.

Another storage technology in which the energy is stored electromagnetically is based on superconducting coils or toroids. A current is inserted into a coil or toroid and is maintained with the only energy loss being the energy needed to refrigerate the coil or toroid. The energy is stored in the magnetic field created by the current and is returned to an external load as current is siphoned from the coil. Once inserted, the current will sustain itself indefinitely. *Superconducting magnetic energy storage (SMES)* systems are commercially available but are expensive and limited in size to the one MW range, although a 20 MWh experimental unit has been constructed with an output capacity of 400 MW. As with nanotube technology they hold significant promise for the future. Estimating the energy density of either an SMES or MCFC system is even more speculative than with other storage systems. If the weight the of structural support and required refrigeration in addition to the superconducting material in an SMES are included in the calculation, then the energy density is very low. Similarly the energy density of an MCFC system is difficult to determine because of the large amount of support equipment required.

Figure 7-20 shows the electricity consumption, excluding generation and distribution losses, in both the world and United States during the period 1980 through 2007. Superimposed on the actual data curves are the exponential approximation of the world data and the linear approximation of the United States data. The exponential approximation of the world data shows a compounded three percent increase per year. By including the necessary excess capacity, this growth implies that in the succeeding years the worldwide generating capacity would need to be increased by 122.9 GW, 126.3 GW, 130.1 GW and so on just to keep pace with demand. If the recent trend continues, by 2040 the worldwide demand would be 43,230 GkWh, more than two and a half times its 2007 demand. The United State has increased consumption at a linear rate of 74.68 GkWh per year and would need to add 18.7 GW to its generating capacity each year to meet its future demands. Its demand in 2040 would be 6,490 GkWh. To add 18.7 GW of capacity would require the construction of over eighteen 1,000 MW generating plants. By taking into account the fact that for wind turbines the ratio of capacity to energy must be even greater than for the average generating plant (3.33 versus 2.2), at least 9,400 3-megawatt wind turbines spread over 3800 square kilometers (1470 square miles) would be needed (assuming 33.3 acres per megawatt—see Chap. 6).

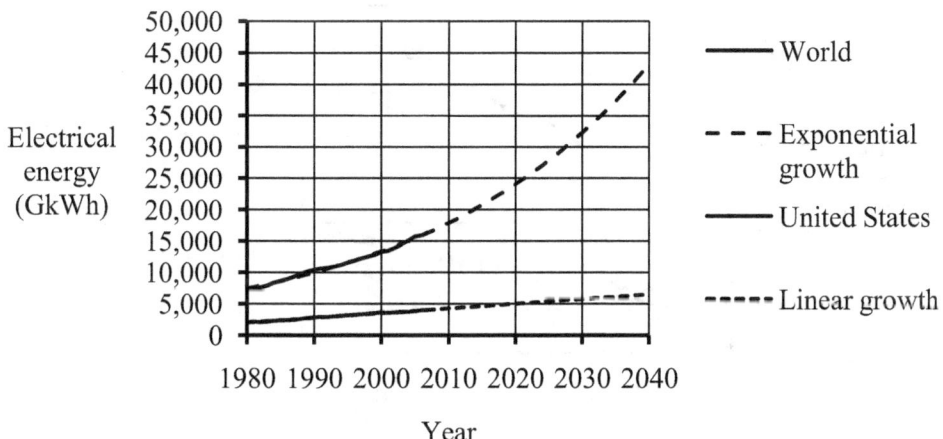

Fig. 7-20 Electricity consumption in GkWh [EIA].

The 1973 through 2009 electricity consumption of three of the four EIA sectors along with the total consumption of all sectors are shown in Fig. 7-21. Because the transportation sector contributes so little of the total and would essentially overlay the horizontal axis, it has not been included. Also, generation and distribution losses have not been included. They would increase the energy amounts shown by a factor of approximately three. Note that the industrial sector has increased its consumption by just 54 percent over the 35-year period shown, but the residential and commercial sectors have increased their respective consumptions by 138 and 204 percent. This is mainly due to the United States shift from heavy industry, such as metal smelting, to office work and the fact that homes are bigger and contain appreciably more air conditioning and electronics. As indicated by Fig. 7-22, not only has consumption increased in the United States, but the consumption per capita has also increased. However, on a per capita basis, the industrial sector's usage has been slightly reduced while the residential sector has increased its usage by 65 percent and the commercial sector has more than doubled its per capita consumption.   The 2007 per capita consumption of electricity in various countries and the world is given in Fig. 7-23. As with total energy, Canada and the United States are by far the greatest users of electricity. The average Canadian and American, including commercial and industrial usage, consumes five to six times as much electricity as the average person in the world. In 2009, the average person in

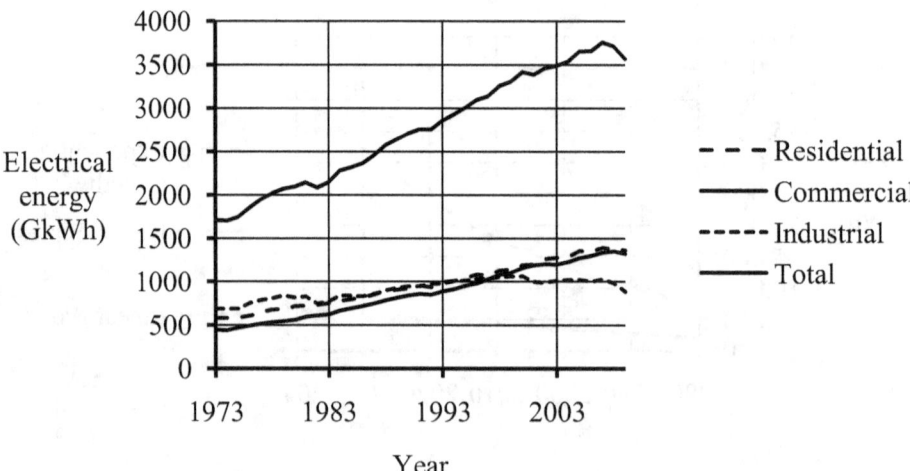

Fig. 7-21 Consumption for sectors and total consumption for these sectors [EIA].

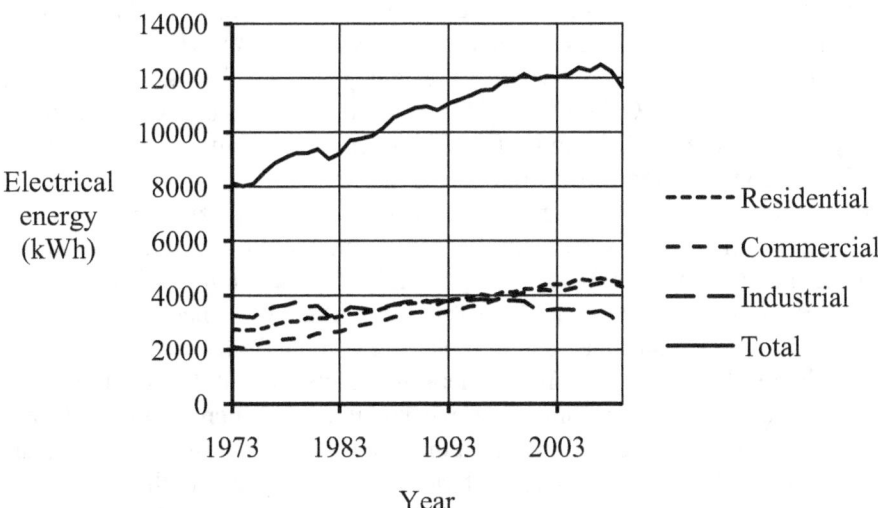

Fig. 7-22 Per capita sector consumption and total per capita consumption [EIA, USCB].

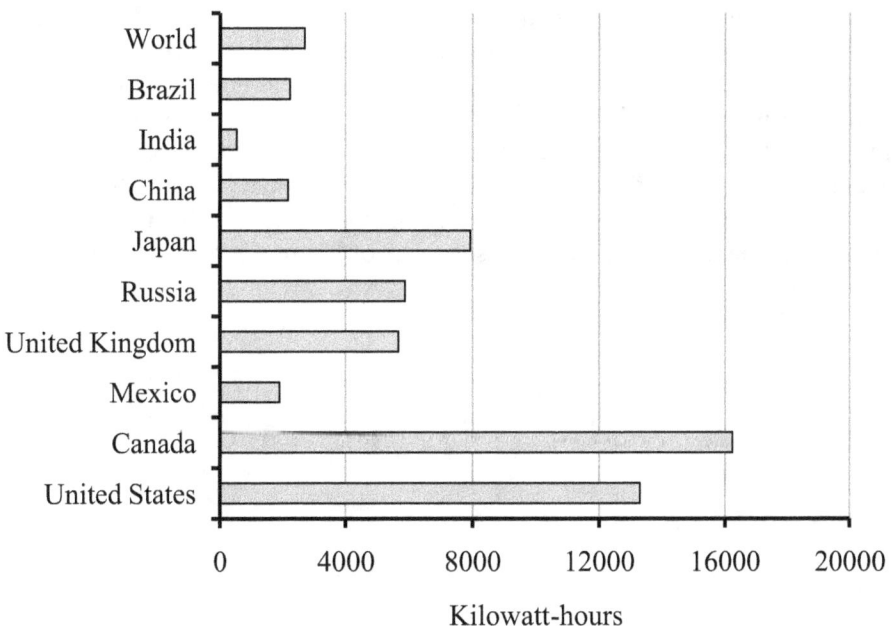

Fig 7-23 2007 per capita electricity usage [EIA].

the United States used 13,302 kWh, while the average person in the world used only 2,694 kWh and the average citizen of China used only 2,171 kWh. However, China is increasing its consumption at an astonishing rate. From 1980 to 2007, China's consumption went from 261 GkWh to 2,835 GWkh and was up by 141 percent in the seven years between 2000 and 2007. Not only is China's consumption increasing, but its consumption growth rate is increasing.

The amount of increased generating capacity does not matter as much as the way in which it is increased. If generating capacity is increased by using renewable energy sources then the environmental damage will be minimal, but if it is increased using fossil fuels the damage could be substantial. Although much of China's recently increased capacity has been due to the Three Gorges Dam Project, there has been a drastic increase in their coal-fired generating plants as well. Coal is the most polluting of the energy sources and is a particularly prominent source of the greenhouse gas carbon dioxide. While the radioactive nuclear waste from nuclear powered generating plants is a problem, it does not contribute to the imminent danger of global warming and may be the best temporary means of solving the world's energy problems. In the longer term, renewable sources must be the solution. Clearly, the best ways of limiting pollution is to limit our increases in demand by conservation and improving the efficiencies of electrical equipment (e.g., using more efficient light bulbs and CHP generating plants where possible).

If our quest to convert to electric vehicles is realized, then the demand for electrical power would increase much more than indicated by the above figures. The transportation sector accounted for 28.5 percent of our 2006 energy requirements and electricity provided very little of the transportation requirements. If over the next three decades, electricity were to provide a modest 20 percent of our transportation energy, then the increase in electricity demand would be in the neighborhood of six to ten percent, depending on the relative efficiencies of the internal combustion engines used today versus those of future electric vehicles.

8

SECTORS

As introduced in Chap. 1, the Energy Information Agency (EIA) divides our energy usage into the four non-overlapping sectors that it refers to as residential, commercial, industrial and transportation. Figure 1-7 gives the amounts and percentages of the 2008 energy consumption by these sectors as 22 percent (21,632 TBtu) for the residential sector, 19 percent (18,541 TBtu) for the commercial sector, 31 percent (31,210 TBtu) for the industrial sector and 28 percent (27,921 TBtu) for the transportation sector. Although the percentages and amounts have remained about the same since 2000, the industrial energy consumption has dropped from 35 percent to 31 percent, the residential and commercial sectors have increased their consumption by approximately one percent each and transportation has increased its percentage by two percent.

The *residential sector* consists of private household dwellings and includes houses, apartments and condominiums. It does not include hotels or other temporary living quarters that are part of the commercial sector. Also, the energy used by the residential sector does not take into account the energy consumed by vehicles, even personal automobiles. All energy consumed by vehicles not restricted to one site (e.g., farm machinery and forklifts are restricted to one site) is accounted for by the transportation sector. How the energy was used within the residential sector in the United States in 2001 is indicated in Fig. 8-1. The amounts in parentheses are in TBtu. The total energy consumed by the residential sector in 2001 was 10,450 TBtu, but this figure includes only the electrical energy entering the residences and does not consider the energy needed to generate and distribute the electricity. These losses add 9,130 TBtu to the total, which then becomes 19,580 TBtu. It is seen from the figure that 71 percent of the 10,450 TBtu used by the residential sector was for the heating and cooling of water and air. However, if the energy required for generating and distributing the electricity were included, this percentage would be reduced and the percentages for lighting, refrigeration and appliances would be increased because they use electricity almost exclusively. A more limited study in

2005 indicated percentages of the total of 11,539 TBtu to be essentially the same as the 2001 percentages.

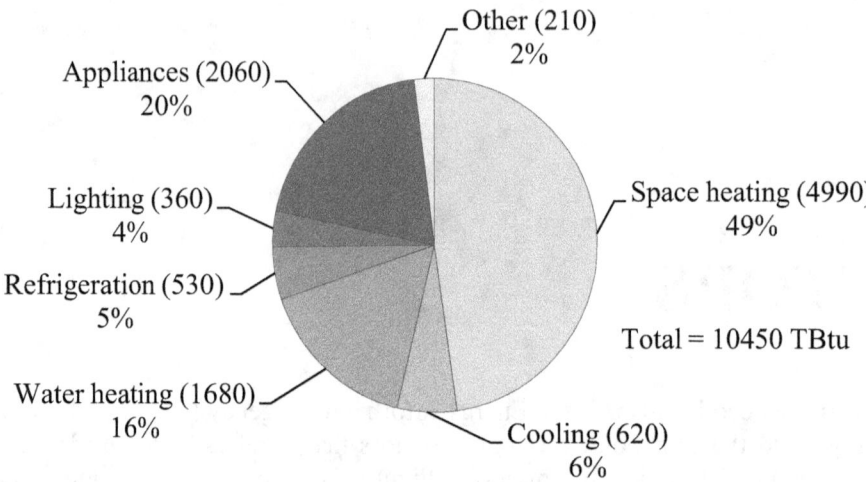

Fig.8-1 2001 residential use of energy in the United States in TBtu [EIA].

Figure 8-2 shows the energy sources used to supply the residential sector in 2001. Again, only the electrical energy entering the residences has been taken into

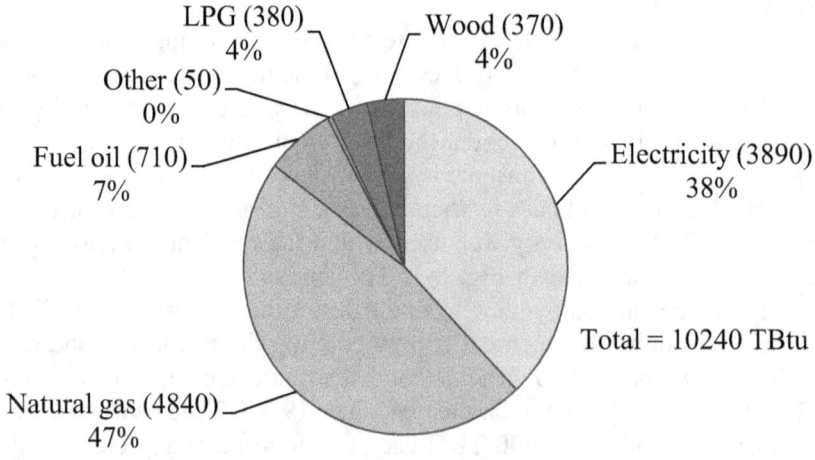

Fig. 8-2 2001 energy sources used by the residential sector in the United States [EIA].

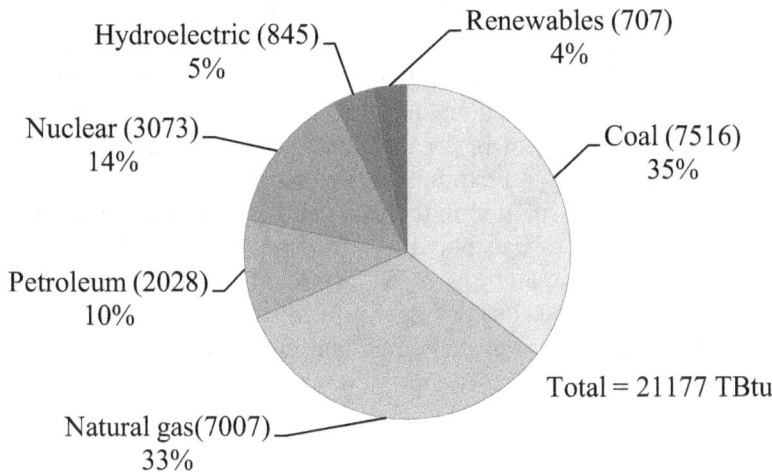

Fig. 8-3 2001 sources used by the residential sector including electrical sources [EIA].

account. If the energy sources needed to generate and distribute the electricity replaced the electricity section of the pie, then sections for coal, uranium and renewables would be required and the remainder of the electricity section would add to the sections for natural gas and fuel oil. This was done for the 2001 data and the results are given in Fig. 8-3. The total consumption including the required generation and distribution of energy was 21,177 TBtu, of which 78 percent was provided by fossil fuels. Petroleum encompasses all petroleum products, including fuel oil and liquid petroleum gas (LPG). Renewables, other than hydroelectric which is given separately, include wood, geothermal, solar and wind energy. (The differences in the totals in Figs. 8-1and 8-2 are due to rounding errors and the fact that, although all of the data was obtained from the EIA website, the original sources of the data differed.)

If the energy required to generate and distribute electricity is taken into account, then lighting would use approximately 5.8 percent of residential energy and space and water heating and cooling would use 40 percent. Although a 4 percent reduction in residential energy might be realized by a 70 percent reduction in lighting through conservation and using fluorescent bulbs and LEDs, approximately 12 percent of the residential energy could be saved by only a 30 percent reduction in the energy needed for heating and cooling. This emphasizes the importance of using sufficient insulation and more efficient means of providing heating and cooling even if the installation costs are relatively high.

Sometimes what is wanted is to keep the heat in a location instead of moving it between locations. Obviously, once a house has been heated or cooled, it is desirable to retain the heat or prevent heat from entering the house. This of course is the purpose of *insulation* such as that depicted in Fig. 8-4. Insulation minimizes the transfer of heat by absorbing very little heat and impedes the movement of heat by both convection and conduction. It may be an evacuated space between containing walls that reflect heat or an inexpensive material that contains air but prevents the air from moving. An example of the former is a vacuum bottle and an example of the latter is the insulation in the walls and ceiling of a house. All cooling systems use insulated pipes to transport the refrigerant between the components in the thermal cycle (see Chap. 6 and Fig. 6-19). Inside a heat exchanger, however, the piping would not be insulated, but designed to transfer as much heat as possible. Raising insulation standards for new homes and subsidizing insulation improvements in older homes offer significant opportunities for reducing residential energy.

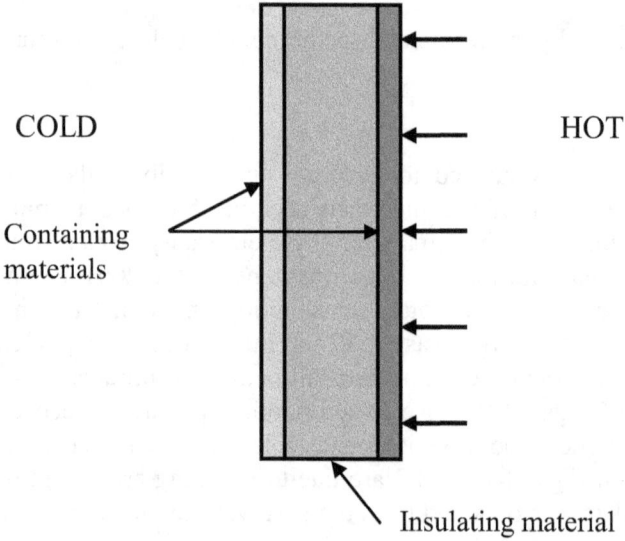

COLD   HOT

Containing
materials

Insulating material

Fig. 8-4 Insulation.

The *commercial sector* consists of service providing facilities, including business, government, religious, education, lodging and other service facilities. As with the residential sector, it excludes all vehicles used for transporting people or goods between sites because energy consumed for these purposes is assigned to the transportation sector. The sources of energy used by the commercial sector in 2009 are given in Fig. 8-5. Again the electricity portion of the pie graph is the electrical energy supplied to the commercial facilities and does not take into account the energy needed to generate and distribute the electricity. Note the dominance of electricity and natural gas as sources for the commercial sector requirements. Also, unlike a hundred years ago when pollution was not considered a serious problem, at present direct consumption of coal is used very little by either the commercial or residential sectors.

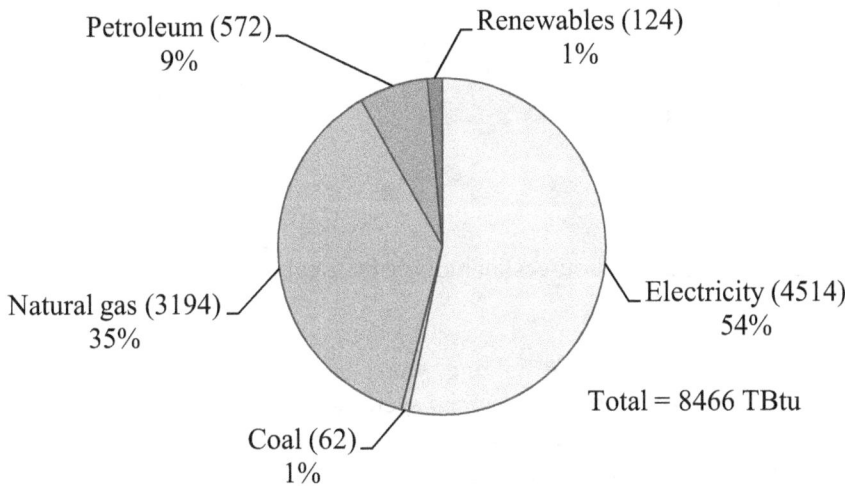

Fig. 8-5 2009 commercial sector energy sources in the United States in TBtu [EIA].

Figure 8-6 shows the breakdown wherein the original sources of the electrical energy, including losses, are added to the other primary sources. As with the residential sector, coal is shown to play a much greater role because coal is used to generate more than half of the electricity. In addition, nuclear and hydroelectric energy now appear as sources and renewable energy provides a slightly greater share of the energy. It is anticipated that renewables, in particular solar energy, will play a much greater role in the future as solar facilitiess are installed on the roofs of businesses to supply both heat and electricity. Not counting electrical generation and distribution losses, the commercial sector consumed 8,466 TBtu in 2009, and with the 9,722 TBtu in electrical losses it consumed 18,144 TBtu.

In Europe, businesses are beginning to take advantage of ground temperature to aid the heating and cooling of buildings. In Austria, a convention center uses 320 eighteen-meter (59.2 ft) borehole heat exchangers (BHEs) and 65 km (40 mi) of polyethylene pipe to supply up to 800 kW of heating and cooling.

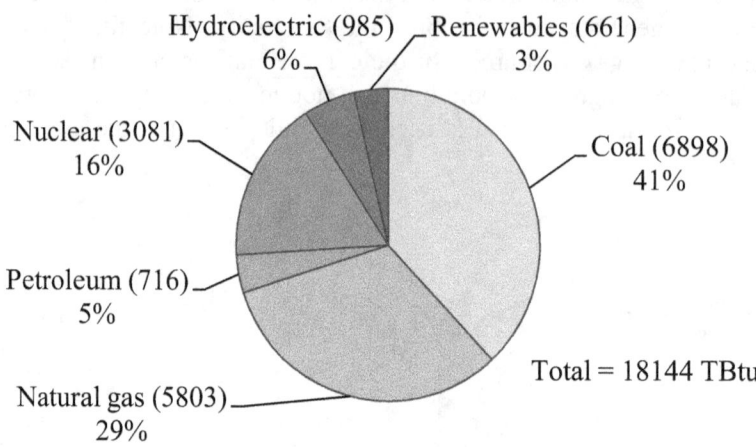

Fig. 8-6 2009 commercial sources including electrical losses [EIA].

It is seen from Fig. 8-7 that lighting, water heating and space heating and cooling are the primary uses of the commercial sector's energy, consuming about 77 percent of the total. Lighting requires 23 percent of the energy with 14 percent of the lighting being provided by incandescent bulbs, 7 percent by compact fluorescent bulbs and 77 percent by tubular fluorescent bulbs. Highly efficient LED lighting is likely to become common in the future and could reduce the total energy requirements of the commercial sector by as much as 20 percent. The fact that temperature regulation of air and water accounts for over half of this sector's energy needs strongly suggests an opportunity for the use of better insulation, combined power and heating (CPH), direct solar and ground source heat pumps (GSHPs). As mentioned in Chap. 7, CPH is already employed in densely populated areas where district heat can be distributed to a group of buildings. Figure 8-7 does not include the energy lost due to the generation and distribution of the electricity consumed. Because the source of essentially all lighting is via electricity, the EIA 2006 data imply that 43 percent of the electricity and (by including the 9,955 TBtu in electrical losses) 34 percent of the energy needed to supply the commercial sector is for lighting. Conservation achieved by simply turning off lighting and electronic equipment when it is not in use and using less heating and cooling when it is not needed offer ways of significantly reducing energy consumption by the commercial sector.

The amounts and percentages of energy consumed in 2003 by the various commercial user types, including electrical losses, are given in Fig. 8-8. "Mercantile" includes all retail stores except grocery stores, which are included in

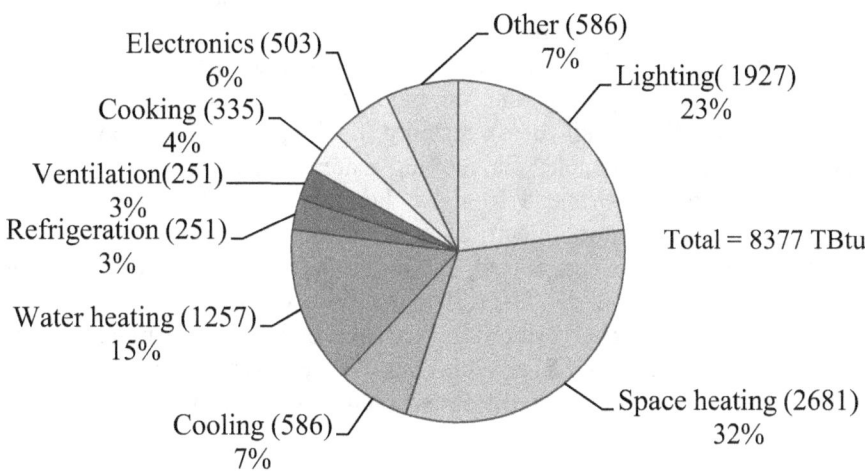

Fig. 8-7 2006 commercial uses of energy in the United States in TBtu [EIA].

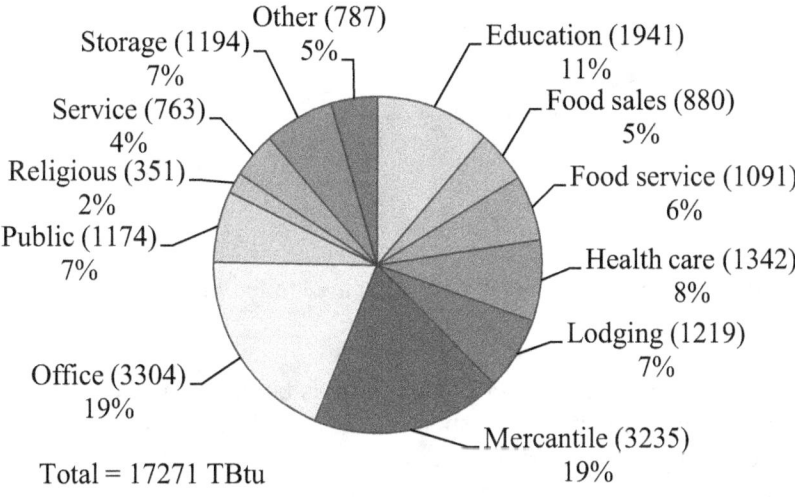

Fig. 8-8 2003 consumption by commercial users in the United States in TBtu [EIA].

the "Food sales" category. "Lodging" consists of temporary facilities for housing people such as hotels and, of course, excludes the more permanent housing accounted for by the residential sector. The "Education", "Office" and "Mercantile" categories required almost half of the energy.

The *industrial sector* consists of all facilities and equipment for producing, processing or assembling goods and encompasses manufacturing, agriculture, forestry, mining (including fossil fuels), refining, fishing, hunting and construction. It also includes the use of vehicles for moving materials on site, but not between nonadjacent sites. Figures 8-9 and 8-10 show the sources of the energy consumed excluding and including the electrical generation and distribution losses, respectively. The total 2009 energy used, excluding the losses, is 21,603 TBtu and, including these losses, is 28,073 TBtu. The increase of 6,470 TBtu due to the inclusion of the electrical losses resulted in an increase in the fossil fuels portion of the chart from 79 percent to 85 percent, but the renewable portion, including hydroelectric, is also slightly increased from seven to eight percent.

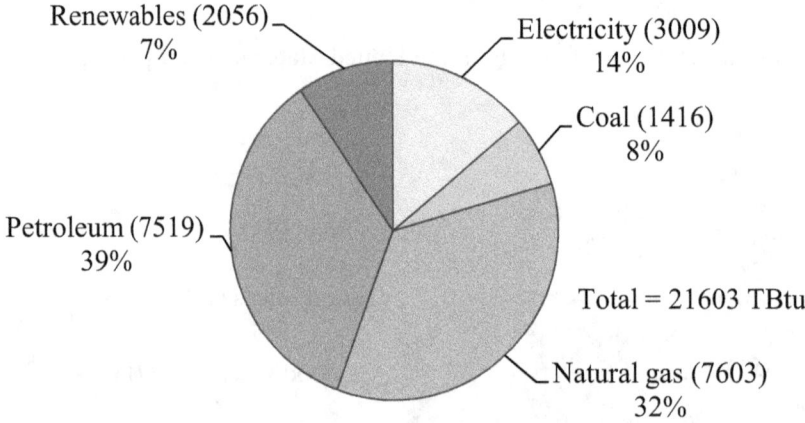

Fig. 8-9 2009 industrial sector consumption in the United States in TBtu[EIA].

In Fig. 8-11, the 2006 energy consumption is broken down according to the type of user. In this figure, "Transportation" and "Machinery" refer to the consumption of energy by the manufacturers of vehicles and other machinery, not their use. "Food/drink" includes the raising and processing of all tobacco, food and drink products and the "Wood" slice of the pie includes lumbering and, except for wood-based paper products, the processing of wood,. This figure includes the losses in generating and distributing the electricity supplied to the industrial sector. It is seen that the production, refining, and processing of fossil fuels and other chemicals consume almost half of the energy. The production of metals and nonmetals, including plastics, consume another fifth of the energy. Much of the energy is needed to fuel boilers or provide process heat. *Process heat* is the heat used to directly raise the temperature of materials during a production process, such as that needed to heat petroleum during refining or to cook packaged foods.

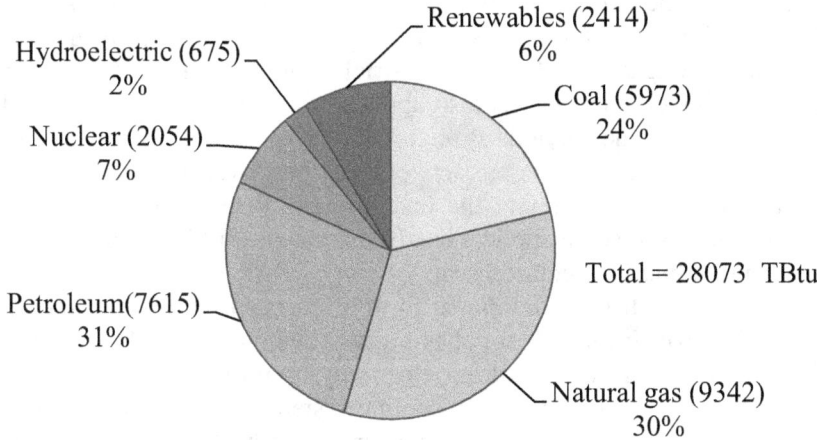

Fig. 8-10 2009 industrial sector usage including electrical losses in TBtu [EIA].

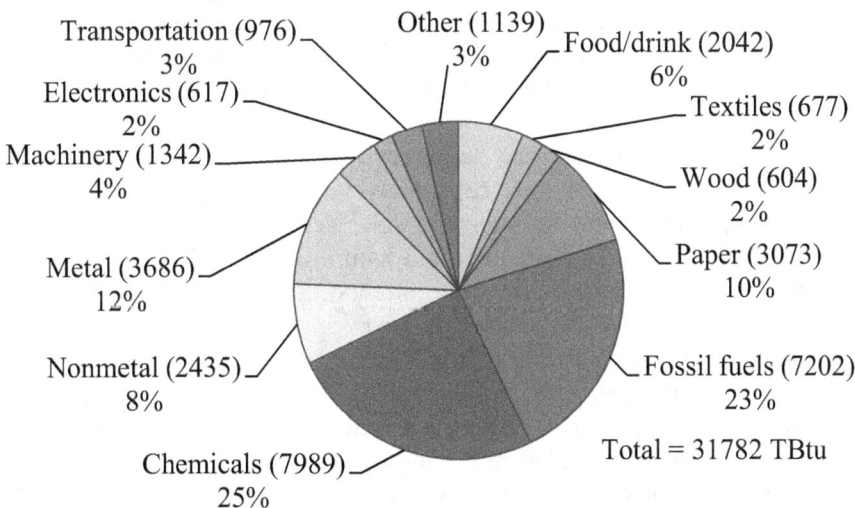

Fig. 8-11 2006 consumption by industry type in the United States in TBtu [EIA].

Agriculture and its associated machinery account for most of the energy consumed by the "Food/drink" industry. Electricity is not needed so much for lighting and heating as it is for driving production machinery, and space heating consumes a much lesser portion of the energy in the industrial sector than in either the residential or commercial sector. Because of the substantial need for supplying heat to support boilers and for processing, the opportunity and convenience for employing CHP is even greater for the industrial sector than for the commercial sector. The use of direct solar heat is also applicable in many cases and could be accomplished using the same methods as are used to drive steam turbines to generate electricity (see Fig. 7-5). In geothermal active areas, heat may be obtained directly from the ground by simply pumping the underground water through radiators or other heat exchangers. For some applications BHEs may be sufficient to provide, or at least assist in providing, the required heat.

It is clear that heat is the dominant form of energy used by all three of the sectors discussed above. It directly supplies process heat and warmth to space and water, and indirectly supplies most of our electricity by providing energy to steam turbines. Ironically, it can even be used in cooling systems, such as gas refrigerators and air conditioners. Unfortunately, a large percentage of heat is presently obtained by burning fossil fuels. Our greatest challenge is to replace this source of heat with renewable sources that are neither polluting nor finite. CHP reduces our dependency on fossil fuels, but it does not eliminate this dependency, nor do GSHPs, BHEs or conservation. Geothermal activity is limited to certain areas. Nuclear fuel is polluting, finite and dangerous for both political and environmental reasons, and is used for generating electricity only. This leaves solar energy as the principal alternative for providing heat directly. Direct solar energy can provide the energy needed to heat air, water and other materials and, by concentrating the sun's rays, can even produce steam for radiators, generating electricity and some industrial processes.

Much of our need for heat can be avoided by generating electricity using hydro-turbines, wind turbines and solar cells. However, suitable hydroelectric sites are limited and, at present, wind turbines and solar cells are expensive to install. However, at the 2007 prices for petroleum and natural gas, wind turbines became competitive with generators powered by these commodities.

The other energy end-use sector, the *transportation sector*, includes all vehicles and other means of transporting goods and people to off-site locations. It includes personal motorcycles, cars, trucks, trains, barges, ships, airplanes and pipelines. It does not include farm or factory machinery, such as tractors or forklifts that are used on-site and are included in the industrial sector. The percentages and amounts of the energy sources for the transportation sector in the United States for 2008 are given in Fig. 8-12. The total energy consumed that year was 27,215 TBtu, 95 percent of which came from petroleum products. Much smaller amounts came from natural gas and biomass and a negligible amount came from electricity. Because 97 percent of this energy originated from petroleum and natural gas, the discussion of the transportation sector essentially revolves around a discussion of these two energy sources, particularly petroleum.

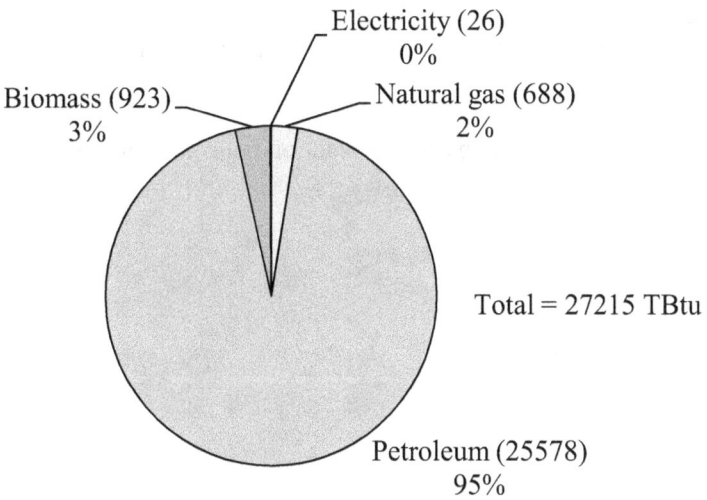

Fig. 8-12 2008 transportation energy sources in the United States in TBtu [EIA].

The Bureau of Transportation Statistics (BTS) has classified the modes of transportation as summarized in Table 8-1. Motorcycles consume a very small percentage of the transportation energy requirements and sometimes the BTS combines passenger cars and motorcycles into one category, although at other times it treats them separately. Similarly, "General aviation" consumes only a small amount of energy, and only the domestic certified, commercial aircraft are considered in the discussions below. The "Light truck" subcategory includes 2-axle, 4-tire vehicles such as pickups, sport utility vehicles (SUVs) and vans that may be used to haul either goods or people. "Passenger cars," or simply cars, are 2-axle, 4-tire vehicles other than light trucks and are primarily for moving people. "Single-unit trucks," hereafter simply referred to as *"trucks"*, are taken to be a means for moving goods short distances while "Combination trucks", more commonly referred to as *"semis"*, are for moving goods between cities. "Intercity bus" includes all vehicles designed to carry more than 10 people and travel between cities, and "Motor bus" includes intra-city transit buses and school buses. "Trains" that are for moving passengers within an urban area are in the "Transit" category, and those for moving freight between cities are in the "Rail" category and constitute the "Class-1 rail" subcategory. Trains that move people between cities are, in the United States,

"Amtrak" trains. At this time, Amtrak carries relatively few passengers and has a negligible effect on energy consumption. This is likely to remain true in the immediate future, but may change in the more distant future. "Water" includes freight hauling boats, barges and domestic ships that travel along our coasts, but excludes international shipping. Although pipelines are stationary, they use energy to transport goods and are, therefore, considered part of the transportation sector.

Table 8-1 Categories assigned by the Bureau of Transportation Statistics.

Air
        Certified carriers
        General aviation
Highway
        Motorcycle
        Passenger car
        Light truck, 2-axle, 4-tire vehicle
        Single-unit truck (truck), single-unit freight vehicle
        Combination truck (semi)
        Intercity bus
Transit
        Commuter train
        Heavy rail (subway and elevated trains)
        Light rail (trolleys, streetcars)
        Motor bus (city buses, school buses)
        Demand responsive (vans, taxis)
Rail
        Class 1 rail (freight trains)
        Amtrak
Water (ships, boats and barges)
Pipeline

These modes of transportation fall into two basic types, those that primarily transport people and those that primarily transport goods. The first of these types consists of the air carriers, both certified and general aviation, motorcycles, cars, light trucks, buses, Amtrak and all subcategories listed under the transit category. The second type consists of trucks (single-unit trucks), semis (combination trucks), freight trains, water transports and pipelines. Air carriers may be used for both, but the energy they use for hauling freight is small compared to the energy they use for transporting people. Light trucks transport both people and goods, but in recent years they have become primarily a passenger vehicle. However, their contribution to moving freight is taken into account in the examples given later. General aviation and motorcycles consume only a small percentage of energy and are not considered in these examples, although the use of motorcycles could reduce energy usage in the future. Also, light rail, demand responsive vehicles, trolley buses and ferry boats consume a very small percentage of the total and are included under "Other transit" in the following discussions in which "Transit" is broken into subcategories.

Figure 8-13 gives the 2006 energy consumption in the United States by transportation mode. Approximately 71 percent of the total 26,661 TBtu was for mainly moving people, and only 29 percent for moving goods. More than half of the energy needed for moving goods was consumed by semis and less than one-tenth was consumed by trains, even though trains hauled 29 percent more freight than semis. As indicated by Fig. 8-12, approximately 97 percent of the energy consumed was supplied by fossil fuels and 95 percent was supplied by petroleum.

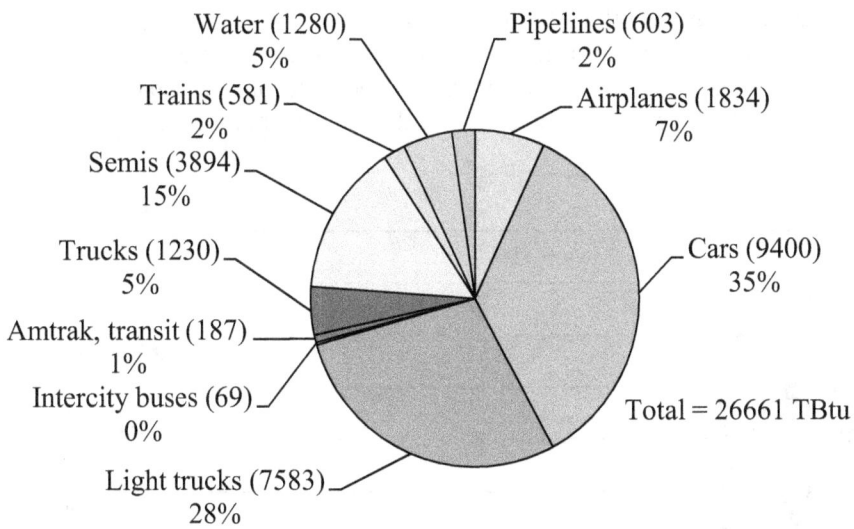

Fig. 8-13 2006 transportation energy usage by mode in the United States [BTS].

Because more than twice as much energy is expended moving people than is spent moving freight, a prominent possibility for reducing our petroleum demand is to transport people by more efficient means. In light of the limited reserves, pollution and political problems associated with petroleum, the necessity of reducing our dependence on petroleum is our most pressing challenge. It is petroleum that is the most limited and the world's most politically sensitive energy source, although natural gas may also produce tensions (as indicated by the disruption of Russian gas supplies to central Europe in 2006, late 2008 and early 2009).

Figure 8-14 shows the 2006 energy consumption, passenger-miles and vehicle-miles of the various modes of passenger transportation as a percentage of the total amounts for these modes. Clearly, cars and light trucks dominate the other means of passenger transport in all three of these measures. Together, they consume 88.9 percent of the energy and account for 85.6 percent of the passenger-miles and 89.2 percent of the vehicle-miles. If airplanes are included with cars and light trucks, then the three modes consume 98.7 percent of the energy (essentially all of which is derived from petroleum) and account for 96.7 percent of the passenger-miles. Intercity buses are responsible for 2.3 percent of the remaining passenger-miles, leaving one percent for all other means of moving people. For energy and vehicle-miles, the percentages for all modes other than cars, light trucks and airplanes are too small to register on this graph. Yet, as shown in Fig. 8-15, it is for cars and light trucks that the *average occupancies* (i.e., passenger-miles per vehicle-mile) are less than two—less than any other mode of transporting people.

Figure 8-16 gives the 2006 average *energy intensity* (i.e., energy consumption per passenger-mile) for the various modes of transportation. The energy required for moving people is primarily dependent on the capacities of the vehicles used and the average occupancies and load factors of these vehicles. A

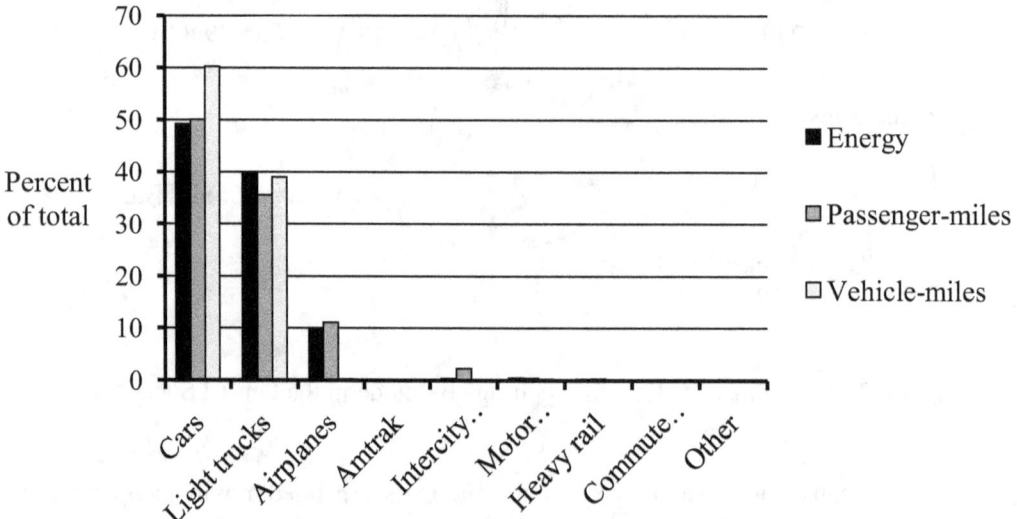

Fig. 8-14 2006 percentages of totals for notable passenger transportation quantities [BTS].

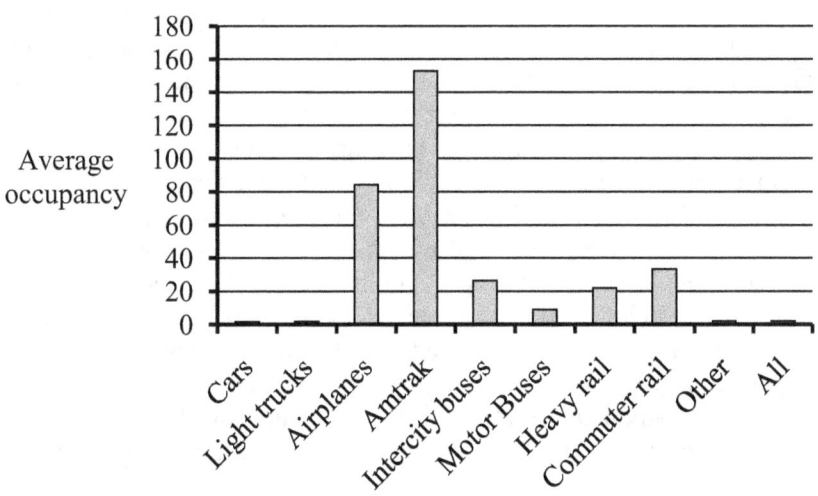

Fig. 8-15 2006 average occupancy of passenger modes of transportation [BTS].

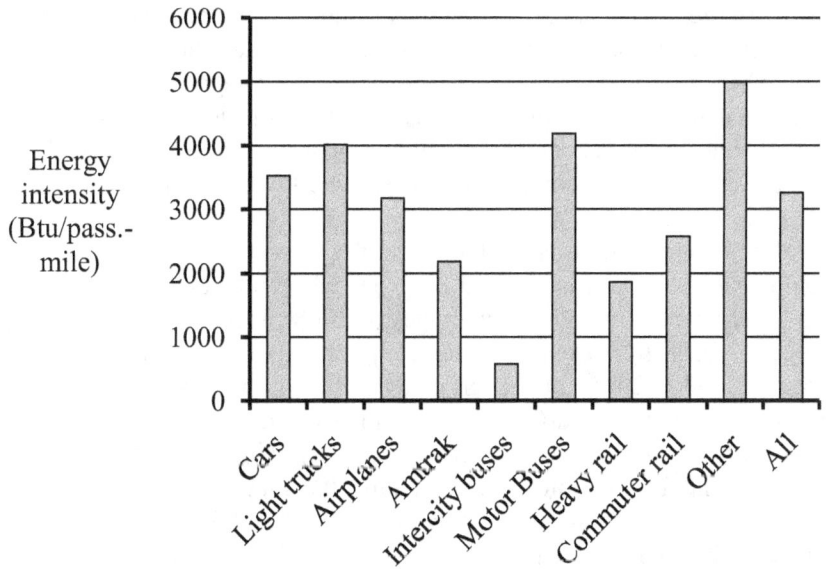

Fig. 8-16 2006 energy intensity for passenger transportation [BTS].

vehicle's capacity is the maximum number of passengers it is designed to carry, and its *load factor* for a trip is the number of passengers it is carrying divided by its capacity. For a given vehicle, the amount of energy per mile is approximately the energy needed for the vehicle when it is empty plus an additional amount for each passenger. The additional amount per passenger is normally small compared to the amount required for the empty vehicle. This is the reason carpooling has so much energy saving potential and an airline's profitability depends so much on its load factor. As much as 33 percent of an airline's costs are fuel costs.

It is apparent from Figs. 8-14, 8-15 and 8-16 that decreasing the energy intensities of cars and light trucks, increasing the average occupancies of cars and light trucks, increasing the load factors of transit systems and aircraft, and shifting the movement of people from cars and light trucks to transit vehicles are the principal means of reducing the energy needed to move people. Decreasing the energy intensities of cars and light trucks can be achieved by increasing either their average occupancies or their *miles per gallon* (*mpg*) values. Reasonably priced cars and light trucks are already being produced with mpg values much higher than those of the current average vehicle, but it takes time to flush out those with lower mpg ratings. Carpooling is the principal way of improving the average occupancy of cars and light trucks. Significant energy savings gained by shifting to transit systems can occur only if the average occupancies of transit vehicles are increased. For example, the average occupancy for motor buses is 8.81 despite the fact that their capacities are normally between 30 and 60 people. There are reasons transit systems find it difficult to increase their occupancies. Although airlines and intercity bus companies can optimize their schedules, transit systems need to operate when the demand is low as well as when it is high. Also, the "Motor bus" subcategory includes school buses that, by the nature of their purpose, have load factors less than 50 percent. Many American cities are either not large enough or not well-structured to justify the high cost of heavy rail or commuter rail, especially those cities west of the Mississippi River.

One of the main premises in the examples presented below is that the average mpg for cars will increase by 0.5 mpg per year and, for light trucks, will increase by 0.4 per year. Although the average for 2008 model cars was 31.2 mpg and the average for 2008 light trucks was 23.4 mpg, the average car in use in 2006 had a mpg rating of only 22.4 and the average light truck in use had a mpg rating of 18. Our hope for increasing the mpg for these vehicles relies on the replacement of our current large internal combustion engines (ICEs) with hybrid, and electric designs. The most mature technology for reducing the mpg is the hybrid design. Table 8-2 summarizes mpg ratings for a few 2009, 2010 and 2011 hybrid models. Many of the hybrids, including the Fusion, Camry and Prius, use nickel-metal-hydride batteries. Plug-in hybrids, such as the Chevrolet Volt that became available in 2011, and all-electric cars (EVs) such as the Nissan Leaf, can charge their batteries from the grid, offering an even greater reduction in their petroleum requirements. However, generating the extra electricity may increase the consumption of coal and natural gas, depending on how we choose to produce our electricity. The Nissan Leaf will have a range of 160 km (100 mi) and the Volt will have a range of 1025 km (640 mi), but its range using battery power alone will be only 64 km (40 mi). Both will use lithium-ion batteries.

Table 8-2 Statistics on a few 2009, 2010 and 2011 model hybrid cars and SUVs.

Car	Year	Seating capacity	Mpg city	Mpg highway
Ford Fusion	2010	5	41	36
Ford Escape	2010	5	43	38
Toyota Prius	2010	5	51	48
Toyota Camry	2011	5	38	34
Toyota Highlander	2009	7	27	25
Honda Civic	2009	5	40	45
Honda Insight	2010	5	40	43
Hyundai Sonata	2011	5	37	39

With any highway vehicle, its important characteristics are its efficiency, ability to accelerate, maximum speed, *range* (i.e., the maximum distance the vehicle can travel before it must refuel or recharge), the time it takes to refuel or recharge, maintenance, initial cost and the weight and space required for its power train and energy storage. Cars and light trucks powered only by ICEs respectively averaged 22 and 18 mpg in 2006. Their ranges are above 400 miles (640 kilometers) and they can be quickly refueled at numerous service stations. Their performance is generally satisfactory, but they tend to be expensive to maintain. The technology for ICE vehicles is quite mature and it is doubtful that cars powered by ICEs alone will be significantly improved in the future. Hybrids get more miles per gallon, perform well and have similar ranges. Their primary maintenance cost is the cost of replacing their battery packs. Generally, the hybrid models with non-hybrid counterparts (e.g., the Toyota Camry and Ford Fusion) are 20 to 25 percent more expensive than the non-hybrids, but government subsidies may make new hybrids more competitive. To demonstrate the differences among cars powered by ICEs, hybrid cars, electrics cars and fuel cell-powered cars, the important features of the various types of cars are shown in Figs. 8-17 through 8-22. As shown in Fig. 8-17, these features for an ICE-powered car include a fuel tank, an ICE and a drive train consisting of a transmission, drive shaft, differential, axles and wheels.

There are two basic hybrid designs and both include a fuel tank, an ICE, battery pack, electric motor/generator and drive train. One is the parallel structure shown in Fig. 8-18 and is the technology employed by the Toyota Prius and Honda Civic. In the parallel structure the ICE, electric motor/generator and transmission are built into a single unit. This structure is sometimes referred to as a power-assist design because the electric motor/generator assists the ICE when extra power is

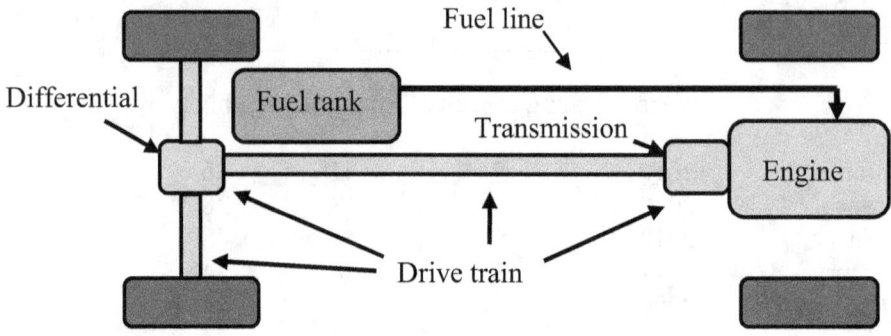

Fig. 8-17 Basic structure of a car powered by an internal combustion engine.

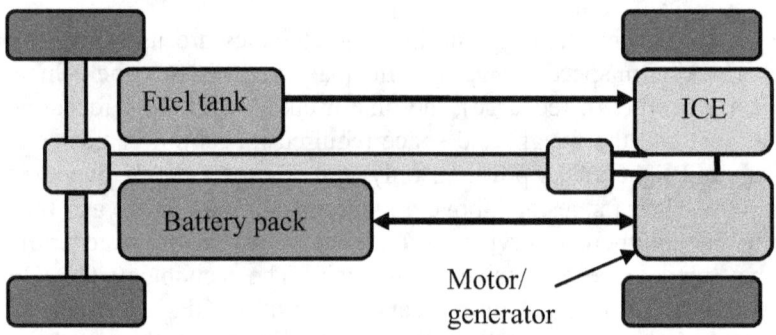

Fig. 8-18 Basic design of a parallel hybrid vehicle.

needed. An electric motor may also be used as a generator depending on whether it supplies torque (i.e., rotational force) as a motor or has a torque applied to it as a generator. The fundamental principle behind a hybrid is to use the electric motor/generator as a generator to recharge the battery pack during braking and assist in powering the vehicle during acceleration. This allows the energy normally lost as heat radiating from the brake pads and disks to be directed to the battery pack. This energy then can be applied to the electric motor/generator as it is needed. Also, hybrid vehicles are designed to turn off the ICE when it would normally be idling and use the electric motor to initiate acceleration. In addition, the electric motor/generator allows the ICE to run more closely to its optimal operating speed.

The other hybrid structure is the serial design diagrammed in Fig. 8-19. In this design, the ICE's sole purpose is to drive an electric generator that is for keeping the battery pack charged. The vehicle's power is provided only by an electric motor/generator that can also charge the battery pack while the vehicle is braking. Because the ICE needs to be operating only when the battery pack requires charging, it can always operate at its optimal speed and can turn off when the battery pack is charged. Also, the ICE does not need to be as large as the one in the parallel design, but the motor/generator must be larger because it alone powers the vehicle. The transmission is a simple one-speed gearbox. The vehicle shown in Fig. 8-19 includes a plug-in feature that permits the battery pack to be charged through a standard 110 or 220 volt wall outlet. The converter is for converting the ac current from the outlet to the dc current required by the battery pack. This plug-in feature may or may not be present on either a serial or parallel design, but is more advantageous in a serial system. Being able to supply energy from an external electrical system saves energy only if the external system is more efficient than the ICE onboard the vehicle. Normally this is the case. An ICE is typically somewhat less than 20 percent efficient while the grid and battery pack may be capable of delivering power with more than 30 percent efficiency.

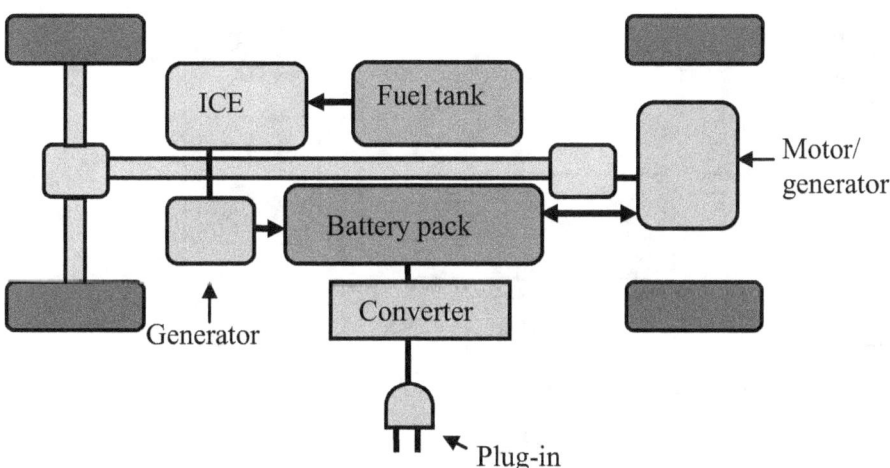

Fig. 8-19 Basic design of a purely serial, plug-in hybrid vehicle.

Because the battery pack in the serial design must supply all of the energy to the motor/generator, it must be capable of discharging rapidly during acceleration. The parallel design alleviates this problem by connecting the ICE directly to the drive train. The requirement for a fast discharge rate causes the battery pack to be more expensive.

Figure 8-20 depicts a purely electric car. The main problem with EVs is that even very small cars have relatively short ranges. As discussed in Chap. 7, batteries have low energy densities and the energy that a vehicle can carry determines its range. The reason hybrid vehicles have fuel tanks is that liquid fuels, especially gasoline and diesel, have high energy densities and, therefore, are able to extend their vehicle ranges to compete with ICE vehicles. A second problem with EVs is that it takes much longer to charge batteries than it takes to pump liquid fuel into a fuel tank. Newly developed batteries, such as nickel-metal-hydride and lithium-ion batteries or the more recent experimental nanotube batteries, have much higher energy densities and may be charged and discharged more quickly than lead acid batteries (the type of batteries used in ICE vehicles), but are much more expensive (see Table 7-1 in Chap. 7). Most 2010 model hybrids will use nickel-metal-hydride or lithium-ion battery packs and the serial hybrid design of the 2011 Chevrolet Volt will have a lithium-ion battery pack large enough to power it up to 40 miles (64 kilometers) on electricity alone. This amount of energy is sufficient for the average daily commute, but for longer drives the Volt must rely on liquid fuel and its ICE. In contrast with a vehicle that uses only an ICE, a contributing factor to the efficiency of an EV is that it does not use fuel while it is stopped.

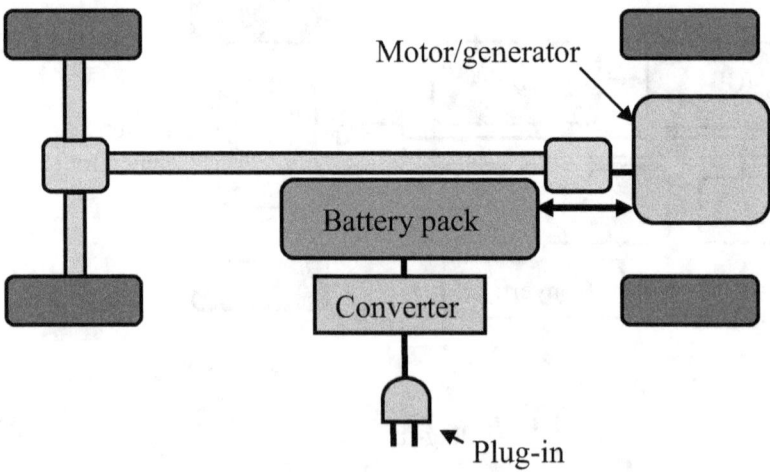

Fig. 8-20 The structure of a purely electric car (EV).

Because of their simpler structure, EVs are more efficient and require less maintenance. Except for the cost of replacing the battery pack, EVs require little maintenance because motor/generators and simple, one-speed gearboxes are low maintenance. If the electricity is generated by wind turbines or solar panels, then no fossil fuels are consumed.

While the concentration has been on using nickel-metal-hydride and lithium-ion batteries in future hybrids, hybrid plug-ins and EVs, there are other possibilities. One such possibility is the battery-capacitor that combines an ultra-capacitor construction with that of a lead-acid battery by splitting the cathode so that material in one half of the cathode is similar to an ultra-capacitor and the other half is similar to that of a lead-acid battery. A normal lead-acid battery is slow in delivering power, but an ultra-capacitor delivers power quickly. Also, a battery-capacitor combination can be charged four times as fast as a standard lead-acid battery, does not have the overheating problem of lithium-ion batteries and is much cheaper than either nickel-metal hydride (NiMH) or lithium-ion batteries.

There has been some experimentation with vehicles that use the compression of air to store the energy produced by braking and an air turbine to assist acceleration. The energy is stored in a tank of compressed air instead of a battery pack. Such a design is simpler than that of an electric hybrid and improves efficiency over that of a purely ICE design, but has not received the attention of an electric hybrid because the amount of stored energy for the available space tends to be less than that of a battery pack and there is no plug-in feature.

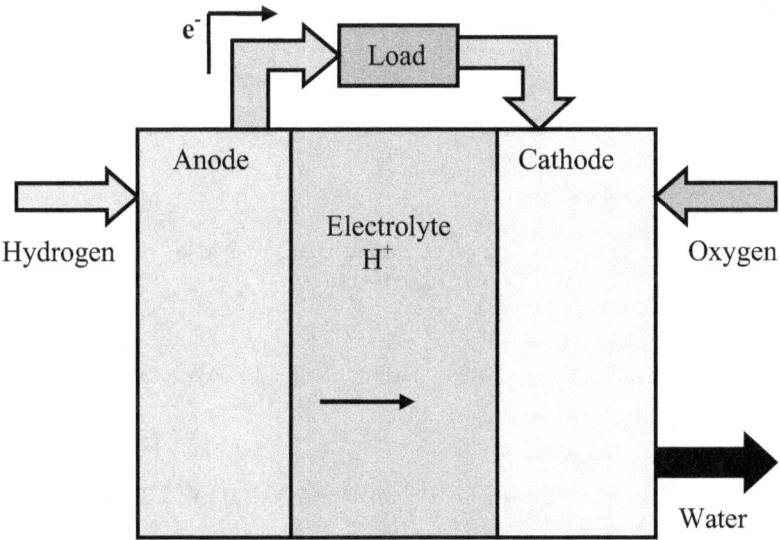

Fig. 8-21 Basic fuel cell design.

The other possibility for powering cars and light trucks is to use fuel cells. The basic fuel cell design is given in Fig. 8-21. Fuel cell vehicles are fueled by hydrogen and oxygen. Oxygen may be obtained from the air, but hydrogen must be produced by the processes discussed in Chap. 6. For the simplified fuel cell shown in Fig. 8-21, the electrolyte allows positively charged hydrogen ions to pass through it, thus causing a voltage to exist between the anode and cathode. This voltage causes the electrons freed from the ions to travel through an external circuit and this electric current may be passed through a load, such as an electric motor, to do useful work. The voltage created by a single cell is no more than the theoretical maximum of 1.23 volts, but a more practical maximum is one volt. Because of losses within the cell, most fuel cells operate between 0.6 and 0.8 volts. For greater voltages the cells must be connected serially to form a stack.

Table 8-3 Characteristics of a few prominent fuel cells.

Type	Temperature range	Anode In→Out	Cathode In→Out	Electrolyte/ ions	Power range	Efficiency range (percent)
Alkaline	60-90 °C 140-200 °F	$H_2 \rightarrow H_2O$	O_2 in	KOH/ OH⁻	0.3-5 kW	60-70
Proton Exchange membrane (PEM)	70-120 °C 160-250 °F	H_2 in	$O_2 \rightarrow H_2O$	Polymer/ H^+	50-250 kW	40-60
Molten carbonate (MCFC)	600-650 °C 1100-1200 °F	$H_2 \rightarrow H_2O$	O_2, CO_2 $\rightarrow H_2O$	Carbonate/ CO_3^{-2}	1-2 MW	50-80
Solid oxide (SOFC)	1000 °C 1830 °F	$H_2 \rightarrow H_2O$	O_2 in	Ceramic/ H^+	100 W-2 MW	40-80
Phosphoric acid (PAFC)	150-200 °C 300-390 °F	H_2 in	$O_2 \rightarrow H_2O$	H_3PO_4/	0.2-11 MW	35-70
Direct Methanol DMFC	60-120 °C 140-250 °F	CH_3OH, $H_2O \rightarrow CO_2$	$O_2 \rightarrow H_2O$	Polymer/ H^+	<1W-100 kW	20-40

Although all fuel cells operate by combining hydrogen and oxygen to form water, they may be constructed from many different designs using different input gases and electrolytes and there may be outputs other than water. Table 8-3 summarizes the characteristics of a few of the more prominent designs. In addition to the electrolyte, fuel cells may require a catalyst. For an alkaline, proton exchange membrane (PEM) or phosphoric acid fuel cell (PAFC) the catalyst is platinum, which increases the cost of the fuel cell substantially. Neither a molten carbonate fuel cell (MCFC) nor a solid oxide fuel cell (SOFC) requires an expensive catalyst because they operate at such high temperatures. Also, they do not have a problem with carbon monoxide poisoning, a problem from which the lower temperature fuel cells suffer. To avoid carbon monoxide poisoning, the input hydrogen to low temperature fuel cells must be highly purified. On the other hand, high operating temperatures cause long startup times and this limits their use in small vehicles, but MCFCs and SOFCs could be used in semis, trains and ships or for stationary applications. For stationary applications, because of the heat they release they may be used in conjunction with steam turbines in a combined cycle arrangement. In such applications, their efficiencies may be over 80 percent.

If a fuel cell is to provide the power for a car or light truck and the hydrogen is not produced on-board the vehicle, then there is the problem of storing it. The energy content of a kilogram of hydrogen is essentially the same as that of a gallon of gasoline, but it takes about 12 cubic meters to store a kilogram of hydrogen at standard atmospheric pressure. Therefore, hydrogen fuel must be compressed or liquefied. If it is compressed to 200 atmospheres, then it still requires about 60 liters to store one kilogram of hydrogen. It is estimated that it takes about three kilograms of hydrogen to drive a small car 800 km (500 mi), so to provide such a range would require a volume of 180 liters (47.6 gal). To use liquefied hydrogen in a car or light truck would require too much potentially dangerous, cryogenic equipment to be practical and the required refrigeration would reduce the overall efficiency.

The direct methanol fuel cell (DMFC) alleviates this problem because the input fuel is methanol instead of hydrogen. Although the fuel density of methanol is less than half that of gasoline or diesel, it is somewhat greater than hydrogen compressed to a pressure suitable for use in small vehicles. However, the anode reaction produces carbon dioxide and an onboard reformer that produces hydrogen from the methanol is required. Ethanol could be used but the efficiency would be somewhat lower. Even gasoline or diesel could be used, but a petroleum-based fuel cell would not reduce the carbon dioxide emissions even though efficiency would be improved. A vehicle design that uses a liquid fuel and reformer is shown in Fig. 8-22. All liquid fuels would require a reformer and there is the question of the amount of fossil fuels used in the production of the methanol or ethanol.

The simulation results of a study by Sheldon S. Williamson and Ali Emadi that appeared in the May, 2005, issue of the *IEEE Transactions on Vehicular Technology* are given in Table 8-4. These results consisted of mpg values, energy usage and well to tank (WTT), tank to wheel (TTW) and well to wheel (WTW)

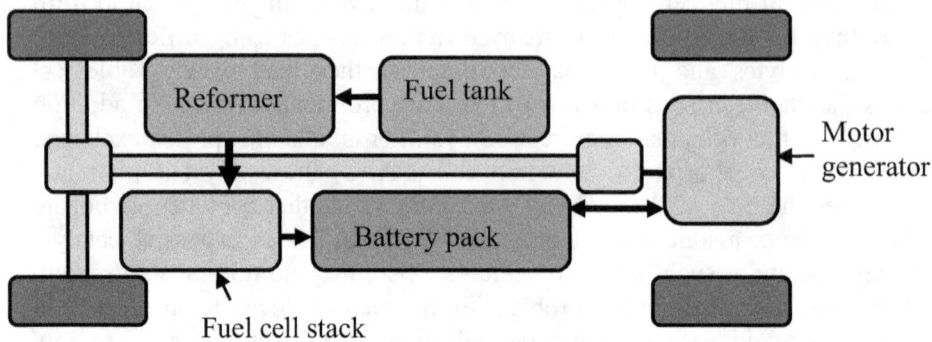

Fig. 8-22 Design of a fuel cell vehicle that uses gasoline or diesel.

Table 8-4 Statistics for medium-sized SUVs using various power trains.

Type of drive	City mpg	Hwy. mpg	WTT eff.	TTW eff.	WTW eff.	Energy Btu/mi
Gasoline	18.1	22.3	88	12.1	10.6	7000
Diesel	22.7	26.3	86	18.2	15.6	5800
Gasoline hybrid	24.5	33.8	88	22.3	19.6	5000
Diesel hybrid	29.6	37.7	86	30.3	25.9	4000
Fuel cell hybrid	21.8	30.3	88	16.2	14.2	6000

efficiencies of five medium-sized SUVs having the same body style. The assumed gross weight was 2,050 kg (4,520 lb). Cars would, of course, show somewhat better values depending on their body characteristics, especially weight. The efficiencies are given as percentages. Also listed are the WTW Btu/mi energy requirements of the given vehicle types. This study did not examine hydrogen fuel cell or electric vehicles. But, if one assumes a PEM fuel cell, a 50 percent loss in the production of hydrogen (see Table 6-4), a 30 percent loss in the compression and distribution of the hydrogen, a 35 percent loss in the fuel cell (see Table 8-3) and a 15 percent loss in the drive train of an SUV, then the efficiency would be

$$(1-0.50)(1-0.30)(1-0.35)(1-0.15) = 0.19,$$

or 19 percent. If an electric SUV does not use *regenerative braking* (the use of braking energy to charge a battery pack), uses a lithium-ion battery pack, there is a 70 percent loss in generating and distributing the electricity, a 15 percent loss in charging and discharging the battery pack, a 20 percent loss in the electric motor and a 15 percent loss in the drive train, then the overall efficiency would be 17 percent. With regenerative braking, the efficiency may be twice this amount during city driving.

Except for diesel hybrids, a May, 2004, article in *Scientific American* by Matthew L. Wald indicates similar results for WTW efficiencies. The Wald article gave the efficiency of a hybrid diesel to be slightly less than 20 percent. In addition, it gave the efficiencies of several fuel cell designs, and the only design that exceeded 20 percent efficiency was a hydrogen fuel cell design for which it assumed the hydrogen was supplied by a steam reforming process. This design had an efficiency of 23 percent. If the hydrogen is produced using electrolysis, then the efficiency drops to less than 10 percent. The gasoline fuel cell design had an efficiency that was about the same as a diesel hybrid and the methanol fuel cell design had an efficiency that was about the same as a gasoline hybrid. No vehicle design specifications other than those for the power train were given. The size and shape of a vehicle is important in determining Btu consumed per mile.

What follows is a series of examples that project the changes in United States petroleum and electricity consumption under different scenarios. The initial time was chosen to be 2006 because the data for that year was the most complete and accurate at the time of writing, although some estimates were needed even for 2006. The 2008 sector energy amounts and percentages given in Fig. 1-7 were essentially the same as those for 2006, although the amounts for 2008 were slightly less. The projections extend to 2040 and take into account the growth in population.

- The population growth rate assumed was 1.03 percent, which was determined from population data extending from 1973 to 2006.

This growth rate predicts a United States population of 422 million in 2040, a value slightly greater than the 406 million predicted by the United States Census Bureau. Population growth in the United States will be heavily dependent on immigration, which tends to be unpredictable.

There are several factors that enter into any prediction and these energy projections are only examples of what could happen under assumptions that seem reasonable based on what is known to be technologically possible and economically feasible. Some discussions of possible improvements to the projections by using future technologies are given following the examples. Figure 8-23 provides a backdrop for the projections that follow by summarizing the 2007 energy usage.

The first projections concern the transportation sector. Clearly, because cars and light trucks consume 63 percent of the transportation energy and roughly 17 percent of the United States total energy, if our energy needs are to be satisfied the energy use by cars and light trucks (i.e., pickups, SUVs and vans) must be addressed. One common observation is the increased number of light trucks that are used primarily for transporting people and the fact that these vehicles use more fuel than cars. In the first example let us assume that:

- The ratio of light trucks to the total number of cars and light trucks that existed in 1980 provided enough light trucks to meet the requirements for moving small amounts of goods and equipment. In 1980 this ratio was 0.18.
- It is possible to estimate the total number of annual replacements using 1995 through 2005 data by subtracting the increase in the number of cars and light trucks from the number of these vehicles sold. The estimate used is six percent.
- The total number of cars and light trucks per capita remains the same as it was in 2006 until 2040.
- The replacement of light trucks is 18 percent of the total replacements.

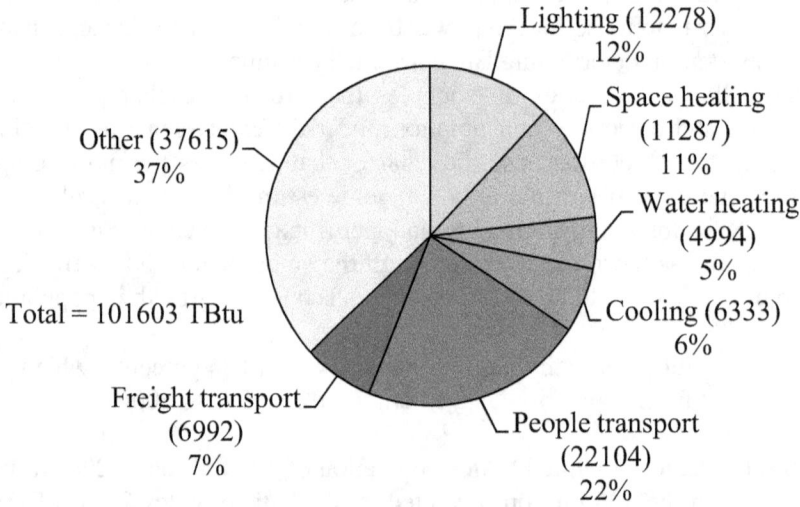

Fig. 8-23 2007 energy consumption by usage in the United States in TBtu [EIA, BTS].

Figure 8-24 gives the BTS data for the numbers of cars and light trucks from 1995 through 2005 and, using these assumptions, the calculated numbers of these vehicles from 2006 through 2040. By 2006 the light trucks to total ratio had reached 0.42, but, under the above assumptions, by 2040 it would be reduced to 0.23.

An estimate is made for the amount of fuel consumed by further assuming that:

• Each year the mileage for cars increases by 0.5 miles per gallon (mpg) from its 2006 value of 22.4 and the mileage for light trucks increases by 0.4 mpg from its 2006 value of 18.0,

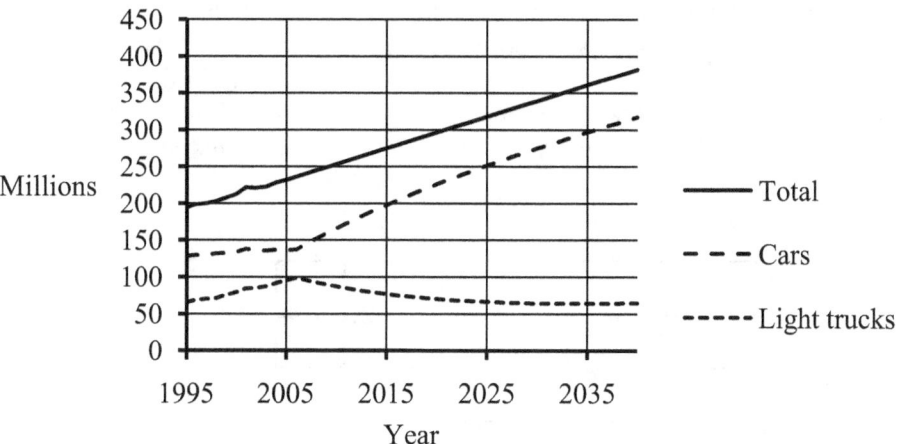

Fig. 8-24 Vehicles resulting from changing the ratio light trucks/total to that of 1980.

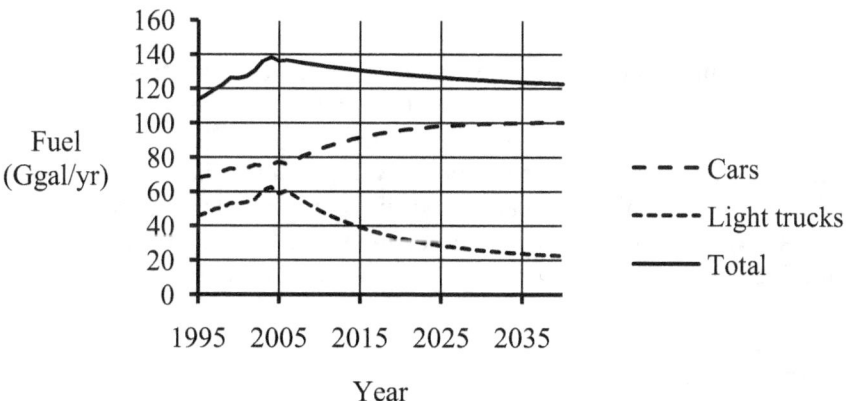

Fig. 8-25 Fuel usage resulting from the ratio change indicated in Fig. 8-24 and improved mpg ratings.

The results obtained by using both the ratio change and the mpg increases are graphed in Fig. 8-25. These mileage assumptions are quite speculative, but are achievable at reasonable costs with the current hybrid technology discussed above and a replacement rate that is slightly greater than the current replacement rate. Figure 8-26 shows the fuel savings due to the ratio change only and does not indicate the savings due to the gains in mpg. Shifting the ratio light trucks/total cars and light trucks to its 1980 value as opposed to holding it constant at its 2006 value would save about 195,000 barrels of gasoline and diesel per day in the year 2040. This amount of savings is very modest compared to the approximately 15 million barrels of gasoline and diesel presently consumed daily in the United States.

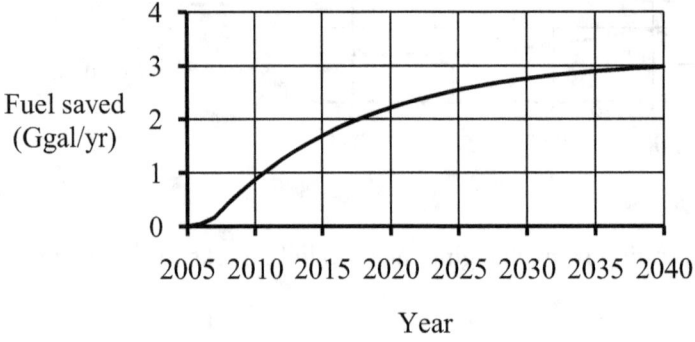

Fuel saved (Ggal/yr)

Year

Fig. 8-26 Annual fuel savings resulting from the ratio change indicated in Fig. 8-24.

Most of the savings indicated by Fig. 8-25 is due to the predicted increases in the mpg. If the population estimate assumed is accurate, it is seen from Fig. 8-25 that, because of our growth in population, our total gasoline and diesel usage will be about the same as that of the year 1998 despite the savings. Fig. 8-27 gives the percentage increases in the numbers of vehicles, vehicle-miles (i.e., miles driven), passenger-miles, passenger-miles per capita and fuel per capita from 1980 to 2004. Clearly, if we are to seriously attack our problems related to petroleum we must change our driving habits, but this does not mean we will have to lower our standard of living.

Figure 8-28 gives the results of a scenario based on the initial year being 2006 and the assumptions that:

- The per capita total vehicle-miles for cars and light trucks returns to its 1980 level by 2030 in equal annual increments and then remains constant until 2040.

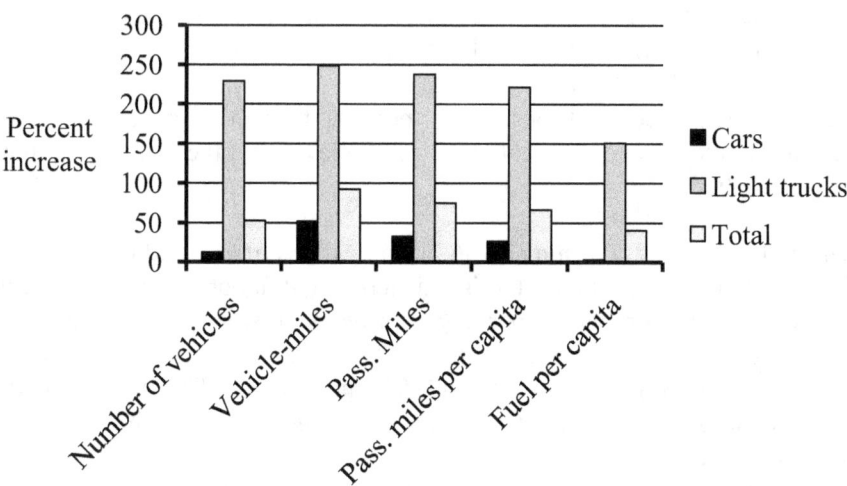

Fig. 8-27 Percent increases from 1980 to 2004 in the United States [BTS].

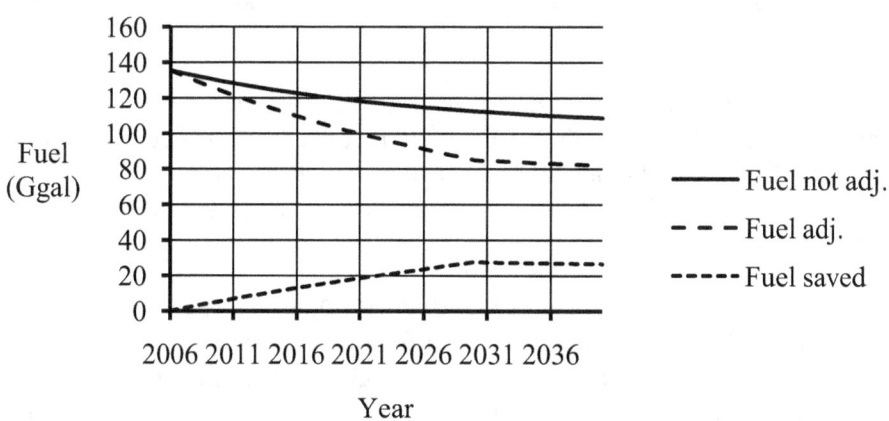

Fig. 8-28 Fuel usage in the United States if driving statistics return to 1980 values by 2030.

- The ratio between cars and light trucks returns to its 1980 level by 2030 in equal annual increments and then remains constant until 2040.

- The average mpg for cars increases by 0.5 mpg per year from its 2006 value of 22.4 and the average mpg for light trucks increases by 0.4 mpg per year from its 2006 value of 18.
- For the unadjusted fuel curve the per capita vehicle-miles and light truck/total ratio remain constant at their 2006 levels but the increases in mpg are still applied.

In 2006, cars and light trucks consumed 136 billion gallons of fuel and the projected usage in 2040 is 82 billion gallons. That is a difference of about 3.5 million barrels per day and is a 1.75 million barrels per day improvement over our consumption if only the improved values of mpg are assumed. The overall savings by 2040 would be almost 40 percent of our 2006 demand for these fuels. The savings curve represents the savings over and above the mpg savings, i.e., it is the difference between the top two curves.

In the United States, only about nine percent of petroleum is refined into jet fuel, but this percentage is sufficient to warrant its examination as a possible reduction in petroleum consumption. The Boeing Company claims that all three versions of its new 787 Dreamliner are 20 percent more efficient than any previous airplanes of comparable size. Passenger capacities of single-class Dreamliners range from 210 to 330. Using the 20 percent gain in efficiency as a basis, the next example assumes that:

- The per capita ton-miles for domestic certified passenger airplanes remains constant from 2006 through 2040, but the fuel per ton-mile decreases by 20 per cent from 2006 through 2040 in equal annual increments.
- The unadjusted fuel per ton-mile does not decrease but the per capita ton-miles remains constant.

The results of the example are given in Fig. 8-29. By 2040 the savings due to this gain in efficiency is 3,807 million gallons of jet fuel per year, or 248,000 barrels per day. However, even after the 20 percent gain in efficiency, the estimated amount of jet fuel used in 2040 is 1,770 million gallons more than the 13,458 million gallons consumed in 2006. Note that this example is based on ton-miles per capita remaining constant. Only 21 percent of the ton-mileage is accounted for by freight; the remainder is due to passengers and their luggage (the BTS assumes 91 kg, or 200 lb, per passenger). If just 15 percent less freight per capita is transported by airplanes that transport freight only, then there would be 480 million gallons per year, or 31,300 barrels per day, in additional savings by 2040.

The above examples concentrated on increasing the efficiencies involved in moving people either by improving the efficiencies of the vehicles or switching to more efficient vehicles or both. But an alternate possibility is to shift some of the personnel transport to other modes of transportation. For a given type of vehicle, the

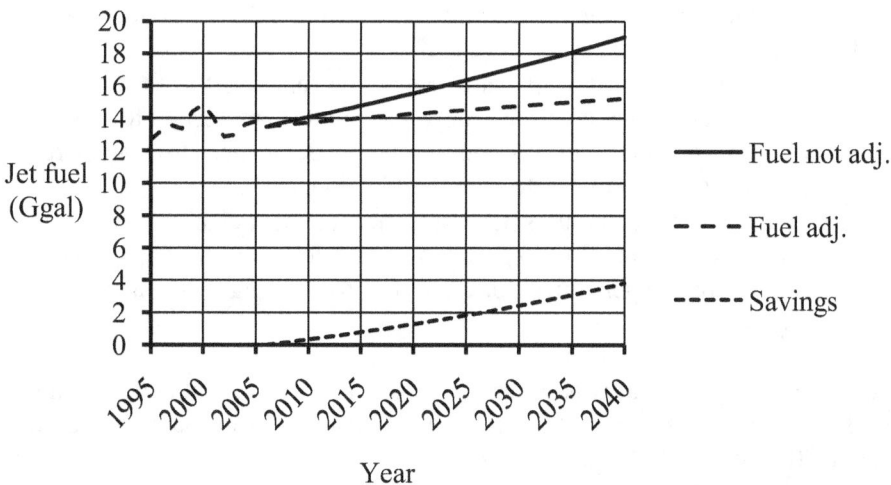

Fig. 8-29 Commercial jet fuel reduction of 20 percent per ton-mile by 2040.

energy per mile is approximately some base amount plus an amount that is proportional to the number of passengers it is carrying. Generally, for a particular load factor, the larger the vehicle the less energy is needed per passenger-mile. However, a small vehicle with a high average load factor may require less energy per passenger-mile than a much larger vehicle with a much lower load factor. This is one of the main reasons Amtrak uses about four times as much energy per passenger-mile as intercity buses (see Fig. 8-16). Also, intercity bus energy consumption per passenger-mile was about 30 percent that of heavy rail (subway trains), 16 percent that of cars and 14 percent that of light trucks.

The overall average number of passenger-miles was dominated by cars and light trucks and could be lowered by shifting more personnel transport to mass transit systems. However, transit systems have a problem with load factor because much of their loads occur when people are going to and getting off from work. In order to gain significant energy savings, the load factors of transit systems would need to be increased substantially. With improved load factors, transit systems have the potential for saving a considerable amount of energy. Let us assume that:

- The annual miles per car amount remains constant at its 2006 value of 12,427.
- The total numbers of urban passenger-miles from 2006 through 2040 for cars is 12,427 times the numbers of cars indicated in Fig. 8-24 times the 2006 percentage of car-miles that are urban car-miles times the 2006 average occupancy of cars.

- The average urban mpg for cars increased from 17.4 in 2006 to 41.4 in 2040 in equal annual increments of 0.706.
- The fuel used by transit vehicles remains constant at its 2006 value from 2006 through 2040 (i.e. no additional vehicle-miles are needed to absorb the additional passengers and no more energy is needed to transport the increased passenger load due to increased load factors).
- The unadjusted fuel amounts take into account the changes in mpg and population as indicated in the assumptions for Fig. 8-28, but not the reduction in passenger-miles for cars due to shifting passenger-miles to transit systems.

The amounts of energy saved by assuming 2, 4, 6, 8 and 10 percent shifts from cars to transit systems is given in Fig. 8-30. For a six percent shift the savings would be 3.65 Ggal in 2040, or 238,000 barrels per day.

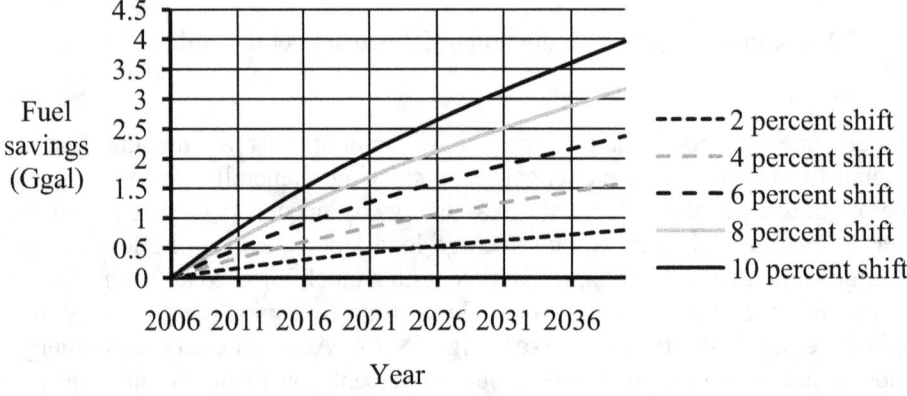

Fig. 8-30 Fuel reductions by shifting urban car passenger-miles to transit systems.

Over longer distances, rail transportation has difficulty competing with airplanes under the Amtrak system of operation in which they must compete with freight traffic for use of the tracks. There are few places where new tracks dedicated to passenger traffic are economically feasible, the eastern seaboard from Washington to Boston being the most obvious. Amtrak already has high-speed trains running in this corridor, but the tracks are old and limit the speed to about half that of the high-speed trains in Europe and Japan. Trains do have one advantage over airplanes in that people would depart and arrive at downtown locations rather than outlying airports.

Other locations for which high-speed rail lines are being proposed are the corridors between San Francisco and San Diego, Chicago and St. Louis, Cleveland and New York, Pittsburgh and New York, Las Vegas and Anaheim and Orlando and Tampa. Recently allocated federal funds are mainly to be spent to support the construction of the San Francisco-San Diego and Orlando-Tampa projects. Both of these lines are rated at over 150 miles per hour. The most recent rail lines in Europe and Japan use magnetic levitation, commonly referred to as *maglev*. In maglev systems the trains ride on a magnetic cushion and are pulled along by a moving magnetic field. Energy is not lost to the friction between the wheels and the tracks. The Las Vegas to Anaheim and proposed new Baltimore to Washington lines are to be maglev systems. The cost of a line varies substantially with the terrain and cost of the right-of-way as well as the type of construction. The cost of the Baltimore-Washington system is estimated to be $132 million per mile while the estimated cost of the Las Vegas-Anaheim system is only $48 million per mile despite the fact that both are to be of maglev construction. A proposed San Francisco-Los Angeles wheels-on-rail system is expected to cost $63 million per mile. (See the article entitled "Revolutionary Rail" by Stuart Brown in the May, 2010, issue of *Scientific American*.)

Although the development of long distance train travel could reduce the energy consumption of intercity travel, the amount of energy saved would be small relative to our overall energy consumption and other approaches to saving energy are likely to be more economically sound. However, Fig. 8-31 does provide some indication of the savings that could result if there were a shift from air travel to train travel and:

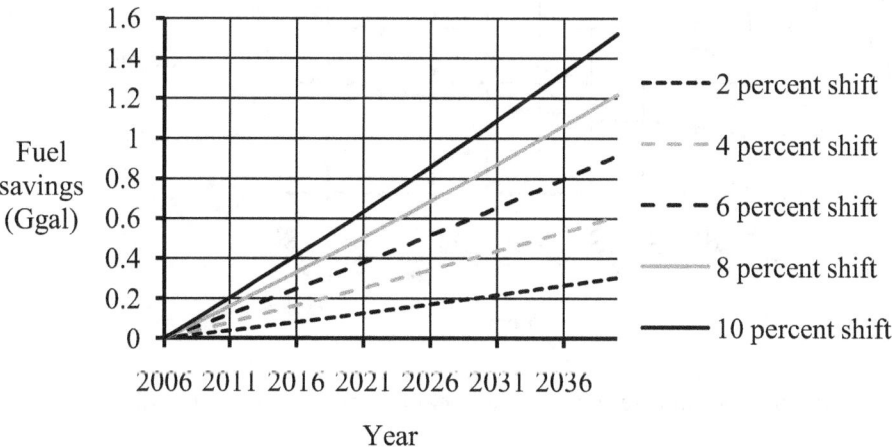

Fig. 8-31 Jet fuel reductions by shifting airline passenger-miles to Amtrak.

- The adjusted fuel amounts indicated in Fig. 8-29 are decreased by 2, 4, 6, 8 and 10 percent from 2006 through 2040 in equal annual increments.

A six percent shift would result in a savings of 0.914 Ggal in 2040, or 60,000 barrels of jet fuel per day, in addition to the savings indicated in Fig. 8-29.

Now let us apply a similar approach to that used to create Fig. 8-25 to analyzing the fuel consumption of trains and trucks. But instead of concentrating on passenger-miles, the emphasis is on ton-miles of intercity freight. In creating Fig. 8-32 it is assumed that:

- The total ton-miles of freight grows at the same rate it grew between 1995 and 2005.
- The ratio of ton-miles hauled by semis to that hauled by both semis and trains is gradually reduced in equal annual increments toward its 1980 value.

Figure 8-32 shows the ton-miles for trains and semis and the total number of ton-miles resulting from these assumptions.

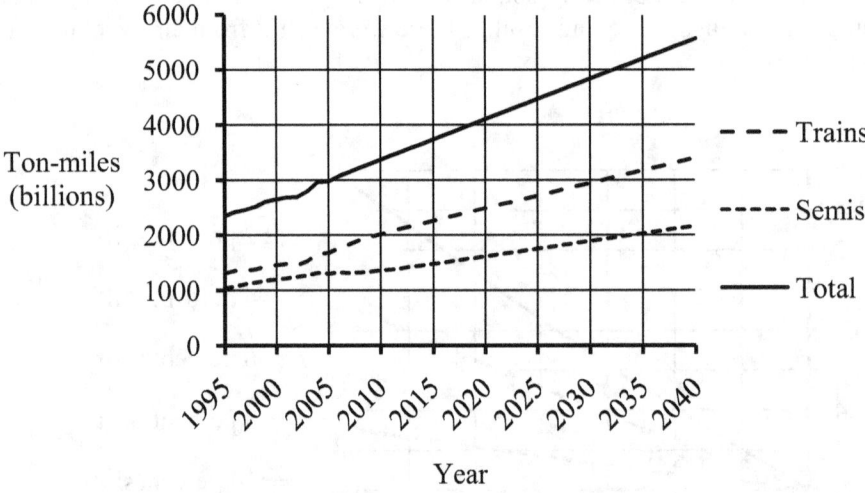

Fig. 8-32 Reducing the ratio semi ton-miles/total ton-miles to that of 1980.

Now let us assume that:

- The amount of diesel fuel per ton-mile of freight hauled by trains is 0.0025 and the amount of diesel fuel per ton-mile of freight hauled by semis is 0.019. These fuel usage rates are the same as those in 2006.

Figure 8-33 shows the amounts of fuel that would result from holding the ratio ton-miles for semis/total ton-miles constant at its 2006 value, shifting the ton-miles from semis to trains as indicated in Fig. 8-32 and the corresponding savings. The amount saved in 2040 would be 3.72 Ggal, or 243,000 barrels per day.

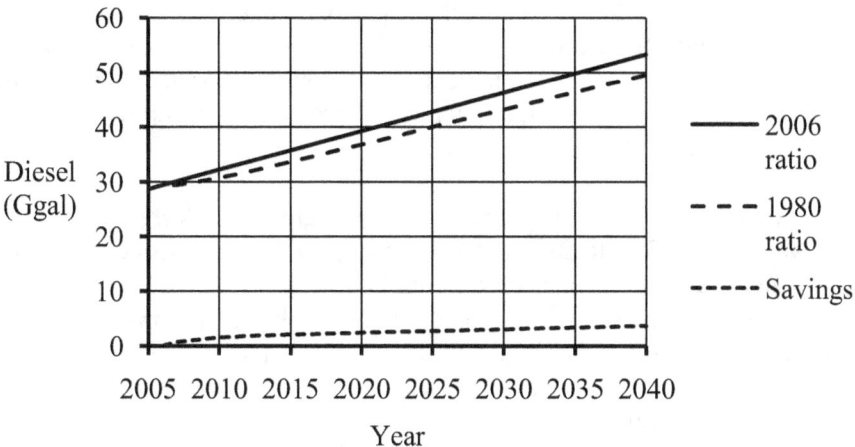

Fig. 8-33 Fuel saved by changing the ratio semi ton-miles/total ton-miles to that of 1980.

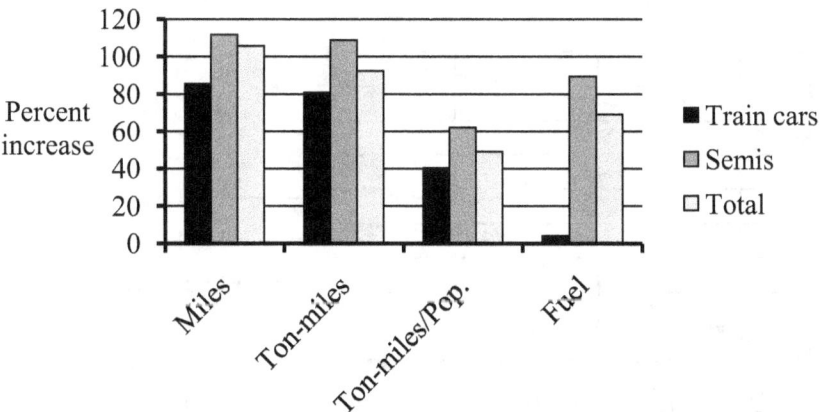

Fig. 8-34 Percent increases from 1980 to 2004.

    In order to significantly reduce the amount of fuel needed for hauling intercity freight, the rate of increase in the amount of freight must be reduced. Figure

8-34 shows the large percentage increases in miles traveled, ton-miles of freight, per capita ton-miles and fuel consumed by freight hauling that occurred from 1980 to 2004. In particular, note that the ton-miles per capita went up almost 50 percent and the ton-miles per capita for semis was 62 percent more than that of 1980. The amount of fuel used by trains increased by only four percent, indicating the improved efficiencies with which trains haul freight. Although, to some degree, the movement of freight indicates the amount of economic activity, the amount of diesel used for freight hauling is a major component of our petroleum usage and must be decreased if there is to be a serious reduction in our petroleum consumption.

    In the next example it is assumed that:

- The total per capita ton-miles for semis and freight trains decreases from its 2006 value of 10,355 to its 1980 value of 6,442 by 2030 in equal annual increments and then remains constant until 2040.
- The ratio of ton-miles for semis to the total number of ton-miles for both semis and freight trains decreases from its 2006 value of 0.436 to its 1980 value of 0.377 in equal annual increments between 2006 and 2030 and then remains constant until 2040.
- The total unadjusted fuel for semis and freight trains per capita remains constant at its 2006 value of 108.9 gallons. For the unadjusted fuel there is no change in the ratio of ton-miles for semis to ton-miles for both semis and trains.

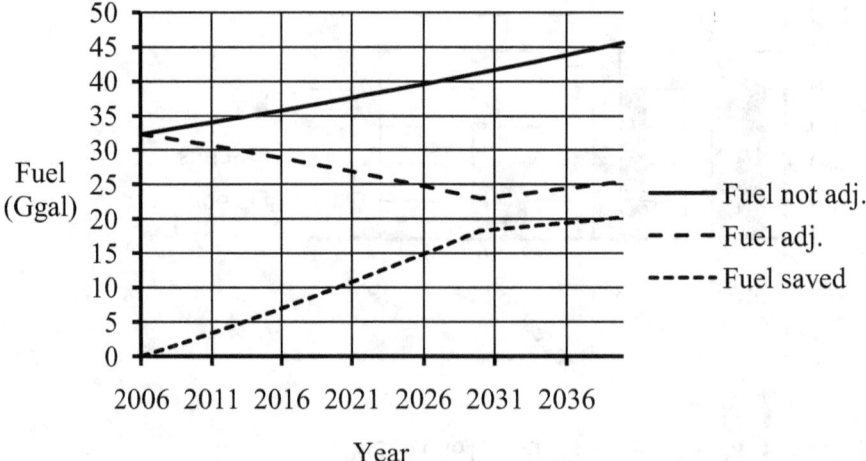

Fig. 8-35 Fuel saved if total/pop. and semi/total ratios are returned to their
        1980 values.

The results are graphed in Fig. 8-35. Using these assumptions, the savings by 2040 would be 20.23 Ggal per year or 1.33 million barrels per day.

Figure 8-36 summarizes the results contained in Figs. 8-28, 8-29, 8-30, 8-31 and 8-35. For Figs. 8-30 and 8-31, the curves corresponding to six percent shifts are used. The three curves labeled "Pass. not adj.", "Pass. adj." and "Pass. savings" are the totals obtained by adding the corresponding data in Figs. 8-28, 8-29, 8-30 and 8-31 and relate to the fuel amounts needed for transporting people. However, only the principal means of transporting people are reflected in these curves and transportation vehicles such as general aviation and ferry boats are not considered. The three curves that similarly refer to the freight hauled by semis and freight trains are from Fig. 8-35 and do not include single-unit trucks or ships that transport goods between ports within the United States. The remaining three curves are the sums of the corresponding passenger and freight curves and indicate the total fuel amounts and savings resulting from the above assumptions. In all cases the data are converted to millions of barrels per day (Mbbl/day). The numbers in parentheses are the projected amounts for the year 2040. Also given in the figure is the total amount of fuel, almost entirely gasoline, diesel and jet fuel, that was consumed by cars, light trucks, semis, most transit vehicles, trains and certified domestic airlines in 2006. Although the unadjusted curves tend to increase in proportion to population, they also tend to decrease because they take into account projected increases in mpg for cars and light trucks. As seen from the "Total not adj." curve, these two effects tend to cancel and the total "unadjusted" consumption is about the same in 2040 as it was in 2006.

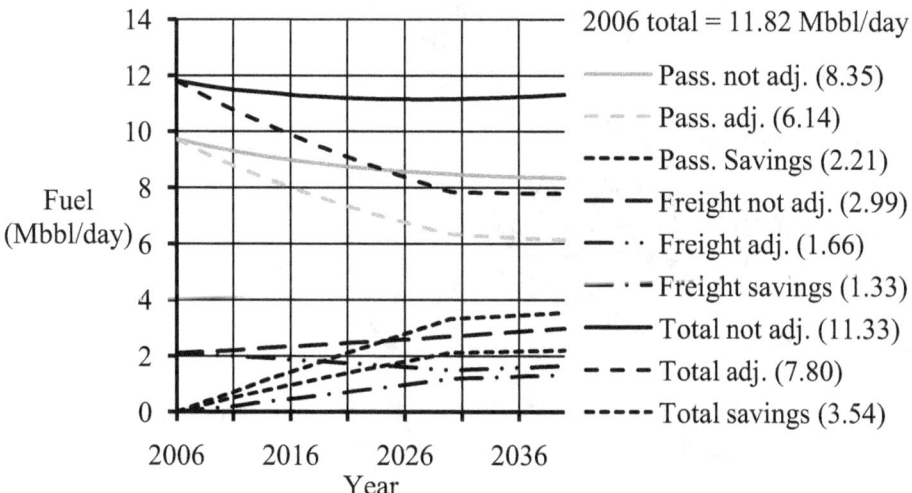

Fig. 8-36 Total fuel amounts computed from Figs. 8-28, 8-29, 8-30, 8-31 and 8-35.

        The concentration in the above examples is on saving the petroleum products gasoline, diesel and jet fuel by only those uses in the transportation sector for which significant gains are most likely to occur. No savings in other transportation areas are considered. In addition, as shown in Fig. 2-20, the transportation sector is responsible for only about 72 percent of our petroleum consumption. Not considered in these examples are the use of gasoline and diesel for driving agricultural and industrial machinery, the use of heating oil and the uses of other byproducts resulting from refining petroleum. Figure 8-37 gives the breakdown of petroleum products consumed in the United States in 2007 in Mbbl/day (the total is essentially the same as those of 2006, the base year assumed in the examples). Approximately 15.1 Mbbl/day of the 20.68 Mbbl/day are gasoline, diesel and jet fuel, of which about 3.3 Mbbl/day are used for purposes not considered in the above examples. Of this 3.3 Mbbl/day, 1.3 Mbbl/day is consumed by general aviation, single-unit trucks, buses, transit vehicles and water transportation. Almost all of the remaining 2.0 Mbbl/day are needed by the industrial sector for agricultural and industrial machinery. Two percent of the petroleum products is residue from the refining process and is used to produce road oil, tar and asphalt. Liquefied petroleum gases are used for heating and as petrochemical feedstock to produce plastics, fertilizer and so on. Still gas is used as refinery fuel and petrochemical feedstock. Included in the "Other" eight percent are aviation gasoline, lubricants, naphtha, waxes, additional petrochemical feedstock and petroleum coke.

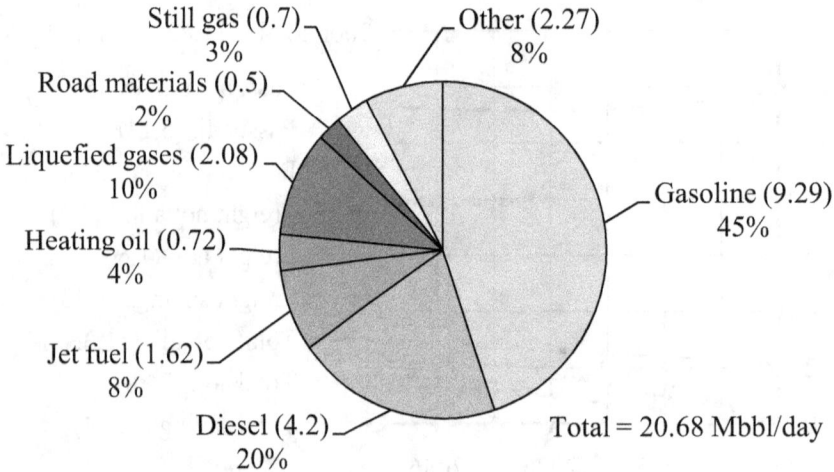

Fig. 8-37 2007 petroleum products produced in the United States in
       Mbbl/day [EIA].

For the products in the examples, Fig. 8-36 shows a decrease of 34 percent from 11.8 Mbbl/day to 7.8 Mbbl/day. Suppose that, despite the rise in population:

- The 3.3 Mbbl/day demand for gasoline, jet fuel and diesel not considered in the above examples is held constant through 2040.
- The remaining 5.6 Mbbl/day for other products is reduced in proportion to that of gasoline, jet fuel and diesel.

The other 3.3 Mbbl/day demand could be held constant through the use of biomass and electricity. The need for heat from petroleum could be reduced in a variety of ways such as the use of CHP, insulation, double-paned windows and biomass. Road materials could be substituted for or gotten from tar sands or other very low-grade forms of crude oil. Coal, natural gas or biomass could be processed to produce most of the other products. With these additional assumptions, the 2040 petroleum requirement would be

$$7.8 + 3.3 + 5.6(7.8+3.3)/15.1 = 15.2 \text{ Mbbl/day.}$$

From 1983 to 2005 the United States petroleum production from oil wells declined at an exponential rate of 2.5 percent annually. Using this rate of decline and the 2005 oil well production of 5.17 Mbbl/day, the projected production in 2040 would be 2.2 Mbbl/day. But this does not include the refinery gain that is typically 6.2 percent nor the liquids captured by processing the output from natural gas wells. The EIA *Annual Energy Outlook* predicts a total United States output of petroleum liquid products of 4.6 Mbbl/day in 2030. If the United States were to produce 3.9 Mbbl/day in 2040 and the above projections were correct, then it would need to import 11.2 Mbbl/day, or 74 percent of its petroleum,. (In 2008 the United States imported 12.94 Mbbl/day.) Petroleum independence would require much more than the steps considered in the above examples. A 2007 EIA model predicts a substantially more pessimistic outlook by projecting a slight increase in petroleum consumption in 2040, but apparently is less optimistic concerning the use of more efficient small vehicles and a shift in the way we haul freight.

The above examples ignore the possible use of electricity, biomass fuels, natural gas and synthetic fuels as substitutes for the liquid petroleum fuels now used almost entirely by the transportation sector. Natural gas and coal-derived synthetic fuels, although they do relieve the depletion of petroleum reserves and the United States dependency on petroleum imports, are nonrenewable and may not reduce pollution. This leaves electricity and biomass fuels as the best alternatives for bringing down our dependency on petroleum.

In 2006, the United States transportation sector consumed 5,377 million gallons of ethanol and lesser amounts of other biomass fuels. Because these fuels have lower volumetric energy densities, they provided less than two percent of our transportation fuel needs. Because the United States gets most of its ethanol from corn and there is only a small net energy gain in producing ethanol from corn, there is serious controversy over our production of ethanol as a fuel source. However, as discussed in Chap. 6, sugar cane yields about eight times as much energy as the energy needed to process it into ethanol. The energy needed to grow the sugar cane is renewable energy from the sun. Unfortunately Hawaii is the only place in the United States where an appreciable amount of sugar cane can be grown. Switchgrass and sugar beets are somewhat better than corn. But if a significant amount of our fuel is to come from ethanol, then the United States must rely on imports from tropical countries such as Brazil where increased ethanol production may cause serious deforestation. The use of E85 would dramatically reduce our dependence on petroleum, but would shift this dependence to that of importing large amounts of ethanol. If by 2040 all gasoline were replaced by E85 and the above savings were realized, the United States would use less than one million barrels of gasoline per day, but our energy dependency on imports would remain and the environmental impact could be severe. Ethanol could also be used as a jet fuel supplement, although this possibility is currently in the research stage.

The use of biodiesel in place of or as a supplement to diesel is also a likely means of decreasing petroleum demand. Biodiesel is presently mainly produced from soy beans, which are readily grown in many parts of the United States. It also results from processing yellow grease and could be produced from algae grown on farms or through the sequestration of carbon dioxide by coal-fired generating plants. Although biodiesel has received less attention than ethanol, it is possible that B30 could become common by 2040. This and the savings considered above would reduce the United States diesel requirements to less than two million barrels per day. The combined production of ethanol and biodiesel could possibly decrease the United States petroleum demands to less than half the 20.69 Mbbl/day it was in 2009. But if the world followed suit, it would do so at the expense of an undesirable shift in agriculture and, perhaps, the environment.

Finally, there is electricity. The roles of electricity, the primary sources used to generate electricity and the direct application of the primary sources are illustrated in Fig. 8-38. The quantities in parentheses are the 2008 energy amounts in quadrillions of Btu (PBtu). This figure is yet another way of showing the energy usage by the United States and, in particular, it shows the role of electricity. The above examples demonstrated how petroleum's 38 PBtu might be reduced by new technologies and shifting the transportation of people and goods to different vehicles or different modes of transportation. Now let us consider a shift from petroleum to biomass and electricity.

Electricity is used very little by the transportation sector at present, but may have the greatest impact on this sector's use of petroleum by 2040. Despite the fact that the transportation sector uses so little electricity, 40 percent of the energy consumed in the United States comes to us via electricity. This percentage will increase with the increase of plug-in hybrids and EVs and so will the energy needed

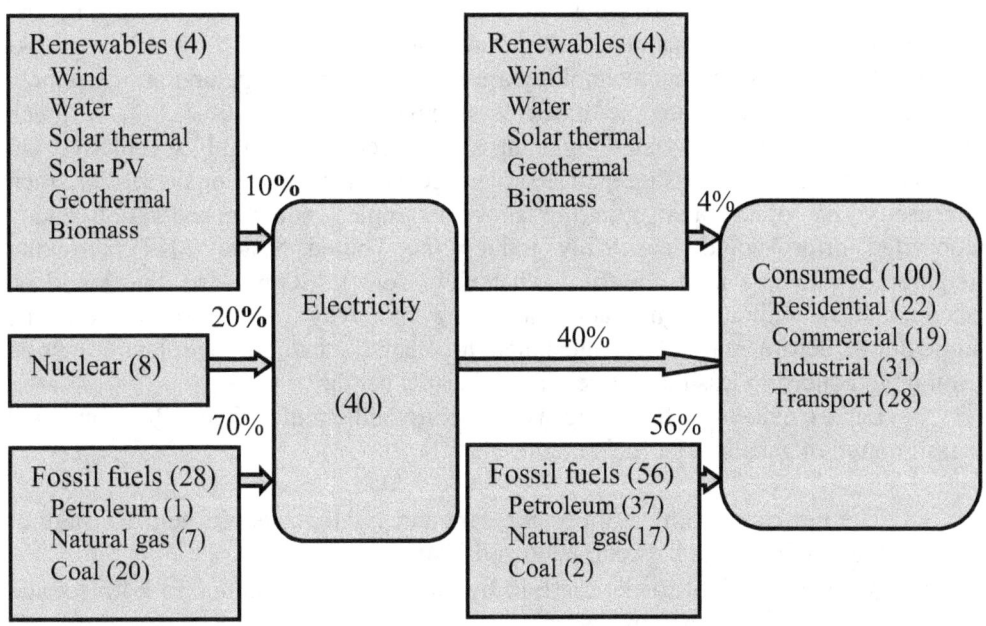

Fig. 8-38 2008 contributions to the United States energy needs in PBtu [EIA].

to produce electricity. Electricity is not a primary energy source, but must be generated from other energy sources including the electromagnetic energy from the sun. Today approximately half of the United States electricity is generated by burning coal and another 17 percent by burning natural gas. Only 1.2 percent comes from burning petroleum. Using electricity instead of petroleum to power vehicles is a matter of replacing petroleum with the energy sources that provide electricity. Because our present methods of generating, distributing and storing electricity incur an average energy loss of about 69 percent, what is there to gain by switching to electricity from fuels that are used directly? There are two basic reasons. First, ICEs are so inefficient that less primary energy is required to provide the necessary electricity than is required by the ICEs. Second, electricity can be readily supplied using non-fossil fuel sources. As discussed in Chaps. 5, 6 and 7, these sources include the nonrenewable source uranium and renewable sources, most of which originate with the electromagnetic energy from the sun.

As indicated above and in Chaps. 6 and 7, the primary deterrent to using electricity to power vehicles that do not travel fixed paths is storage. However, most of these vehicles are small, i.e. cars and light trucks, and most of the miles traveled by these vehicles are traveled in short round trips of less than 80 kilometers (50 miles). Current technology allows lithium-ion battery packs to store enough energy to travel 64 kilometers (40 miles—the Chevrolet Volt) without recharging and battery packs based on nanotechnology are able to store much more energy. At present, batteries using nanotechnology are in the research stage and are extremely expensive, but should become much less expensive by 2040. Just how much petroleum may be saved by 2040 by using electric vehicles is highly speculative. On the other hand, if decreasing petroleum usage is made a national priority, then electricity, the other savings in the above examples, some conservation and a concerted effort could reasonably reduce the United States 2040 petroleum requirements to less than one-third what they were in 2006. Also, this could be accomplished without significantly affecting our way of life or without the agricultural detrimental effects attributed to ethanol, and our petroleum imports would be reduced to less than three million barrels per day.

Let us make the following assumptions and predict their effect on 2040 consumption of gasoline, jet fuel and diesel:

- The per capita vehicle-miles for cars and light trucks are held constant at their 2006 values between 2006 and 2040.
- All cars and light trucks use E40 by 2040 and the reduction to E40 is made in equal annual increments.
- Two-thirds of the energy required for urban car-miles is supplied by electricity.
- All light trucks have the same efficiency gains in terms of mpg as indicated for Fig. 8-28.
- All cars have the same efficiency gains in terms of mpg as indicated for Fig. 8-28, but for the city car-miles two-thirds of the energy is actually supplied to electric cars and plug-in hybrids by electricity.
- The vehicle-miles ratio light trucks/total is returned to its 1980 value as it is in Fig. 8-28.
- The passenger-miles for certified domestic passenger air carriers are held constant at their 2006 value between the years 2006 and 2040 as indicated for Fig. 8-29.
- Certified domestic air carriers have the same efficiency gain in terms of passenger-miles per gallon as indicated for Fig. 8-29.
- All certified domestic air carriers use jet fuel that has 20 percent of its energy derived from biomass.
- The total ton-miles per capita for semis and trains are held constant at their 2006 value between 2006 and 2040.
- The ton-miles ratio semis/total is returned to its 1980 value as it is in Fig. 8-35.
- All trucks and trains use B30 by 2040 and the reduction to B30 is made in equal annual increments.

Figure 8-39 provides the results of this example. The total amount of fuel needed by cars, light trucks, certified domestic air carriers, semis and trains in the year 2040 would be 81.4 Ggal or 5.31 Mbbl/day.

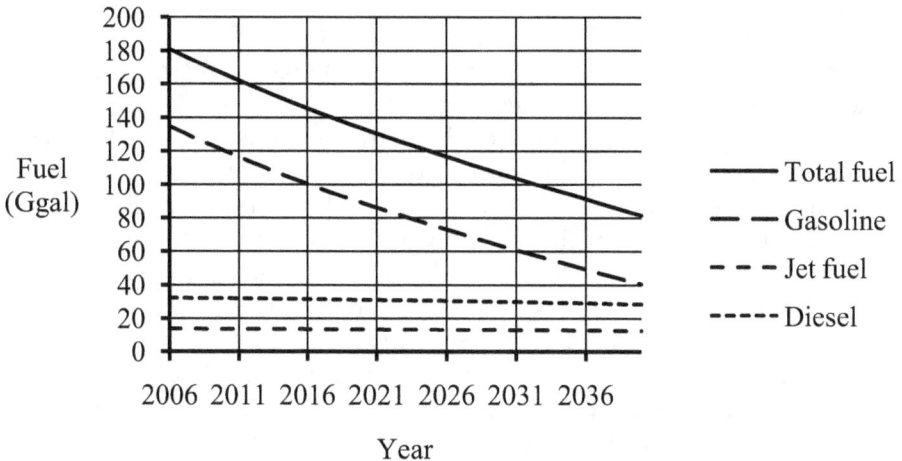

Fig. 8-39 The additional effects of electricity and biomass on petroleum consumption.

Once again, suppose that:

- The 3.3 Mbbl/day demand for gasoline, jet fuel and diesel not considered in the above examples is held constant through 2040.
- The remaining 5.6 Mbbl/day for other products is reduced in proportion to that of gasoline, jet fuel and diesel.

Then the total petroleum requirements would be

$$5.3 + 3.3 + 5.6(5.3+3.3)/15.1 = 11.8 \text{ Mbbl/day},$$

and our imports would be 7.9 Mbbl/day, 67 percent of our consumption. Although this reduction would be a significant improvement over our present level of imports, a reduction in worldwide production of petroleum could still pose a serious problem to the United States.

The primary barrier to reducing petroleum usage is population growth, which is expected to increase our population by over 40 percent between 2006 and 2040. The results shown in Fig. 8-36 were based on the per capita vehicle-miles, passenger-miles and ton-miles remaining constant at their 2006 values. In Fig. 8-39, as in Figs. 8-28 and 8-35, the per capita vehicle-miles and ton-miles were returned to their 1980 values. This allowed the effects of population growth to be moderated. By 2050 the United States population is expected to level off, but immigration and other factors make any such projections very speculative. Even if this is true, the world's oil supply may be in serious decline and much more will need to be done than is suggested by these examples.

Much of the gain in Fig. 8-39 as compared to Fig. 8-36 is attributed to the shift from petroleum-derived energy to electricity and biomass-derived energy. If it takes ten percent more ethanol to get the same mpg as gasoline because ethanol contains less energy, then this shift would require the production of 1.93 Mbbl/day of ethanol. By assuming the mean production per acre implied by Table 6.2 (760 gallons per acre), it would take a sugar cane field larger than the state of Iowa to meet this level of demand. If the well-to-wheel efficiency of electricity were 1.6 times that of a gasoline-driven ICE, then the additional electricity demand in 2040 would be approximately 956 GkWh, a 26 percent increase in the amount of electricity generated in 2006. This is the approximate output of 120,000 3-megawatt, suitably located wind turbines. If the land requirement is 33.3 acres per megawatt (the average amount assumed in Chap. 6), then the turbines would be spread over roughly 48,500 square kilometers (18,700 square miles). However, if, as indicated in Chap.6, only five percent of the area is needed for the towers and access roads, then 46,000 square kilometers (17,800 square miles) still could be used for farming or other purposes and the wind farm would use only 2,500 square kilometers (900 square miles). If solar PV is used, an area of 6,800 square kilometers (2,600 square miles) would be required. Neither wind turbines nor solar panels need to be placed on farmland. A 20 percent mixture of jet fuel and biomass fuel would require the production of 0.2 Mbbl/day of biomass fuel and the biodiesel needed for the B30 would require the production of 0.8 Mbbl/day of biodiesel. Assuming an algae field that yields 5,000 gallons per acre per year (see Table 6-4), the area needed to produce the amount of biodiesel and biofuel indicated in the example would be 12,400 square kilometers (4,800 square miles).

Although electricity is not an EIA sector, it provides much of the energy to the residential, commercial and industrial sectors and the way it is generated will determine the greatest changes in the consumption of our primary energy sources. How much electricity is needed in the years to come will depend on not only its expanded usage by the transportation sector, but also on how much of it is needed for other uses. Figure 7-20 indicates electricity consumption increases at a linear rate of 74.68 GkWh per year, a rate that was seen to be quite accurate between 1980 and 2005. Suppose that:

- Electricity consumption between 2006 and 2040, excluding transportation, continues to increase at the rate established by the 1980 through 2005 data.

- The transportation usage of electricity by hybrid plug-ins as indicated in Fig. 8-39 is added to the other consumption.

The resulting projection is shown in Fig. 8-40. If the added electricity due to the added transportation usage is included, this projection implies that electricity consumption will almost double from 3,666 GkWh in 2006 to 7,161 GkWh in 2040. Excluding transportation, the electricity consumption would be 6,205 GkWh.

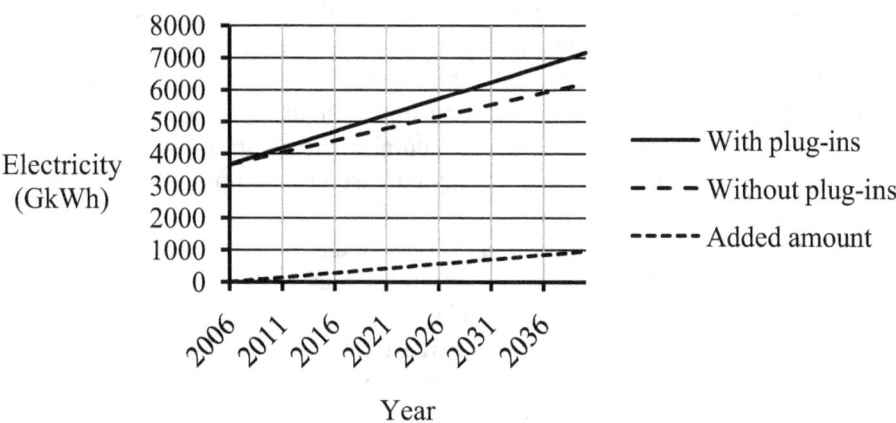

Fig. 8-40 Electricity usage with and without the plug-in hybrids assumed in Fig. 8-39.

Some contend that no more than 30 percent of the electricity supply should come from wind, solar PV and solar thermal because of their variability, but others think that up to 70 percent could be generated from these sources. If 50 percent of our electricity in 2040 were to originate with these sources and wind turbine output averages 30 percent of capacity, then the generation of half of the 7,161 GkWh would require approximately 450,000 3-megawatt wind turbines placed over 182,000 square kilometers (70,200 square miles), an area the size of North Dakota. However, 95 percent of the land could be used for other purposes. If solar PV is used, an area of 25,500 square kilometers (9,900 square miles) would be required, but if put on buildings as opposed to land, then more of the land could be used for other purposes.

The above examples have made a lot of assumptions and the projections may not be particularly accurate, but they do demonstrate the difficult problems we face. The projections are reasonable and are comparable to the projections made by others, however. These examples were primarily designed to point out the available choices and suggest the consequences of those choices. They emphasize the importance of replacing nonrenewable fossil fuels with renewable sources and the necessary shift to electricity, which permits flexibility in choosing the primary energy source used. They also emphasize the effects of population growth, which may affect the world as a whole more than it affects the United States.

The assumptions made in the above examples are realistic and technologically possible, but to achieve the projected outcome will be expensive and time-consuming. If there is no change in the way we generate electricity, then the consumption of the fossil fuels to produce electricity and its associated effects would almost double. Shifting from fossil fuels, particularly coal, to renewables would require the construction of massive numbers of wind turbines or solar facilities. The introduction of much greater amounts of electricity into the transportation sector and boosting the percentage of electricity generated by renewables to 50 percent would require an effort equivalent to that of building our interstate highway system. The increase in the use of biofuels to replace petroleum may require the setting aside of large sections of our farmland and the, possibly unreasonable, importation of ethanol from tropical countries.

In addition, the changes suggested by these examples will require a change in the American mindset, but not necessarily our standard of living. We must use smaller, more efficient cars, better means of moving people and goods, set higher standards for the construction of buildings, increase our use of alternative means for providing space and water heating and continue making ever more efficient appliances and lighting products. It is not necessary that we drastically alter the amount of energy we consume, but it is necessary that we drastically change the way we obtain that energy. Thirty years is enough time to make the changes, but the sooner we start the easier it will be to do the things needed to avoid the undesirable, perhaps severe, economic and environmental consequences. It is undeniable that whatever we do will be expensive and our energy will cost more. To do nothing for the next few years could be the most expensive of all. It is not just a matter of lessening the burden on our descendants, what we do now will significantly affect anyone under 60 years old that lives an average lifespan.

9

POLLUTION

The two most pressing concerns related to energy usage are the finiteness of nonrenewable sources and the pollution that results from energy usage. There is no one definition of pollution. The legal definition is quite broad and is "the wrongful contamination of the atmosphere, water or soil to the material injury of the right of an individual". However, in this book the even broader definition of *pollution* is a composite of the definitions found in other sources and is the undesirable contamination of the environment in which it is located. The contaminants are called *pollutants*. Although we normally think of pollution as chemical or nuclear in nature, visual and aural pollution are also considered. Loud noises or unsightly scenes may not cause physical harm, but they certainly may degrade the quality of life. The physical presence of energy-related equipment requires a certain amount of land and is often an impediment to animals. This impediment is sometimes dangerous, as are wind farms to migrating birds. Extremely intense electromagnetic waves may also be harmful to the cells of humans and other animals and there is some concern about high-voltage transmission lines. The intense magnetic fields that would surround any future superconducting transmission of energy may likewise present problems.

Pollution due to human activity began with the arrival of humans and grew as our population expanded. Serious pollution problems began to multiply with the advent of the industrial revolution in the late 18th century, but their seriousness was not officially recognized until the 20th century. It was not until the 21st century that the majority of scientists agreed that human activity is probably a major contributing factor to the current period of global warming. Although not all pollution is caused by our consumption of energy, it is the undesirable side effects of energy usage that are the focus of this chapter.

As indicated by the legal definition, pollution may be divided into soil pollution, water pollution and air pollution. The three are interrelated by the contaminant transfers between them. Air pollution may be absorbed by soil or surface water and, conversely, the contaminants in soil or water may diffuse into the air. One of the most serious pollution problems is the leaching of soil pollution into subsurface water reservoirs. If it were not for the transfer of contaminants from soil to water or air, soil pollution would tend to be localized. But air and water are able to quickly spread contaminants over a wide area. It should also be noted that the amount, density and location of a substance determines whether or not it is a pollutant. Plants depend on a certain amount of carbon dioxide being in the atmosphere, but too much carbon dioxide may cause the weather to change in a way that is detrimental to the plants in the area. Certain amounts of what are normally considered pollutants occur naturally and are considered safe, but high concentrations in certain locations may be quite dangerous. Mercury and lead are in a number of products that are considered safe when used properly, but if the same concentrations were located elsewhere (e.g., in a landfill) they may be considered dangerous. A discarded, rusting battleship may be unsightly in a beautiful harbor, but may be beneficial when used in the ocean to build a reef. Most substances could be considered pollutants depending on the circumstances. Some activities may result in multiple types of pollution. Littering is a form of visual pollution, but may also cause the contamination of the surrounding soil or water.

Pollution may result from any of the stages involved in the production, storing, transporting, processing or usage of energy sources. It may also be caused by the manufacture of the equipment needed to transport or store energy or to convert the energy sources into useful work, or the production of the plastics and metals needed to manufacture this equipment. The principal causes of energy related pollution are accidental leakage, the intentional disposal of harmful wastes, the burning of fossil fuels, road construction, the creation or concentration of radioactive materials and the presence of energy-related equipment, especially if it is loud or unsightly. The most notable sources of aural pollution are trucks, trains, airplanes, construction machinery and industrial machinery. Even wind farms can emit enough noise to be annoying to those who live nearby. Visual pollution includes that caused by intense or excessive lighting, power lines, landfills, junkyards and mining, especially strip mining. As mentioned previously, high levels of electromagnetic energy may damage animal cells. However, it is chemical pollutants that present the most danger and most pressing problems that must be resolved in the near future. The more notable chemical pollutants along with their most detrimental locations, sources and effects are listed in Table 9-1.

The primary concern with regard to carbon dioxide is its buildup in the Earth's atmosphere where it traps the sun's radiation and raises the Earth's temperature. Trapping the sun's radiation is known as the *greenhouse effect* and *global warming* is the rise of the average temperature of the Earth's atmosphere as a whole. For millions of years before the industrial revolution there was a balance between the carbon dioxide breathed out by animals and the carbon dioxide consumed by plants. However, over the last two centuries there has been a spike in the amount of carbon dioxide in the atmosphere caused by the burning of fossil fuels and waste materials containing carbon and the increased demand for cement. The

Table 9-1 Principal chemical pollutants.

Pollutant	Locations	Sources	Effects
Carbon dioxide	Air	Burning hydrocarbons, wastes, cement	Global warming
Carbon monoxide	Air	Burning hydrocarbons	Suffocation, organ damage
Methane	Air	Natural gas leaks from wells, refineries, vehicles, pipelines, storage tanks	Global warming, asphyxiation, fire or explosion
Ozone	Air	Burning fossil fuels, vapors from gasoline and solvents	Lung damage, asthma, allergies, plant damage
Liquid petroleum products	Soil, water, air	Waste disposal, leaks from vehicles, processing, storage	Cancer and other health problems, explosion, fire
Sulfur oxides	Air	Burning hydrocarbons	Acid rain
Nitrogen oxide	Air	Burning hydrocarbons	Ozone, global warming
Nitrogen	Water	Fertilizers, animal wastes	Oxygen depletion by algae
Phosphorus	Water	Fertilizers	Oxygen depletion by algae
Pesticides, herbicides	Soil, water, air	Agricultural usage	Cancer, mental problems, birth defects
Mercury, lead	Soil, water, air	Mining, refining, wastes, combustion, additives	Cancer, nerve, mental and other health problems
Arsenic	Soil, water	Semiconductor processing, waste disposal	Serious health problems and fatal poisoning
Radioactive materials	Soil, water, air	Uranium mining, refining, waste recycling, disposal	Cancer and other serious health problems
Particulate matter	Air	Mining, burning diesel	Respiratory problems

carbon dioxide from cement is related to energy usage only through the construction of roadways, runways and other transportation facilities. A detailed discussion of the greenhouse effect and global warming is reserved for the latter portion of this chapter. Carbon dioxide disperses quickly into the atmosphere and is not considered an immediate threat to animal life. For example, working around melting carbon dioxide ice poses no hazard. It is uncertain what an increased concentration in ocean water would have on marine life. It is toxic only when it occurs in extreme concentrations.

On the other hand, carbon monoxide has a relatively small effect on the atmosphere, but is very toxic when even partially confined. It is dangerous to leave a car idling in even an open garage. One year, after the Macy's Parade in New York City, several people became sick and some were hospitalized when too many cars at once tried to exit a parking garage at the same time. Even small densities of carbon monoxide can be harmful if breathed over an extended period of time. Inhaling carbon monoxide may cause sickening, permanent organ damage or death. It may collect in the lungs and cause death by suffocation. Like carbon dioxide, carbon monoxide is produced by burning fossil fuels and waste containing carbon. It also occurs naturally in coal mines and is a major hazard for miners. A very small amount of carbon monoxide occurs naturally in the atmosphere.

Methane is also a greenhouse gas, but, for reasons discussed later, it is not as immediate a threat to the world's atmosphere as carbon dioxide. The primary cause of excess methane in the atmosphere is leakage. There are varying amounts of natural gas leakage in all natural gas production, transportation, storage, refining and usage stages. Not all methane in the atmosphere is related to energy usage. The digestive systems of animals and decaying organic material produce a significant percentage of the methane in the atmosphere. Like carbon dioxide, it may be deadly when confined or semi-confined. It becomes deadly when its concentration exceeds 30 percent, 300,000 *parts per million by volume (ppmv)*. At such levels, it reduces the oxygen entering the lungs of animals and causes death by asphyxiation. There is not adequate evidence indicating that the inhalation of much lower densities of methane has long term harmful effects on animals. Buildings that use any fossil fuel for heat should have alarms that detect dangerous levels of both methane and carbon monoxide gas. As is well known, years ago miners used canaries to warn them when the concentration of either carbon monoxide or methane became too great. As are all hydrocarbon gases and vapors, methane is flammable and can explode when its density is as little as five percent (50,000 ppmv). Open flames should never be allowed near where natural gas or methane is stored. There are similar problems with ethane, butane and propane, but they are not mined directly as is natural gas and their usage is much less. Butane and propane are used for heating and propane is used as vehicle fuel. Propane, in particular, is heavier and does not disperse in air as quickly as methane and is more prone to exploding in unconfined areas.

Whether or not ozone is harmful depends on where it is located in the atmosphere. Ground level ozone is responsible for a number of respiratory ailments and damages plant life, but the ozone that occurs at the top of the atmosphere is needed to shield the Earth from receiving too much ultraviolet radiation. Among other things, too much ultraviolet radiation causes skin cancer, plant damage and plankton reduction. In the latter half of the 20th century our use of chlorofluorocarbons and bromoflourocarbons in cooling systems and aerosol cans caused these chemicals to leak into the air where they were broken down to produce the chlorine and bromine that ultimately rose into the stratosphere. This presence of chlorine and bromine resulted in a depletion of ozone in the polar regions and produced what is referred to as *ozone holes*. A worldwide effort to reduce the usage of chlorofluorocarbons and bromoflourocarbons has, at least temporarily, kept the ozone holes from growing larger than they were a decade ago. Near to the ground, the burning of fossil fuels by vehicles, power plants and industry produce nitrogen oxides. Along with volatile organic compound vapors (VOCs), such as those emitted by gasoline and solvents, these oxides are turned into ozone by heat and sunlight. Because of summer's higher temperatures, ozone is known as the summer pollutant. Ozone may cause allergies, asthma or permanent lung damage and interferes with the growth of plant life.

Liquid petroleum products may be pollutants even when they are not burned. Their leakage and disposal pollute the soil and water and their vapors pollute the air. The liquid hydrocarbons that leak into the soil seep into the groundwater and some of this leakage eventually finds its way into lakes, waterways and oceans. Figure 9-1 gives the percentages of the various causes of direct petroleum pollution in North American waters in the 1990's. This data was obtained by the EIA from the National Research Council. Note that 96 percent of this pollution is due to natural seepage and user spills. Although spills from ships and barges get the most publicity, they account for only two percent of the total. The spills from ships have declined since the year 2000, because beginning in the 1990's, ocean-going tankers have been constructed with double-hull designs. However, it is the spills by seagoing vessels and oil platforms that cause the most immediate environmental damage. The sea life in the area of these spills is often devastated. Other sources of petroleum-related spills are sludge ponds around drilling sites, pipeline leakage and leakage around oil rig sea platforms. User related pollution is mainly due to accidental gasoline, diesel or oil spills, leakage from storage tanks and disposal of various petroleum products. All underground storage tanks need to be replaced by tanks with double-lining. A survey by the Government Accountability Office estimates that $12 billion is needed to clean up gasoline leaks from service station storage tanks alone. Liquid petroleum products are carcinogens and can cause a variety of health problems if ingested, and their vapors are explosive.

The worst petroleum spill occurred when the drilling rig known as Deepwater Horizon exploded in the Gulf of Mexico on April 20, 2010. The explosion caused the pipe of the deepwater well to break a mile below the surface,

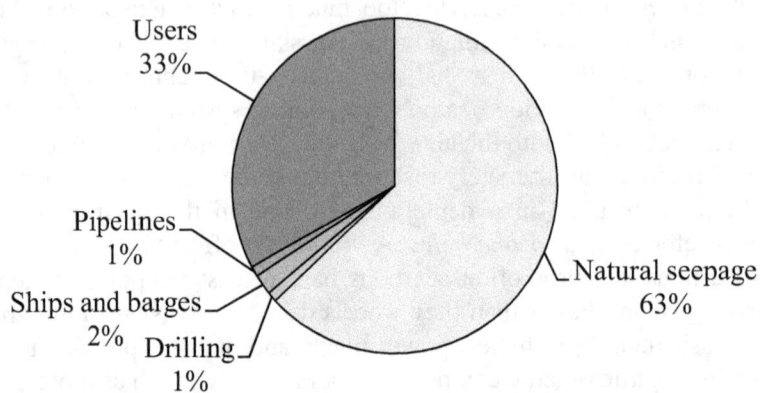

Fig. 9-1 1990-1999 petroleum spills in North American waters.

just above the ocean floor. The cost of the cleanup and losses to the fishing and tourist industries has been billions of dollars. The true cost to the environment may never be known. How long it will take for the environment to recover is speculative, as is the total amount of oil that spilled into the ocean. It is thought that around five million barrels escaped into the ocean during the spill's 87-day lifetime.

The oil sand area of the Athabasca Valley in Alberta, Canada, is a prime example of how our quest for petroleum can pollute the land, water and air. Almost 20 percent of the United States petroleum imports are from Canada and about half of these imports are from this area. An article in the March, 2009, issue of *National Geographic* entitled "The Canadian Oil Boom" by Robert Kunzig notes that the oil sand area extends over 3,512 km² (1,356 mi²) and contains 1.7 trillion barrels of potential petroleum products. About 173 billion barrels of petroleum can be economically recovered from these deposits at today's price of oil and as much as 300 billion barrels may eventually be extracted from the area. It is estimated that 20 percent of the 173 Gbbls will be recovered by surface mining and the other 80 percent will be produced by in situ extraction. Surface mining requires the clearing of the forest and the removal of two short tons of surface soil and two short tons of oil sand in order to ultimately recover one barrel of oil. The sludge ponds created by this surface mining already cover about 130 km² (50 mi²). The University of Waterloo estimates that 45,000 gallons of contaminated water per day could be seeping into the Athabasca River. For oil sand more than 60 m (200 ft) under the surface, in situ mining is employed. In situ mining involves installing vertical and horizontal pipes and using one set of pipes to steam heat the oil sand and another to extract liquid bitumen. Both surface and in situ mining initially produce bitumen that must be sent to refineries to be processed into petroleum.

   Although in-situ mining may be less environmentally damaging to the land, it is less energy efficient and causes more air pollution. Surface mining requires the energy needed to operate huge shovels and trucks. In situ mining requires the energy needed to create vast amounts of steam and may damage underground water reservoirs, but does offer an opportunity for the use of CHP in generating electricity. Both mining technologies require large amounts of water and extensive energy-consuming processing before the bitumen is retrieved and refined into petroleum. Including mining and processing as well as the final consumption, petroleum obtained from oil wells produces 58 kg (128 lb) of carbon dioxide per barrel, that from surface mining produces 165 kg (364 lb) of carbon dioxide per barrel and that from in situ mining produces 176 kg (388 lb) of carbon dioxide per barrel. If continued to completion, the Athabasca oil sands area will be among the world's greatest localized environmental disasters. Extracting petroleum from the oil shale in northwest Colorado will be even more inefficient and result in even more carbon dioxide being introduced into the atmosphere per barrel of petroleum ultimately consumed. What is happening in Alberta portends the future of the oil shale area if large scale development of oil shale deposits were to become a reality.
   Sulfur and nitrogen oxides are due primarily to the burning of hydrocarbons. Sulfur dioxide, when mixed with rain clouds, reacts with water vapor to produce sulfuric acid which falls as acid rain. This rain pollutes lakes and rivers and is hazardous to animal and plant life. Nitrogen oxides are converted into ozone by heat and sunlight. They are also greenhouse gases, but, at present, play a minor role in global warming compared to carbon dioxide and methane. The principal

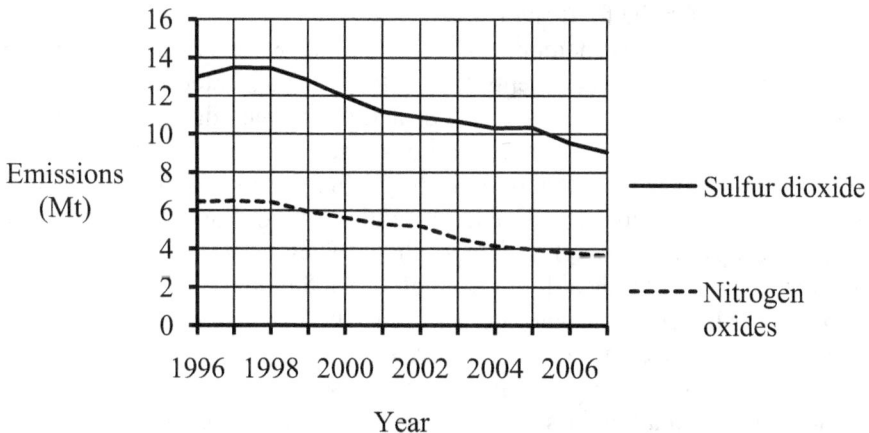

Fig. 9-2 Emissions from conventional and CHP power plants in the United States [EIA].

contributor to these oxides is electrical power plants. Serious attention has been given to reducing these pollutants and, beginning in 1972, several laws have been passed to reduce their entry into the air. From Fig. 9-2 it is seen that these laws are having a dramatic effect on the amount of sulfur dioxide emitted by the power plants in the United States. The amount emitted has dropped from 13 million tonnes (14.3 million short tons) in 1996 to 9 million tonnes (9.9 million short tons) in 2007. During that same period, the nitrogen oxides being released by power plants dropped from 6.5 million tonnes (7.1 million short tons) to 3.65 million tonnes (4 million short tons). Additional reductions of nitrogen oxides were obtained by requiring all vehicles to have catalytic converters. The sulfur oxides are eliminated from petroleum products by removing the sulfur during the refining process.

Both nitrogen and phosphorus are contained in fertilizers and nitrogen is in animal wastes. Some portions of these elements are absorbed by the soil, seep into the groundwater and eventually end up in lakes, rivers and oceans or are washed into surface water directly. There, they tend to deplete the oxygen needed by underwater life. Their connection to energy is through the use of fertilizers to grow energy crops and farm wastes that are used as fuel. Corn, the primary source of ethanol in the United States, especially requires a considerable amount of fertilizer. Like nitrogen and phosphorus, pesticides and herbicides are energy-related pollutants only to the extent they are used to raise energy crops. However, unlike nitrogen and phosphorus they are considered pollutants when they are airborne or in the soil as well as when they are dissolved in water. They may be airborne during their application and contaminate the soil directly or through plant waste. They enter lakes, rivers and oceans in the same manner as nitrogen and phosphorus. Initially, the air and soil contamination is a direct threat to pesticide/herbicide and agricultural workers only. But, once in our water systems, they become a threat to all animal and plant life and may pass upwards through the food chain.

Heavy metals, such as mercury and lead, and semimetals, such as arsenic, are dangerous even in very small concentrations. Heavy metals enter our air, soil and water through many of our energy related activities, including coal mining, petroleum refining, combustion of hydrocarbons and other materials and waste disposal. Even though their concentrations in fossil fuels are small, small amounts may be devastating to humans and animals. As with pesticides and herbicides they are readily passed upwards through the food chain. They cause cancer and affect nervous systems and mental health in general. One famous case is that of an Alamogordo, New Mexico, man who ate a pig that he had fed seed grain treated with a mercury pesticide. He lay in a coma for years.

Lead was used as a gasoline additive to increase octane until it was found to cause child learning difficulties in smoggy urban areas. It was replaced by MTBE (metyl-tertiary-butyl ether) that has since been banned in some states because leaky storage tanks have allowed it to seep into the groundwater. However, MTBE also causes fuel to burn more completely, thereby reducing the amount of pollutants being exhausted. One option is to use ethanol to improve octane ratings and reduce pollutants. Arsenic is, of course, a lethal poison. Arsenic and other semimetals are related to energy usage because they are employed in the manufacture of the semiconductors that are contained in solar panels. However, because arsenic is used only during the manufacture of semiconductors and not during their use, it is potentially dangerous only in the vicinity where the semiconductors are produced. In addition, semiconductor manufacturing requires large amounts of water that must be cleansed of these pollutants.

Radioactive pollutants are essentially due to the mining, refining, reprocessing and disposal of the uranium, plutonium and other radioactive materials used by nuclear generating plants. These pollutants are clearly lethal and cause many health problems, most notably cancer. Even low doses of radioactivity are dangerous if they are received over a long enough period of time. People who work in the nuclear industry must wear badges that measure the accumulated dosages they have received. It is the accumulation of radioactivity exposure that indicates when someone is in danger of suffering permanent damage.

The Environmental Protection Agency's (EPA's) definition of *particulate matter emissions* is "all finely divided solid or liquid material, other than uncombined water, emitted to the ambient air as measured by applicable reference methods." It is the extraction of solid matter that is of interest here. Solid particulate matter is mostly caused by burning diesel in vehicle engines and coal in electrical generating plants. If the particulate matter were simply uncontaminated dust, the effects could be limited to mild respiratory problems. But the particulates emitted from burning diesel or coal may contain a number of the pollutants in Table 9-1 and cause the problems associated with those pollutants.

The Environmental Protection Agency (EPA) defines *pollution prevention* as "reducing or eliminating waste at the source by modifying production processes, promoting the use of non-toxic or less-toxic substances, implementing conservation techniques, and reusing materials rather than putting them into the waste stream." Although "implementing conservation techniques and reusing materials rather than putting them in the waste stream" is the best way of avoiding pollution, no matter how much these approaches are practiced, modern life necessitates the creation of some pollution. The pollution that is produced must be minimized and it is the "modifying production processes" and "promoting the use of non-toxic or less-toxic substances" that is given most of our attention in this chapter. The modifications involve diluting the pollutants to an acceptable level before they are released into the environment, concentrating and storing the pollutants and chemically changing them into something less hazardous. Promoting the use of non-toxic or less toxic means of obtaining energy is best accomplished by using renewable energy sources.

The alternatives to pollution prevention are to simply release pollutants into the air, soil or a large body of water or to store it where it no longer poses a problem. The latter is called *sequestration*. Before the 20th century, pollutants were released and we depended on dispersion to dilute them until they were judged not harmful. If dilution is done in such a way that it is not harmful to the local environment, then

certain amounts of pollution may be acceptable. But an acceptable amount is one
that does not accumulate and eventually cause an environmental problem. As the
20th century progressed, we began to realize that our atmosphere and water resources
could not absorb ever increasing quantities of pollutants. As the world's population
grew and standards of living improved, there was an increasing demand for energy
and products that resulted in a sharp increase in pollutants. Indiscriminately pouring
pollutants into our environment is no longer an option. As discussed in Chap. 5, the
sequestration of radioactive materials has faced many challenges. Sequestration of
gaseous pollutants is limited by the number of suitable containers, the volumes of
the containers and the ability of the containers to permanently retain the pollutants.
If massive amounts of pollutants are produced (e.g., carbon dioxide from coal-fired
generating plants), then the gas must be compressed and the container must be
extremely large. Normally, underground caverns or exhausted petroleum reservoirs
are used. Solid pollutants may be sequestered in landfills, but there is danger that
they will be washed into the groundwater or, over time, react with other chemicals
and release gaseous pollutants.

   There is a variety of pollution prevention equipment presently being
employed. Pieces of equipment called scrubbers are the usual means of removing
oxides from the gaseous outputs of power and other industrial plants. A *scrubber*

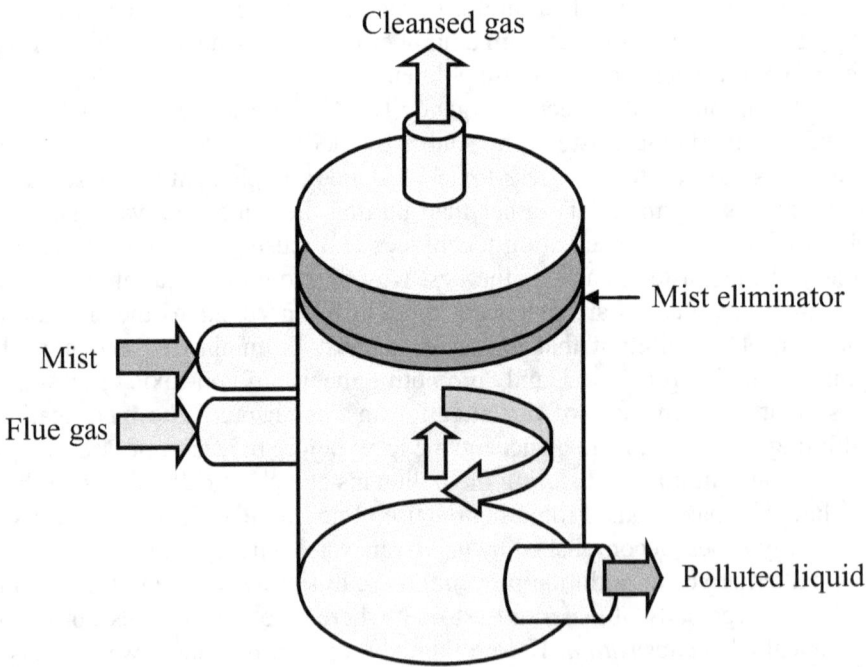

Fig. 9.3 Simplified diagram of a cyclonic scrubber.

uses a liquid mist to remove small solid particles and absorb oxides from the gas. It then uses a mist eliminator to gather the mist into droplets that can be extracted as a liquid from the scrubber while the gas escapes through a separate vent. The liquid mist is normally an amine (e.g., ethanolamine, C_2H_7NO) or a limestone and water solution. There are several scrubber designs. Figure 9-3 provides a simplified diagram of a cyclonic scrubber which causes the mist to swirl downwards and then pass upwards through the mist eliminator. Depending on its design, a scrubber may be large or small and removes 50 to 98 percent of contaminants in the flue gas. One of the design characteristics is the pressure drop between the input streams and the output streams. Greater pressure drops are associated with more thorough cleansing of the flue gas and greater cost because the scrubber must be more complex. The energy needed to operate a scrubber increases with pressure drop and flow volume, but large volume scrubbers tend to require less energy and remove more contaminants per cubic meter of flow.

One of the phrases that entered our vocabulary in the last few years is "clean coal." There is no such thing as clean coal. What is meant by the phrase is the cleaning of the flue gases that result from burning coal. This cleaning is most commonly done by a process called *amine scrubbing* which involves passing the flue gas through a filter and an amine and water solution that filters out the ash and absorbs both sulfur dioxide and carbon dioxide. The solution is then separated into water, sulfur compounds and carbon dioxide by heating and then cooling the solution. The carbon dioxide can then be used for some purpose or sequestered. A simplified diagram of a coal-fired generating plant that uses an amine scrubber and is operating in Germany is given in Fig. 9-4. The tank is transported to an underground reservoir where the carbon dioxide is sequestered over 600 m (2000 ft) below ground under a cap rock seal that traps the carbon dioxide so that it cannot seep back to the surface.

One serious disadvantage of scrubbers is that they consume a lot of water. According to the National Energy Technology Laboratory report entitled "Estimating the Needs to Future Thermoelectric Generation requirements—2009 Update," a 550 MW steam-turbine plant loses approximately 0.5 gallons of water per kilowatt-hour if no carbon capturing equipment is used, but loses about 0.9 gallons per kilowatt-hour if such equipment is employed. For a combined-cycle plant, these losses are 0.3 versus 0.5 gallons per kilowatt hour, respectively. As pointed out in Chap. 7, water losses are critical in arid regions.

Catalytic converters, which are also used to muffle sound, are installed on all vehicles and some stationary devices that burn hydrocarbons. They use a metal catalyst to aid the following reactions:

$$2NO_x + O_2 \rightarrow (x+1)O_2 + N_2,$$
$$2CO + O_2 \rightarrow 2CO_2,$$
$$\text{Hydrocarbons} + O_2 \rightarrow \text{carbon dioxide} + \text{water},$$

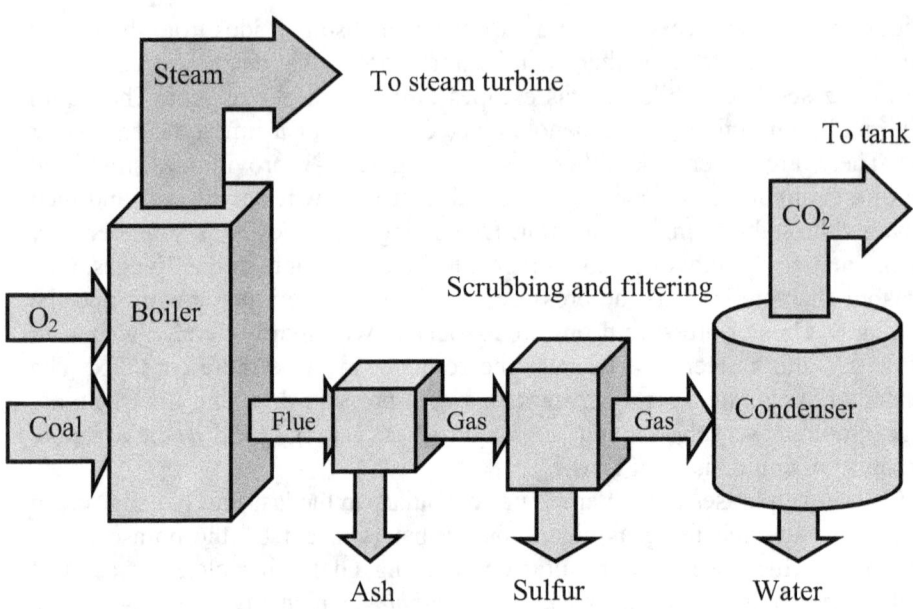

Fig. 9-4 Simplified diagram of an amine scrubbing system.

where NO_x represents a nitrogen oxide. The most effective catalysts are the precious metals platinum and palladium which increase the cost substantially. The reduction of nitrogen oxides by the first reaction, in turn, reduces the amount of ozone in the lower atmosphere. Although the converter does reduce the amounts being exhausted into the lower atmosphere of some carbon monoxide and hydrocarbons left unburned by the initial combustion, the latter two reactions do produce the greenhouse gas carbon dioxide. Modern catalytic converters are called *three-way converters* because of the three chemical reactions they produce. They have become increasingly more complex and require computer control because the first reaction needs a lesser air to fuel ratio than the third reaction. Modern catalytic converters may remove over 99 percent of the pollutants. Large devices based on the design of catalytic converters used in vehicles may be used in industrial and power plant operations.

There are basically three devices for removing solid particulates from gas streams: cyclonic separators, electrostatic precipitators, and baghouses. Cyclonic separators are similar to cyclonic spray scrubbers without the mist and mist eliminator. It operates like a cyclone inside a cylindrical tank by spinning the gas around. The heavier particulates are spun outward and drop downward while the cleansed gas goes upward in the center of the cyclone. An electrostatic precipitator charges the particulates in the entering gas and uses an electric field between plates inside the precipitator to extract the particles from the gas stream. Some electrostatic precipitators use a tubular geometry instead of plates. The collected particles are usually removed by vibrating the plates or tubes. Baghouses, which are large filtering systems, contain bags that filter the incoming gas and a means of continually removing the captured particulates. There are several baghouse designs and they mainly differ in the way they remove the captured particles.

It is seen from Fig. 9-4 that combinations of the equipment types discussed above are used to eliminate multiple types of pollutants. Also, variations of the types discussed above and more sophisticated boilers are sometimes installed to increase the efficiency in removing the pollutants. One such variation is the addition of a limestone dust stream into the boiler to absorb sulfur dioxide from the flue gas stream directly. The dust is later removed by filtering or an electrostatic precipitator. Such a technique is known as *dry flue gas desulfurization*. Another variation is to add to the flue gas stream a natural gas reburner or catalytic converter to reduce the amount of nitrogen oxides emitted. It is possible for combinations including baghouses to extract as much as 90 percent of the mercury entering coal-fired plants. However, these combinations may be very expensive to purchase and maintain and it may be prohibitively expensive to install the proper combinations in some existing plants. The EIA estimates that just the equipment needed to remove 80 percent of the nitrogen oxides from a coal-fired plant would cost over $70 per kilowatt of the plant's capacity.

Of the methods considered above, catalytic converters, reburning, and burning more thoroughly by injecting additional oxygen achieve a much reduced toxicity in their outputs. Gaseous outputs, such as carbon dioxide, that are not acceptably diluted or are output in excessive quantities may be sequestered. Scrubbers, electrostatic precipitators, baghouses and other filters and limestone injection all concentrate the pollutants into liquids or solids while releasing relatively clean gaseous outputs to the environment. Processes that use these types of equipment have not completely solved the problem. The concentrated pollutants still must be changed chemically or disposed of in such a way that they offer no further danger to the environment. Sludge ponds are often used to separate the solid matter from the liquid output of scrubbers, but there are still the problems of cleansing the liquid and disposing of the solid matter by sequestration or changing it chemically.

Sequestering solid waste usually means putting it into landfills and is not a way of solving the problem, but is simply a way of delaying the problem. There is considerable controversy over what to do about the land-filling of toxic waste. These wastes may be changed chemically by high temperature burning, sunlight, putting them in molten sodium carbonate or feeding them to microorganisms custom-designed according to the type of waste. Some could be packaged in corrosion-resistant containers as is done with radioactive waste and some, such as lead and mercury, could be separated out as useful byproducts. Steven Marcus interviewed several experts and, in an article published in the September 4, 2009, issue of the *New York Times*, quoted a National Academy of Science report as saying "there exists some technology or combination of technologies capable of dealing with every hazardous waste so as to eliminate concern over future hazards." The problem is cost. This article also indicates that it costs 15 to 50 cents per pound to adequately incinerate toxic waste and may cost up to 60 cents per pound to chemically treat it. It costs only one to three cents per pound to put it into a landfill, but no one knows what future health and legal costs may arise from toxic landfill waste.

Although various bills may be submitted to Congress in the near future, for the most part the federal government seems content to monitor the land-filling of hazardous materials and occasionally clean up some of the worst cases. The Toxic Release Inventory (TRI) program is charged with monitoring these landfills, and manufacturers or processors that output more than small amounts of hazardous wastes must report their outputs to the TRI. The Superfund program, which is administered by the EPA, is responsible for the cleanup of the nation's hazardous landfills. This program inspects waste sites, determines their level of toxicity, places the hazardous sites on a national priority list, cleans up sites according to their priority, and, where possible, collects compensation from the companies responsible. The Superfund program was created by the Comprehensive Environmental Response, Compensation and Liability Act of 1980. Associated with the Superfund program is a trust fund by the same name that was funded by a tax on the industries that create hazardous landfills. However, that tax was suspended in 1995 when the money in the trust fund reached six billion dollars. By 2003 all of the money in the trust fund had been spent and the federal government did not renew the tax collections. To pay for cleanups the Superfund program must now rely on the government's general revenue or collecting compensation from those who created the landfills. *Brownfields* is the term given to areas that are potentially dangerous enough that they cannot be developed. As the number of brownfields grows, it is increasingly clear that the practice of landfilling our hazardous wastes cannot continue indefinitely. Although putting this waste into landfills is initially less expensive than the alternatives, the long-term costs of these landfills are likely to increase the total costs considerably.

Having briefly discussed the spectrum of pollution problems related to energy let us now discuss in greater detail the elephant in the room—greenhouse gases and their relationship to global warming. The fireman's theory states that the hottest fire should be attacked first and, in the case of pollution, the most pressing problem is global warming. In the past there has been a lot of debate as to how much the greenhouse gases resulting from human activities contribute to global warming, but now an overwhelming number of experts agree that the methane and carbon dioxide due to our activities, particularly carbon dioxide, are major players in the current warming of Earth.

The EPA defines *global warming* as "an increase in the near surface temperature of the Earth." Global warming is caused by certain gases, called *greenhouse gases* (*GHGs*), that permit incoming solar radiation to pass through the Earth's atmosphere, but block some of the outgoing radiation from the Earth's surface and lower atmosphere from escaping into outer space. It is the lower frequencies that tend to be blocked and cause the temperature to increase. Although the phenomena of trapping heat by the GHGs has always been a natural phenomena that has made animal and plant life possible, the present concern is that human, or *anthropogenic*, activities are overloading our atmosphere with GHGs, thereby causing the temperature to rise considerably above its recent natural limits.

Before we begin our discussion of global warming let us examine the medium in which the warming is occurring. Our *atmosphere* is the mixture of gases that surround the Earth. It is divided into layers according to their elevation above sea level and their temperature ranges. These layers are summarized in Table 9-2. Each layer begins at the top of the layer just below it and extends upward to the elevation given in the table. Within each layer the temperature varies more or less linearly from the temperature at its lower limit (shown on the left) to its temperature at its upper limit (shown on the right). The troposphere, the level in which we live and the part of the atmosphere we are most interested in, ranges from sea level up to between 7 km (4.4 mi) at the poles to 17 km (10.6 mi) at the equator. (International jet aircraft typically fly 11.5 km (7 mi) above sea level.) The atmospheric pressure is approximately one atmosphere (14.7 pounds per square inch) at sea level and decreases exponentially with elevation. Temperatures high in the thermosphere are measured according to radiation, not the movement of molecules, and may be quite high.

Table 9-2 Atmospheric layers.

Name	Upper elevation in kilometers	Upper elevation in miles	Temperature range °C	Temperature range °F
Troposphere	7-17	4.4-10.6	10- -75	50- -103
Stratosphere	50-55	31-33	-75- -5	-103- 23
Mesosphere	80-90	50-53	-5- -100	23- -148
Thermosphere	No limit	No limit	> -100	> -148

Figure 9-5 gives the average composition of the troposphere (the air we breathe) at sea level. The numbers in parentheses are in parts per thousand by volume while the numbers in the adjacent table labeled "Other" are in parts per million by volume (ppmv). There are, of course, variations in this composition depending on elevation and location, e.g., city air contains much greater percentages of carbon dioxide and other pollutants. The compounds listed separately under "Other" are the GHGs. Despite the fact that almost all of our air consists of nitrogen, oxygen and argon, it is the GHGs that mainly determine the temperature and other environmental characteristics of the troposphere. The compounds included under the label "Remainder" are neon, helium, krypton, hydrogen, xenon, iodine, carbon monoxide, ammonia and chlorofluorocarbons (CFCs). Although the CFCs are GHGs, they were entirely manmade and, now that they are banned, should decrease to the point of being harmless within the next 50 years. As indicated before, the primary danger of CFCs is their destruction of ozone layer, not their contribution to global warming.

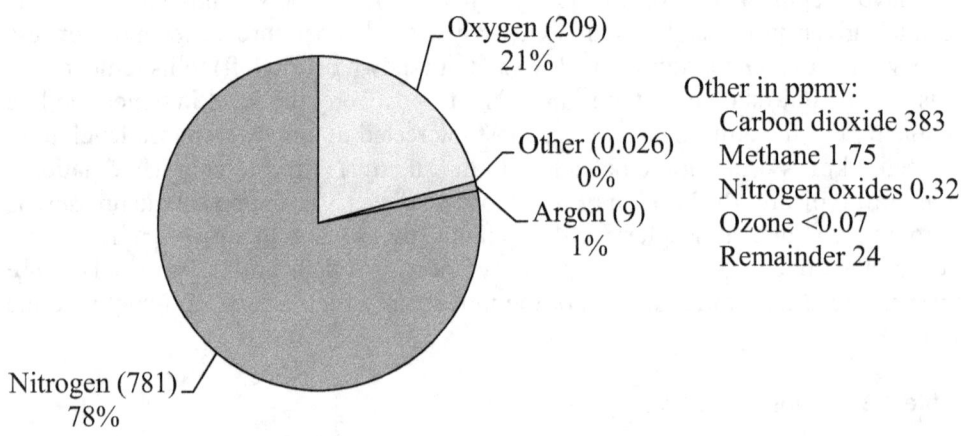

Fig. 9-5 Average composition of clean dry air at sea level.

The composition given in Fig. 9-5 is that of clean dry air. Not considered in the figure is the presence of water vapor, which is a GHG, or particulates. The densities of water vapor and particulates vary drastically depending on elevation and location. On average, water vapor accounts for about 0.4 percent of our atmosphere, but may locally account for as much as 70 percent of the greenhouse effect. Particulates caused by anthropogenic activities are from smokestacks and diesel vehicle exhausts and tend to be regional. They actually cool the Earth some, but may carry pollutants including lead and mercury.

Climate is the long term meteorological conditions, including temperature, precipitation, wind, humidity and pressure, that prevail in a particular region or are averaged over the Earth as a whole, and *climate change* is a significant shift in these conditions. This is in contrast to weather which concerns short term meteorological conditions. Because our interest is in global warming, the discussion here concentrates on the climate change of the Earth as a whole. More precisely, this discussion concerns the greenhouse effect, i.e. the trapping of heat by the GHGs.

Because atmospheric compounds spread over the globe in a matter of months or less and our interest is in what happens over decades or more, studies that are strictly regional are of little interest.

Just how do GHGs differ from other gases and why do they cause heat to be trapped in the troposphere? All compounds absorb and emit certain electromagnetic frequencies according to their natural frequencies of vibration. The sun's radiation is made up of a broad spectrum of frequencies. When it strikes the Earth's atmosphere, the dominant gases allow almost all frequencies to pass, but some of the low frequencies are blocked by the GHGs. Some frequencies that reach the Earth's surface are radiated back into the troposphere. But the Earth's surface causes a frequency shift and the frequencies that are radiated back tend to be the lower frequencies that are absorbed by the GHGs. The GHGs reradiate them in all directions, thus resulting in much of their energy to be directed back toward the Earth. For hundreds of millions of years the percentage of GHGs in the atmosphere was such that this greenhouse effect permitted diverse life to evolve. However, since the industrial revolution, the concentrations of the GHGs, particularly those of methane and carbon dioxide, are greater than they have been for millions of years.

A well established law of physics, called the *Stefan-Boltzmann Law*, states that the absolute temperature in degrees Kelvin of an object is proportional to the fourth root of the ratio of the rate of energy incident on the object to the emissivity of the object. The rate of energy arriving from the sun is nearly constant and, if the Earth had no atmosphere, the emissivity of the Earth is such that its temperature would be about -18 °C (0 °F). However, because of the greenhouse effect on the atmosphere the actual average temperature of the Earth is about 15 °C (59 °F). Also, because the conditions of the atmosphere tend to change the Earth's effective emissivity, a change in atmospheric conditions causes the temperature to change. In particular, if the effective emissivity is decreased the temperature is increased.

Actually, there are several factors that cause the temperature of the troposphere to vary. Although the above discussion is a simplistic explanation, it does demonstrate the basic principal behind that of global warming. The energy output by the sun varies, as does the tilt of the Earth's axis and the Earth's distance from the sun. The Earth moves around the sun in an elliptical orbit and the eccentricity of this orbit changes cyclically with an average period of approximately 100,000 years. The change in the Earth's tilt has an average period of about 41,000 years and its precession has a period of about 23,000 years. These cycles are referred to as *Milankovitch cycles* after the Serbian civil engineer and geophysicist Milutan Milankovitch. All three affect the Earth's climate. Also, the reflectivity of the atmosphere depends on other contents of the atmosphere as well as the GHGs. Particulates in the atmosphere, such as aerosols and volcanic dust, reflect much of the sun's radiation so that it never reaches the lower troposphere. Large volcanic eruptions have blanketed the entire Earth and caused the years following these eruptions to be much colder than normal. In addition, the radioactive elements deposited in the Earth at the time it was formed act as a natural nuclear reactor whose heat, less than one watt per square meter, adds to that received from the sun. Natural combinations of these variations have caused the Earth's climate to cycle between ice ages and warming periods. For the last half million years, the period of these cycles has been affected most by the eccentricity of the Earth's orbit and has been about 100,000 years. The current worry is that an extraordinary increase GHGs will detrimentally change the pattern of these cycles that have gone on for what is believed to be millions of years.

Figure 9-6 illustrates the current transfer of energy surrounding the Earth in watts per square meter (W/m^2). Of the incoming solar radiation, on average 77 W/m^2 is reflected off the atmosphere, 168 W/m^2 reaches the Earth and is absorbed and 30 W/m^2 is reflected off the Earth. The radiation from the Earth's surface is 390 W/m^2 and 102 W/m^2 is carried back into the atmosphere by *thermals*, the convection of air, and *evapotranspiration*, the evaporation of water and convection of water vapor. Of the energy transfer from the Earth to the atmosphere, 324 W/m^2 is radiated back to the Earth. GHGs tend to increase the amounts radiated back to Earth and reradiated back toward the atmosphere. Note that the total energy received from space is balanced by the total energy reentering space and the total amount absorbed by the Earth is balanced by the amount leaving the Earth. Otherwise, the Earth would continually heat up and the temperature would continually rise.

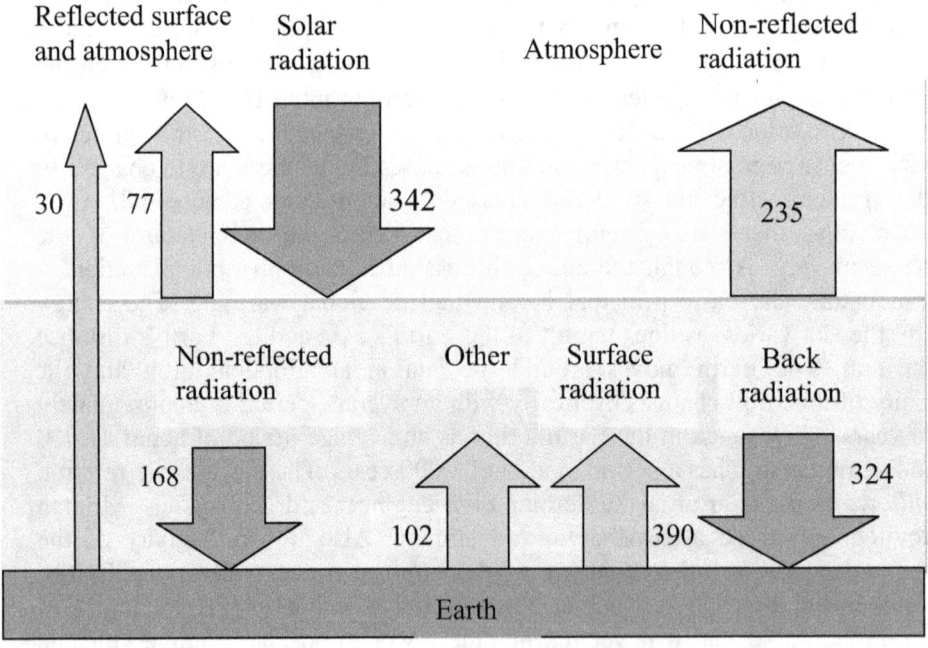

Fig.9-6 Energy transfer surrounding the Earth in W/m^2.

The amounts given in Fig. 9-6 represent worldwide averages over recent years. The 30 W/m^2 of reflected energy from the surface depends on location with the reflected energy near the poles being much greater due to the high reflectivity of ice and snow. On the other hand the back radiation over the oceans is greater because of the greater humidity and the fact that water vapor absorbs more low frequency energy than any other GHG. The contribution of water vapor to the greenhouse effect is over 70 percent where the humidity is high and may be half that where the humidity is low. The variations in the other GHGs are much more limited. The contribution of carbon dioxide is as much as 26 percent where the contribution of water vapor is low and as little as 9 percent where the humidity is high. Likewise, the contribution of methane varies from 4 to 9 percent and that of ozone varies from 3 to 7 percent. The effect of carbon dioxide on back radiation is almost three times that of methane and about four times that of ozone despite the fact that each molecule of methane or ozone absorbs and releases several times as much low frequency energy as carbon dioxide. The reason is that the atmospheric carbon dioxide density is hundreds of times greater than that of either of these gases. Therefore, our principal concern regarding global warming is carbon dioxide and it is mainly carbon dioxide that has been released into the atmosphere by the burning of fossil fuels at increasing rates since the beginning of the industrial revolution.

The reservoirs and movements of carbon are depicted in Fig. 9-7. The amounts of carbon in the various reservoirs is shown in bold type and given in gigatonnes (Gt) and the fluxes between the reservoirs are shown in regular type and given in gigatonnes per year (Gt/yr). There is some disagreement over the amounts shown, but the relative amounts are reasonably accurate. Most of the 720 Gt of carbon in the atmosphere is contained in carbon dioxide as is the fluxes between the atmosphere and other reservoirs. Note that it requires three and two-thirds Gt of carbon dioxide to contain one Gt of carbon because carbon dioxide contains two oxygen atoms and only one carbon atom. Oxygen is four-thirds as heavy as carbon. As a consequence, the atmosphere includes almost 2,640 Gt of carbon dioxide.

Of the 720 Gts of carbon in the atmosphere, 116 Gt/yr is absorbed by plants via photosynthesis, which outputs the oxygen in the carbon dioxide to the atmosphere. The oxygen is breathed by animals which breathe out carbon dioxide and the decay of plants and animals contributes to the atmosphere additional carbon containing gases, much of it methane. A total of 112 Gt/yr is transferred back to the atmosphere. The 4 Gt/yr difference is locked into the soil and detritus. In addition, 0.2 Gt/yr of carbon settles to the ocean floor in the form of the shells and bones of dead sea animals. Some of this 4.2 Gt/yr put into the Earth is ejected into the atmosphere through volcanic activity and other natural phenomena (not shown) and, after millions of years, some of it becomes fossil fuels. Just how much carbon is currently locked into fossil fuels is not known but is estimated to be around 10,000 Gt. At present temperatures and concentrations of carbon dioxide, the ocean absorbs 91 Gt/yr and outputs 89 Gt/yr. The deep ocean is by far the largest reservoir of carbon and may lose or gain carbon over extended periods of time. Human activity puts about 7.9 Gt/yr of carbon into the atmosphere.

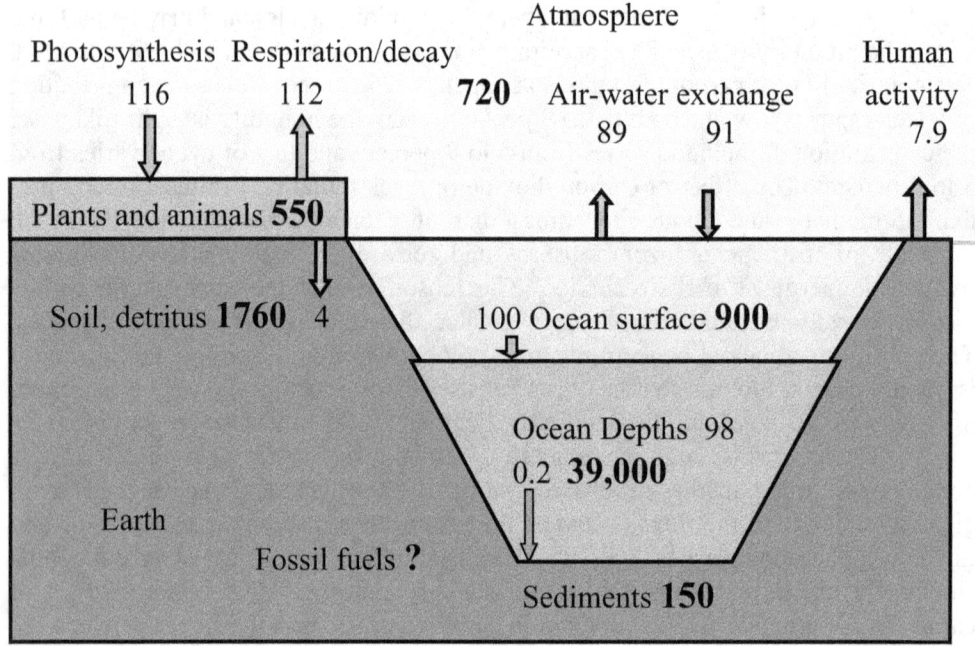

Fig. 9-7 Carbon storage and movement—reservoirs in Gt and fluxes in Gt/yr.

Because there are proposals to mitigate the buildup of carbon dioxide in the atmosphere by sequestering the carbon in vegetation, Table 9-3 is included to provide more detailed data on the carbon stored in the Earth's vegetation and soil. The data was obtained from the Intergovernmental Panel on Climate Change (IPCC), which cites a 1998 report by the German Advisory Council on Global Change as the original source of the data. *Biomes* are areas with similar climatic and geographical conditions. *Boreal forests* are those located near the Arctic Circle, *tropical areas* are those located near the equator and *temperate areas* are the areas in between. Tropical forests are a particularly important means of storing carbon while cropland vegetation stores relatively little carbon. Note that it is the soil in wetlands that stores the most carbon. Wetlands and former wetlands are the sources of peat. The differences between the carbon amounts appearing in Fig. 9-7 and those in Table 9-3 are due to the uncertainties in the separate estimates, the difference in the times the estimates were made and the fact that animals were included in Fig. 9-7 but not included in Table 9-3.

Table 9-3 1998 worldwide vegetation and soil carbon content by biome [IPCC].

Biome	Vegetation Gt	Soil Gt	Total Gt	Area Mkm^2(Mmi^2)	Vegetation t/km^2(t/mi^2)	Soil t/km^2(t/mi^2)
Tropical forest	212	216	428	17.6 (6.80)	12,045 (31,198)	12,273 (31,786)
Temperate forest	59	100	159	10.4 (4.01)	5,673 (14,693)	9,615 (24,904)
Boreal forest	88	471	559	13.7 (5.29)	6,423 (16,636)	34,380 (89,043)
Tropical savanna	66	264	330	22.5 (8.69)	2,933 (7,597)	11,733 (30,389)
Temperate grassland	9	295	304	12.5 (4.83)	720 (1,865)	23,600 (61,124)
Desert/semi-desert	8	191	199	45.5 (17.6)	176 (455)	4198 (10,872)
Tundra	6	121	127	9.5 (3.67)	632 (1,636)	12,737 (32,988)
Wetland	15	225	240	3.5 (1.35)	4,286 (11,100)	64,286 (166,500)
Cropland	3	128	131	16.0 (6.18)	188 (486)	8000 (20,720)
Total	466	2011	2477	151.2 (58.4)	3,082 (7,982)	13,300 (34,438)

Many of the Earth's elements and compounds move around in cycles. Of interest here are the cycles involving the movements of the GHGs between the atmosphere and the solid and liquid parts of the Earth. For the GHGs, the *cycle time* is the average time the GHG spends in the atmosphere. Water vapor, the most effective greenhouse gas, has a cycle time of one week. Carbon dioxide has a cycle time of about five years and for methane the cycle time is about 12 years. Normally, the cycle time is a measure of how quickly the atmosphere can recover from an excessive, but temporary, influx of a substance. But for carbon dioxide the cycle time is misleading in this regard because a substantial amount of carbon dioxide is being continually added to the atmosphere. Roughly half the 7.9 Gt/yr of carbon dioxide put into the atmosphere by human activity is retained, thus causing the amount of carbon dioxide in the atmosphere to increase between 3 and 4 Gt/yr. Therein lies the problem.

The seriousness of this problem is open to debate; so let us now consider the evidence that has been collected. Figure 9-8 shows the difference between the annually measured temperatures from 1880 to 2008 and the average temperature between 1951 and 1980. It is assumed that since 1880 the measuring instruments and the number and diversity of direct measurements have provided a degree of accuracy of ±0.1 °C (±0.18 °F). The amount of temperature rise during this 128-year period was approximately 0.8 °C (1.44 °F). This rise in temperature not only has been seen from land surface instrument measurements, but has correlated with borehole, ocean layer, balloon and satellite measurements. Also, the steep rise since 1970 is evidenced by the melting of ice sheets, glacial retreat, unusual weather patterns and changes in species behavior. The instances of unusual weather include increases in the number and intensity of hurricanes and abnormal heat waves and precipitation patterns. The changes in species behavior include bird migrations that go farther north by 6 km/decade (3.75 mi/decade) and some animals moving to higher elevations by 6 m/decade (20 ft/decade). Plant growth and other spring events are occurring earlier by 3 days/decade.

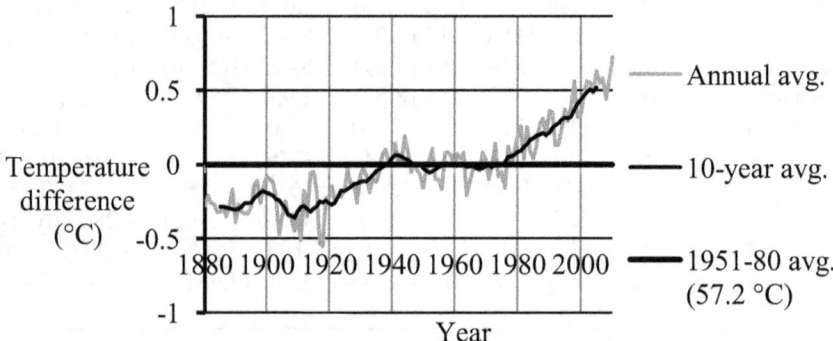

Fig. 9-8 Global average temperature difference from 1951 to 1980 average [NASA].

Figure 9-9 gives the estimated temperature values over the last 1,000 years. The figure was created from data obtained from several sources. The latter part of the curve is from the data shown in Fig. 9-8. The temperature rise shown in Fig. 9-8 may not seem like much, but when compared to earlier data, the record since 1880 shows a very steep rise. Since 1970 the rise appears almost vertical. It is this rapidity of the change that worries scientists most. In the history of estimated temperatures, it is believed that never before has there been such a rapid rise in temperature.

Because reasonably accurate and diversified temperature measurements have been made in only the last 160 years or so, other means, called *proxies,* were used to estimate temperatures before 1850. Thermometers have existed since the late 16th century, but their accuracy is questionable and their measurements were not widely distributed. The principal proxies are tree rings, fossils and ice cores from Greenland and Antarctica. The width and density of tree rings provide annual growth data that can be used to determine year by year temperature and precipitation, but the accuracy of tree ring measurements alone is not very high and the oldest trees are only a few thousand years old. To reach back millions of years, fossil and ice core proxies are required. Both fossils and ice cores rely on isotope ratios or changes in isotope amounts to produce temperature estimates.

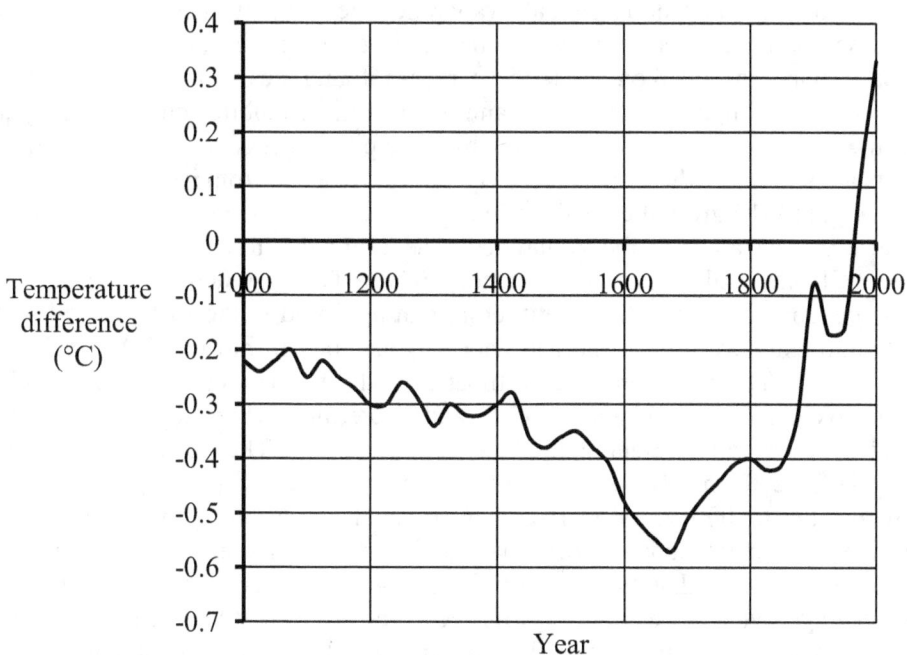

Fig. 9-9 Estimated temperatures for the last 1,000 years.

The most abundant hydrogen contains one proton and one electron, but deuterium also contains a neutron. Therefore, deuterium is heavier and the deuterium that evaporates from the oceans tends to precipitate out of the atmosphere more readily. The ice sheets covering Greenland and Antarctica consist of water that was evaporated from the oceans and then deposited in these areas over several eons. The ice in ice sheets is deposited in annual layers just as trees grow in annual rings. Higher temperatures cause the deuterium to be more mobile than cooler temperatures and, therefore, more likely to migrate as far as the polar regions before being precipitated. The net result is that during warmer periods the amounts of deuterium in the ice layers are more than during cooler periods. Both the thicknesses of the layers and the amounts of deuterium provide measures of annual temperatures. Although pressure has caused the deeper layers (the older layers) to be less distinct, it is believed that layers up to one million years old may provide satisfactory temperature data.

Similarly, the oxygen isotope ^{18}O is heavier than the more abundant ^{16}O and is less likely to evaporate. This causes the ratio of ^{18}O to ^{16}O in the oceans to be greater in the years when there is less water in the oceans. The sediment in the ocean floor also consists of annual layers and sea animal fossils have retained the oxygen isotope ratios, thus indicating annual evaporation, temperature and sea level. The oxygen isotope ratio variation may also be detected in ice cores, but ice cores that date back hundreds of thousands of years are found near Earth's poles only. Hydrogen is much lighter than oxygen and is much more mobile and more likely to reach the poles before it is deposited by precipitation. As a result, deuterium amounts are stronger indicators of the temperatures obtained from ice cores.

Figure 9-10 gives the results of a study made of the ice core taken from the Russian Vostok station in the Antarctica. The original reference was the article entitled "Climate and Atmospheric History of the Past 420,000 Years from the Vostok Ice Core, Antarctica" by Petit, et al, that appeared in the *Nature* issue 399, pp. 429-436. However, the numerical data was obtained from the NOAA website, *www.noaa.gov*. The figure includes estimates of the temperature, carbon dioxide, methane, oxygen isotope changes, $\delta^{18}O$, and deuterium, ^2H, changes. All of the graphs extend from the present to 450,000 years in the past. The ice core extended to a depth of 3270 m (10,730 ft). The curves indicate four major ice ages approximately 100,000 years apart and interglacial periods in between. The four major ice ages correspond to the changes in the Earth's eccentricity, thereby reinforcing the belief that this eccentricity is the main driving force behind the natural ice age cycle. Although very old ice cores are available in Greenland and Antarctica only, scientists believe they provide the best evidence of past global temperatures. The temperature curve indicates the differences between the temperatures at the time and the average of recent temperatures in degrees Celsius. The measurement of gas densities is primarily based on ice core data. As snow falls and is compacted, air bubbles are trapped in the ice. The densities of the various gases in these bubbles can be measured with a good degree of accuracy. Although other proxy data has been collected, it is the ice core data that has most strongly identified temperature changes with the changes in GHG densities.

Although it is not evident in the graphs in Fig. 9-10, detailed analysis of the data indicate that in the past the rises in temperature precede the rises in carbon dioxide and methane. (Keep in mind that time increases from right to left.) It is believed that a pronounced rise in temperature triggers the release of carbon dioxide and methane from the land and oceans. The increase in carbon dioxide and methane, in turn, causes a further increase in temperature and so on until the land and ocean can no longer supply the gases required to further increase the temperature. Just what causes the decreases after the peak is attained is not fully understood, but the subsequent decreases are more gradual than the steep rises. This observation is contrary to the present situation in which a rise in the densities of carbon dioxide and methane are believed to be causing the current rise in temperature.

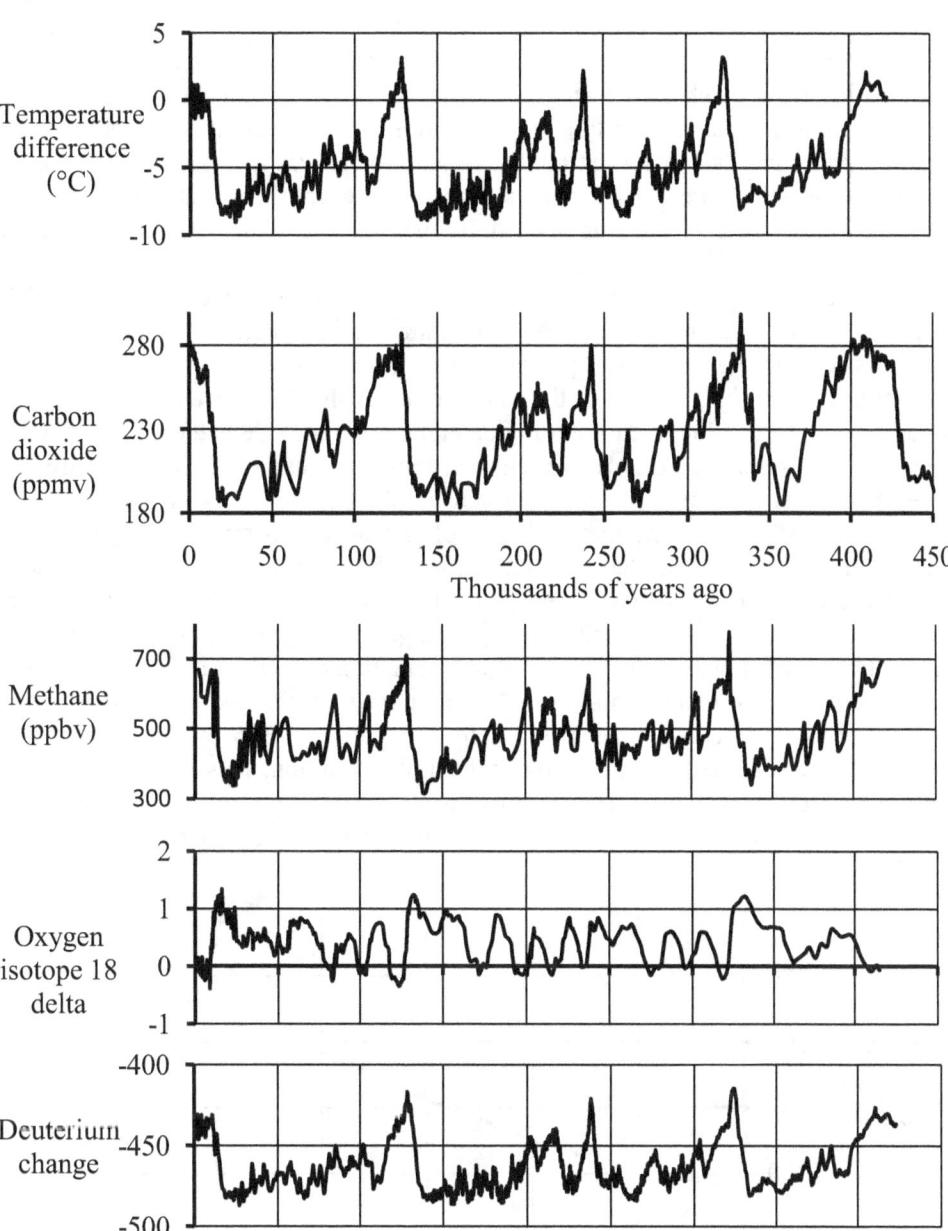

Fig. 9-10 Estimated temperature, carbon dioxide, methane, oxygen
isotope variations and deuterium variations [NOAA].

There also exists proxy data, primarily from fossilized plants and animals, that indicate atmospheric conditions millions of years in the past. These data are less accurate but provide evidence that in the distant past the Earth has experienced periods that were much colder and much warmer than those of the last one million years. It is known that during the age of the dinosaurs the Earth was much warmer near the poles and the concentration of carbon dioxide was much greater than in the period for which there are ice core records. However, the distant past is not as relevant to our present concerns over global warming as understanding the climate changes over the last half million years.

If one accepts, as most scientists do, the hypothesis that an increase in GHG densities causes an increase in global temperature as indicated by the above figures, then it is logical to examine our present situation in light of the known increases in GHGs since the beginning of the industrial revolution. Because most of the greenhouse effect is due to carbon dioxide and carbon dioxide is the dominant GHG released by burning fossil fuels, we will concentrate on carbon dioxide. Figure 9-11 gives the carbon dioxide density recorded at Mauna Loa, Hawaii, since 1958 in ppmv. Mauna Loa is on the remote island of Hawaii that is sparsely populated and there is little locally generated anthropogenic carbon dioxide. This graph is known as the Keeling curve and was taken from the National Oceanic and Atmospheric Administration (NOAA) website, *www.noaa.gov*. It gives the monthly average concentrations that demonstrate the annual fluctuation caused by the growing season and pattern of fossil fuel consumption. During the 61 year period from 1958 to 2009 the carbon dioxide density has risen from 316 ppmv to 386 ppmv, a 22 percent increase. Since the early years of the industrial revolution, it is estimated that the carbon dioxide density has exponentially risen from 280 ppmv to 386 ppmv, a 38 percent increase, and is continuing its exponentially increasing pattern.

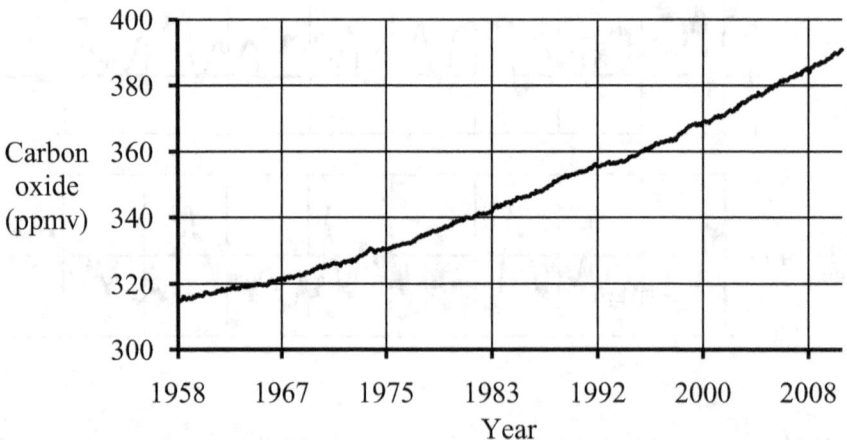

Fig. 9-11 Carbon dioxide monthly average density [NOAA].

The question is: how much of the increase in carbon dioxide is the result of human activity? The answer to this question is not straightforward, but depends on the assumptions made regarding how much of the human induced carbon dioxide is absorbed by the Earth's plant life, soil and oceans. Although some carbon dioxide is generated by the production of cement and other activities, according to the EIA more than 98 percent of the anthropogenic carbon dioxide is due to burning fossil fuels. The increase in carbon density due to injecting one gigatonne of carbon into the atmosphere has been estimated to be about 0.5 ppmv. Using this figure and the EIA amounts of carbon dioxide put into the atmosphere by burning fossil fuels, ignoring the contributions of anthropogenic sources other than fossil fuels and assuming that one half of the anthropogenic carbon dioxide is absorbed by the Earth's plant life, soil and oceans (a widely used estimate), one arrives at the graph depicted in Fig. 9-12. The solid curve is simply a portion of the NOAA curve in Fig. 9-11, the curve made up of long dashes is the density of carbon dioxide that would have resulted if none of the anthropogenic carbon dioxide were absorbed and the remaining curve assumes half of the anthropogenic carbon dioxide was absorbed. This figure indicates that over 85 percent of the increase in carbon dioxide in our atmosphere is caused by burning fossil fuels. There is debate over the correctness of this percentage, but most scientists believe that a high percentage of the increase in atmospheric carbon dioxide is due to human activities.

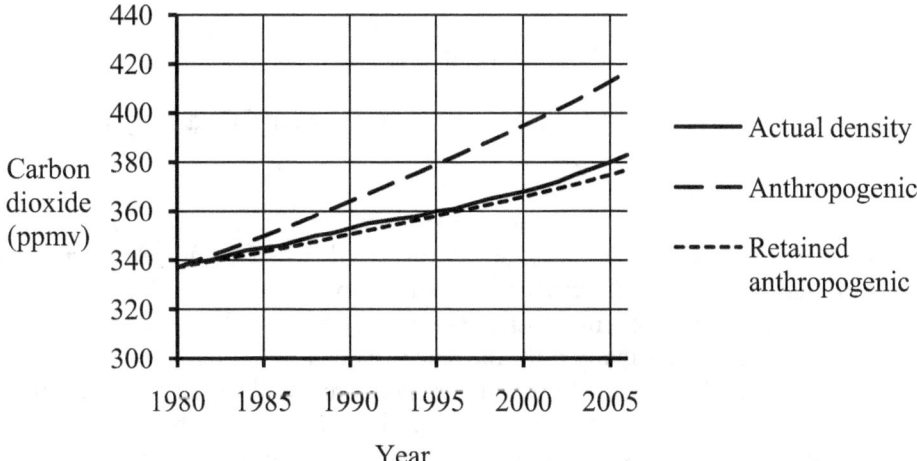

Fig. 9-12 Carbon dioxide densities—actual, anthropogenic if not absorbed and anthropogenic if absorbed.

Figure 9-13 shows how changes in atmospheric carbon dioxide are related to recent changes in temperature. It was obtained by computing the ratios of the changes in temperature in degrees Celsius to the changes in carbon dioxide density in ppmv from 1965 through 2007. Superimposed on the ratio curve is the linear least-squares fit of the ratio data. The least-squares curve fit is almost flat, indicating that there has been an almost constant relationship between changes in carbon dioxide densities and changes in temperature during this 42 year span. The average change was roughly 0.01 °C (0.018 °F) per one ppmv. If there is a 0.5 ppmv increase in carbon dioxide density for each gigatonne increase in atmospheric carbon dioxide, then each gigatonne increase would cause the temperature to increase by 0.005 °C (0.009 °F).

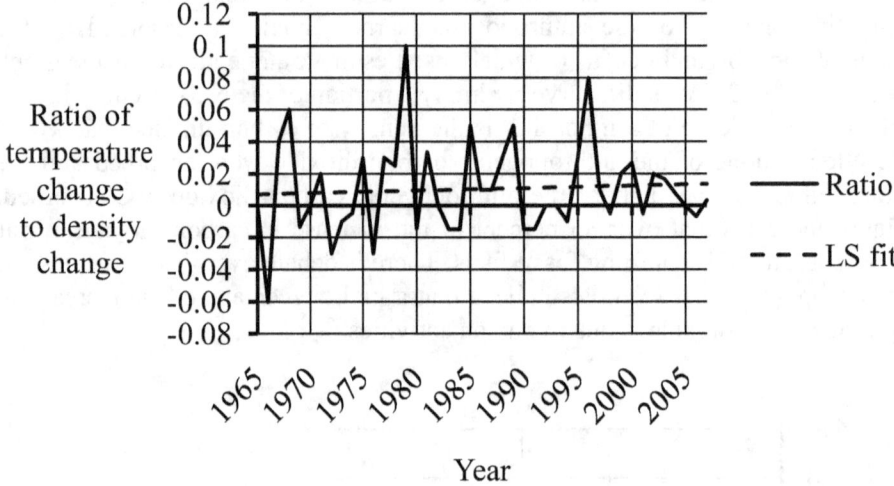

Fig. 9-13 Ratio of change in temperature in °C to change in CO^2 density in ppmv.

Just how much carbon dioxide is produced when a fuel is burned is indicated by its *carbon dioxide intensity*, which is typically expressed in kilograms of carbon dioxide emitted per million Btu (kg/MBtu) generated. These values vary within a particular fuel, especially with fuels such as coal or wood, which include a wide range of material qualities. Some of the more frequently used fuels and their carbon dioxide emission intensities are given in Table 9-4. These values are representative and, for fuels such as those derived from petroleum and natural gas, are quite accurate. Except for the last two values, which were gotten from Wikipedia, these values were obtained from the EIA. Although the biomass examples in the last three lines emit considerable carbon dioxide when they are burned, it is assumed a more or less equal amount of carbon dioxide is absorbed by their regrowth.

To produce the same amount of energy, coal emits a third more carbon dioxide than gasoline and 80 percent more carbon dioxide than natural gas. This is why so much attention is given to the consumption of coal, particularly the consumption by the United States and China. The primary use of coal is to generate electricity. Much less carbon dioxide would be released into the atmosphere if natural gas were used to generate electricity instead of coal. Even better would be to use renewables that emit no carbon dioxide while generating electricity, the use to which they are best suited.

Table 9-4 Carbon dioxide emission intensity of various fuels [EIA, WIKI].

Fuel	Emission intensity in kg/MBtu
Coal	95
Gasoline	71
Jet fuel	71
Diesel/fuel oil	73
LPG	63
Natural gas	53
Methanol/ethanol	66
Wood/wood waste	89
Dry plant material	148
Bagasse	143

Figure 9-14 shows the amounts of carbon dioxide resulting from the consumption of fossil fuels in some of the developed, developing and undeveloped regions of the world in 1980 and 2008. The United States and China, the two countries with the largest coal reserves, accounted for 41 percent of the world's carbon dioxide emissions. Both countries use coal extensively to generate electricity. The differences in the dark and light bars show that while Northern Europe had a 31 percent decrease in emissions from 1980 to 2008, the United States had a 22 percent increase and China had a whopping 348 percent increase.

Figure 9-15 paints a much different picture. While Fig. 9-14 shows China emitting more carbon dioxide than the United States, Fig. 9-15 shows that the average person in the United States emits almost four times as much as the average person in China. On a per capita basis, Americans emit two and one half times as much carbon dioxide as Northern Europeans, twice as much as the Japanese and almost 15 times as much as Indians. Canadians fare only slightly better than Americans.

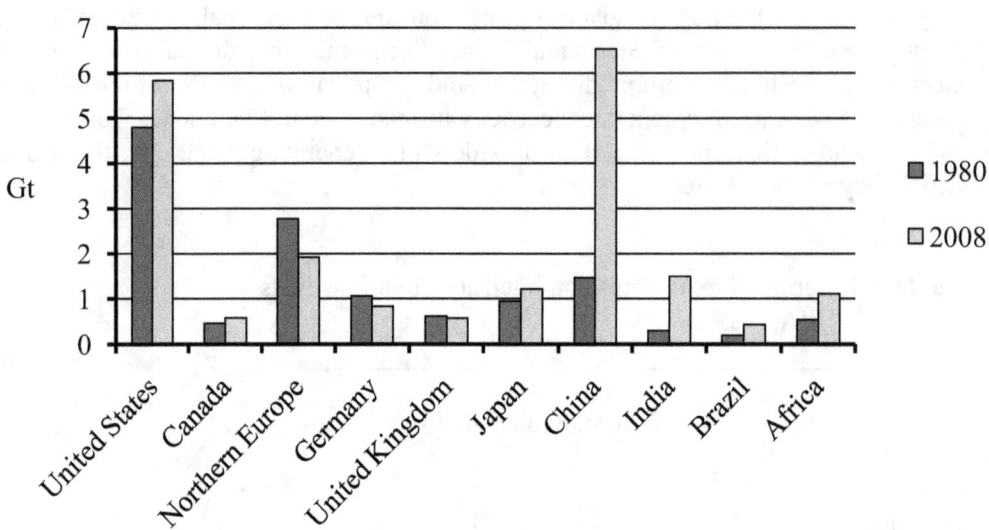

Fig. 9-14 1980 and 2008 carbon dioxide emissions in Gt [EIA].

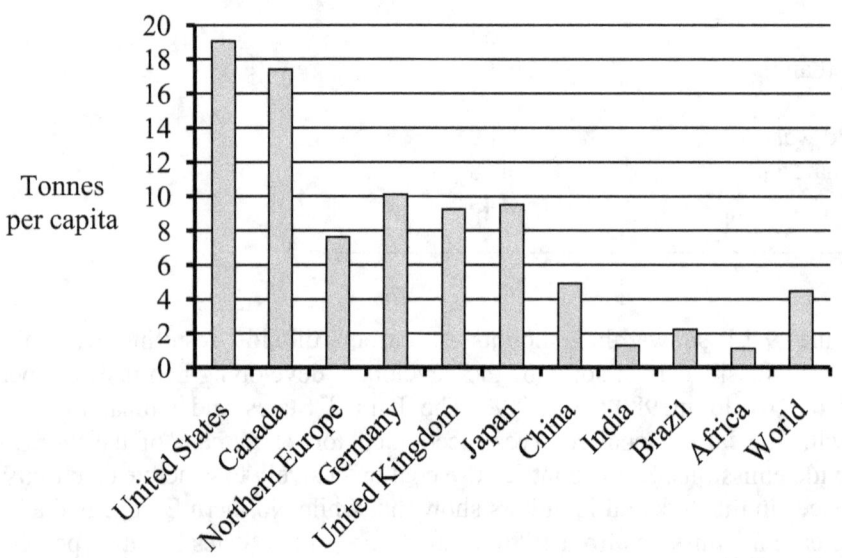

Fig. 9-15 2008 carbon dioxide emissions per capita [EIA, UN].

The carbon dioxide emissions caused by burning fossil fuels for the world, China and the United States in Gt for the years 1980 through 2008 are given in Fig. 9-16. In 2008, the world emitted 30.4 Gt of carbon dioxide, of which 6.5 Gt were emitted by China and 5.8 Gt were emitted by the United States. The discussions of Figs. 9-12 and 9-13 imply that, if the addition of 30.4 Gt of carbon dioxide continued to be added to the atmosphere each year and half of it were retained, then by 2100 the global temperature would rise 7.1 °C (12.8 °F). However, because this simplified computation assumes a constant rate of absorption relative to the amount of carbon dioxide in the atmosphere, the temperature increase is higher than that predicted by most computer models. But even half this increase could be catastrophic in terms of weather patterns and sea rise. Most scientists believe that the oceans will rise between 0.2 and 1 m (0.7 to 3.3 ft), but admit the sea level rise could be much worse.

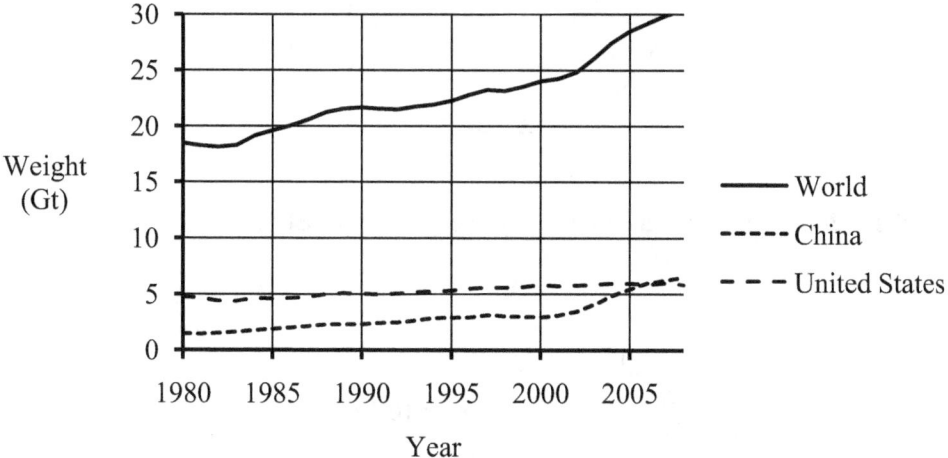

Fig. 9-16 Carbon dioxide emissions caused by burning fossil fuels [EIA].

The percentages of Chinese energy consumption by source in 2006 are given in Fig. 9-17. The percentage of China's energy that comes from coal is much higher than that of the United States and other developed countries—70 percent versus 23 percent for the United States. If the Chinese percentage of coal as an energy source continues as China's rapid development continues, then China's contribution to the emission of carbon dioxide would be enormous by 2050. Although the completion of the mammoth Three Gorges Dam since 2006 would cause the percentage of hydroelectric power to be higher and the percentage of coal to be lower, the percentage changes would be small. China does have plans to increase their use of renewable energy sources considerably and presently produce more solar panels than the United States. But both China and the United States are continuing to build coal-fired generating plants at what many consider a dangerous rate.

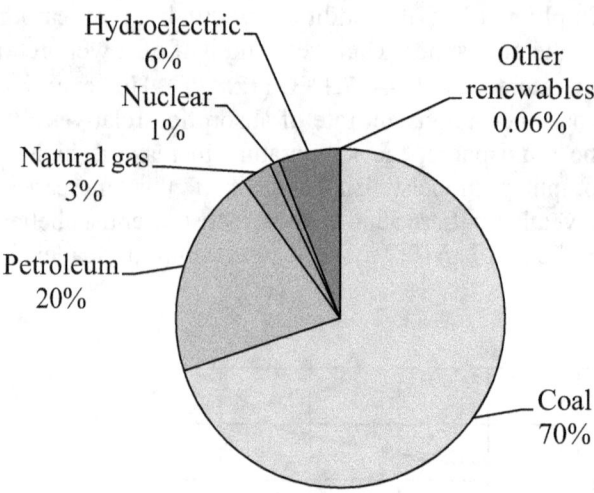

Fig. 9-17 2006 Chinese consumption by energy source [EIA].

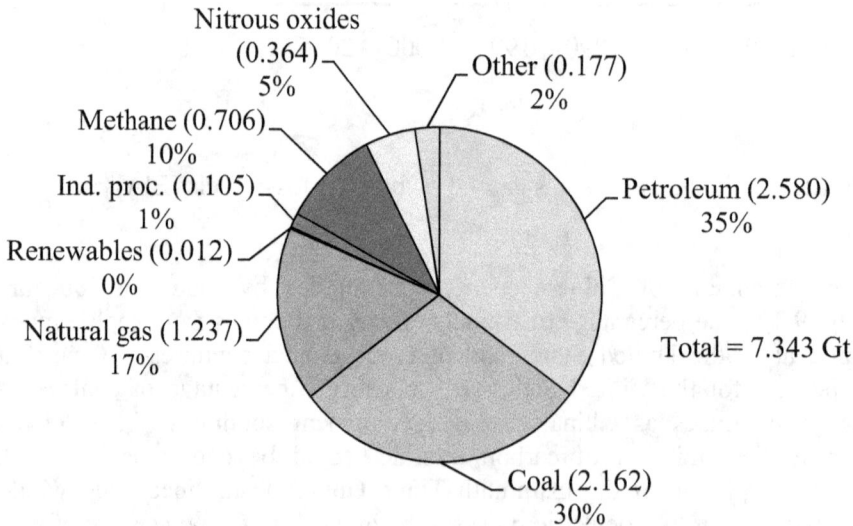

Fig. 9-18 2007 carbon dioxide equivalents emitted in the United States [EIA].

Despite the fact that most of the atmospheric GHG is carbon dioxide, by no means is the greenhouse effect totally due to carbon dioxide. Figure 9-18 shows the carbon dioxide equivalent amounts of the various GHGs emitted into the atmosphere by the United States in 2007. Only carbon dioxide is broken down by its sources and is associated with all sections of the pie except the "Methane", "Nitrous oxides" and "Other" sections. The "Methane", "Nitrous oxides" and "Other" sections give the carbon dioxide equivalent emissions of their respective gases from all sources. The "Industrial processes" section represents the direct carbon dioxide emissions from industrial processes, emissions not caused by fossil fuels. GHGs other than carbon dioxide are lumped together regardless of their source (i.e., petroleum, coal, etc.). Approximately 83 percent of the greenhouse effect is caused by carbon dioxide and over 82 percent of the anthropogenic greenhouse effect is caused by energy-related carbon dioxide emissions, almost all of which are emitted by burning fossil fuels. The total amount of the 2007 emissions was equivalent to 7.36 Gt of carbon dioxide with 5.99 Gt being energy-related carbon dioxide emissions. The carbon dioxide from renewables was due to burning waste to generate electricity. The 0.105 Gt of industrial process direct carbon dioxide emissions are mainly caused by the processing of cement and other stone and ash products. The 1.26 Gt of carbon dioxide equivalent emissions of other gases include emissions from industrial processes, agriculture, waste management (mainly landfills) and so on.

Figures 9-19 and 9-20 illustrate the 2007 equivalent carbon dioxide emissions of the GHGs in the United States by sector. Figure 9-19 gives the amounts going directly into the atmosphere and indirectly into the atmosphere through the generation of the electricity used by the various sectors. Figure 9-20 separates out the amounts caused by the generation of electricity. In Fig.9-20, the amounts indicated for the sectors are the emissions of all GHGs that are output directly by the

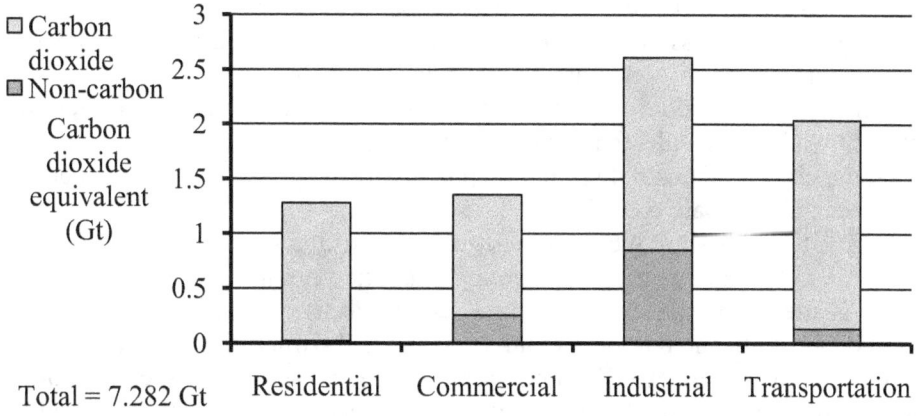

Fig. 9-19 2007 GHG direct and indirect emissions in the United States by sectors [EIA].

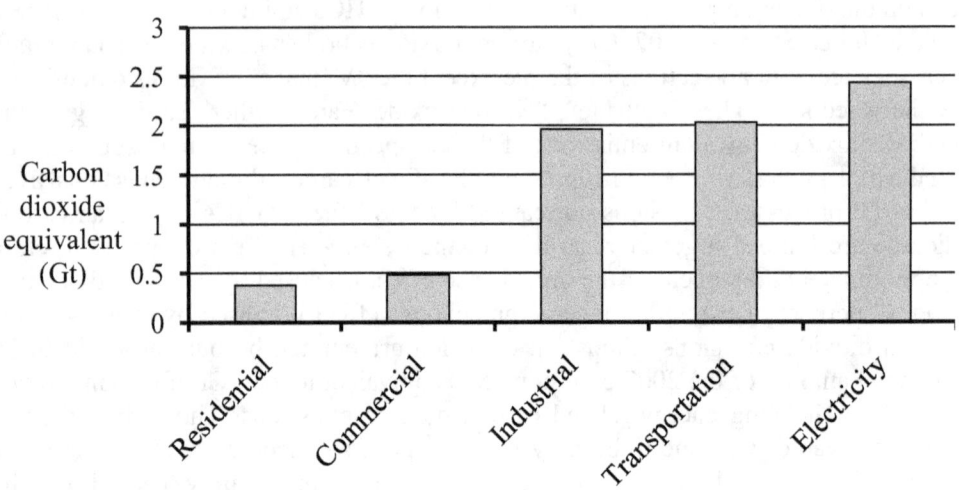

Fig. 9-20 2007 GHG direct emissions in the United States by sectors and
electricity [EIA].

sectors only. The difference in the total amounts given in Figs. 9-18 and 9-20 is
accounted for by the emissions from international air and water transportation.
About two-thirds of the non-carbon dioxide GHGs are emitted by the industrial
sector. While the GHG emissions by the residential, commercial and transportation
sectors are almost entirely energy related, much of those by the industrial sector are
not. The GHG output by the residential sector is essentially carbon dioxide. By
separating out the emissions caused by the generation of electricity, the emissions
from the transportation sector are nearly unchanged, but the emissions from the
residential sector are reduced by 71 percent. The transportation sector uses very little
electricity while the residential sector is a heavy user of electricity. Note that both
the commercial and industrial sectors output significantly less emissions when only
their direct emissions are considered.

The transportation sector, the sector that emits the second greatest amount of
GHGs, currently gets its energy by burning petroleum products. But the automotive
industry is attempting to move toward the use of electricity to power cars and light
trucks. If a significant portion of our cars and light trucks become powered by
electricity, then the direct GHG emissions by the transportation sector would be
substantially reduced, but those caused by the generation of electricity would be
increased. Just how much the GHG emissions from electricity generation would
increase would depend on the energy sources used. If these sources are renewable or
nuclear power is used, then the savings in emissions would be considerable. But if
we continue to generate electricity in the same way we have in the past, then there

may be no net savings because of our heavy reliance on coal. In fact, in some parts
of the United States, the emissions would increase.

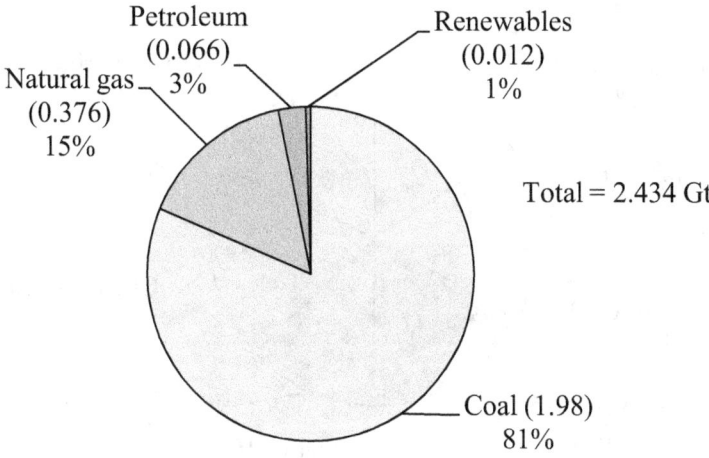

Fig. 9-21 2007 electricity generation emissions in the United States by
energy source [EIA].

One third of all GHG emissions in the United States are caused by the
generation of electricity and much of them are due to coal-fired power plants. Figure
9-21 shows the carbon dioxide equivalent emissions of the fuels used to generate
electricity. Note that, although coal is used to generate only half our electricity, it is
responsible for 81 percent of the GHGs emitted by electric power plants. All but
four percent of the greenhouse effect caused by generating electricity is due to
carbon dioxide. Nuclear plants contribute to the output of GHGs only through the
processing of nuclear fuel.

From Fig. 9-20 it is seen that the industrial sector emits almost 30 percent of
the carbon dioxide. Figure 9-22 breaks down the industrial sector's carbon dioxide
emissions by industrial groups. GHGs other than carbon dioxide are not included in
the figure. The "Petroleum" group consists of oil refineries and other industries
involved in the processing of petroleum. All sections of the pie other than the section
labeled "Other industrial" represent manufacturing or refining groups. Altogether,
manufacturing and refining are responsible for 84 percent of the carbon dioxide
emitted by the industrial sector. "Other industrial" includes agriculture, forestry,
fishing, construction and mining. The data used to construct Fig. 9-22 are from the
EIA sponsored *Manufacturing Energy Consumption Survey* of 2002.

Finally, Fig. 9-23 indicates the methane and nitrous oxide emissions by
activity. In the United States, all non-carbon dioxide GHG emissions account for

only 17 percent of the GHG emissions and energy-related non-carbon dioxide emissions contribute only four percent of the total, much of which is caused by natural gas leakage.

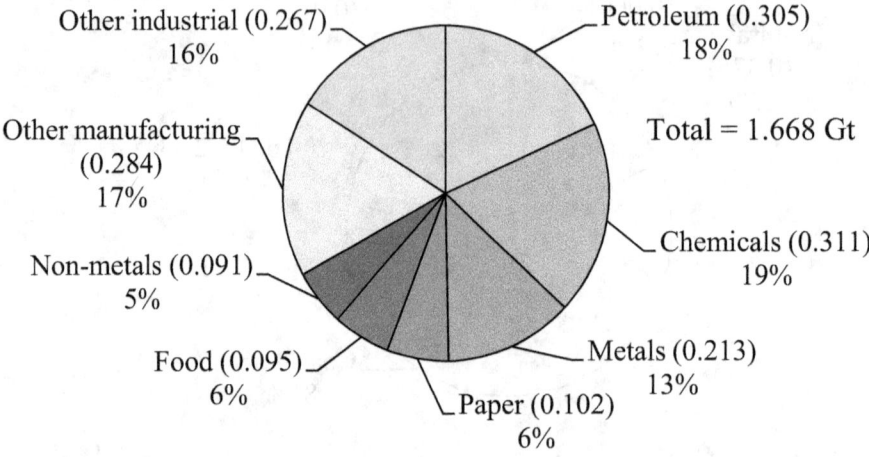

Fig. 9-22 2002 carbon dioxide emissions in the United States by industry [EIA].

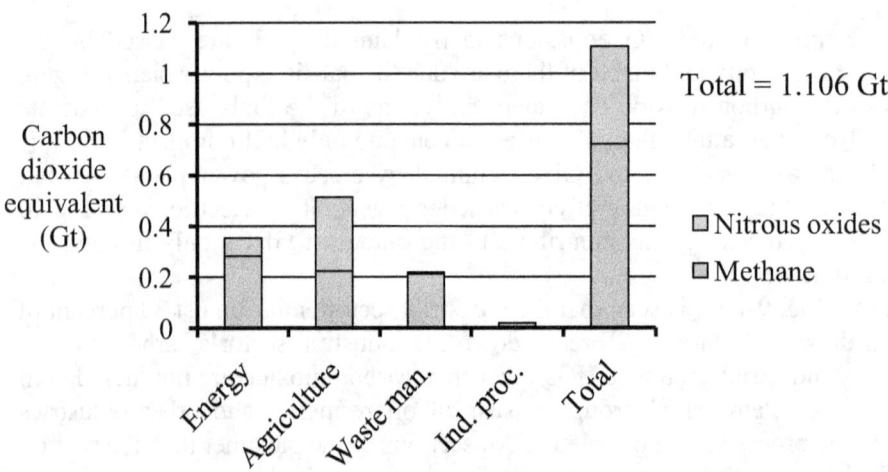

Fig. 9-23 2007 non-carbon GHG emissions in the United States by activity[EIA].

An historic record of the GHGs output to the atmosphere by the United States from 1980 through 2007 is provided in Fig. 9-24. Almost all of the increase in the GHGs since 1980 has resulted from the increase in carbon dioxide. The slight increases in nitrous oxides and other GHGs are offset by the decrease in methane. All GHGs other than carbon dioxide contribute only one sixth of the total greenhouse effect.

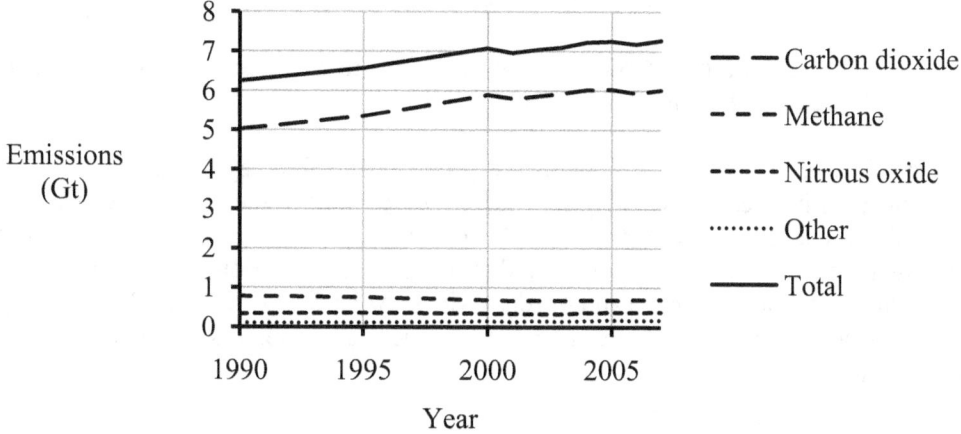

Fig. 9-24 History of carbon dioxide equivalent GHG emissions in the United States [EIA].

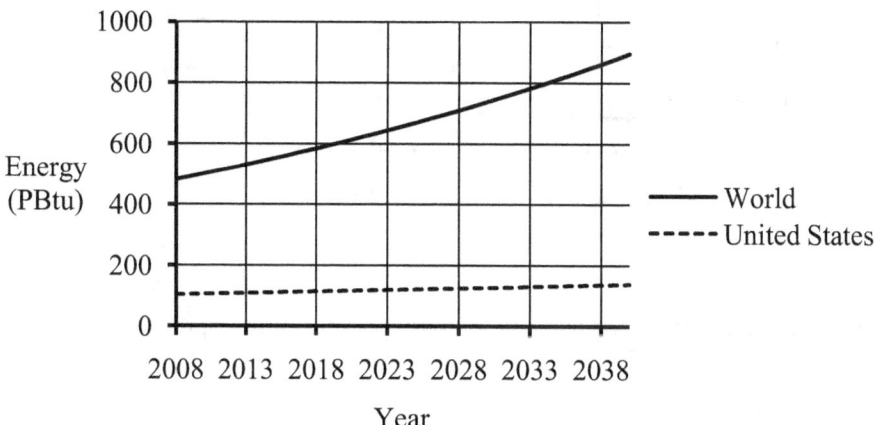

Fig. 9-25 Energy projection based on the growth indicated in Fig. 1-8.

Figure 9-25 gives projections for the energy growths of both the world and United States in peta-Btu (PBtu). It is based on the growth rates computed from the

1980 to 2008 EIA data and is an extension of Fig. 1-8. From the world's data the projected growth is exponential with a rate of 1.95 percent per year, and from the United States data the growth is linear with a rate of 1.094 PBtu per year.

The corresponding carbon dioxide equivalent emissions projections are provided in Fig. 9-26. These projections assume that:

- The 2008 ratio of carbon dioxide emissions to energy consumption amounts are constant from 2008 to 2040.

Figure 9-26 indicates a doubling in the greenhouse effect by 2040 if the world continues on its present trend. Approximately 88 percent of the greenhouse effect is caused by energy-related activities. The other 12 percent is attributed to waste management, agriculture and direct industrial process emissions and GHGs other than carbon dioxide, methane and nitrous oxides. At best, this trend would be a risky course to follow and emphasizes the need for strict international agreements limiting GHG emissions.

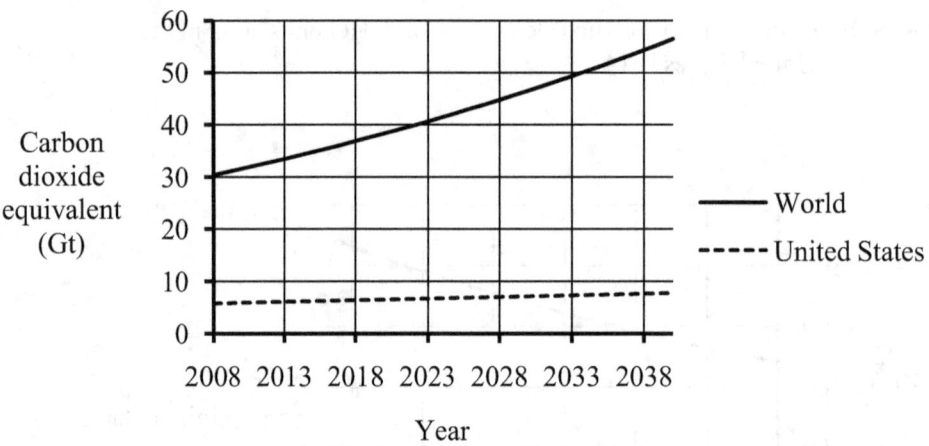

Fig. 9-26 Emissions projection based on the energy consumption projection in Fig. 9-25.

Although the United States has limited means for reducing the emissions of other countries, let us now consider some scenarios for reducing carbon dioxide emissions in the United States. Because transportation is the second greatest producer of carbon dioxide, ways in which we could decrease our carbon dioxide emissions by reducing our petroleum consumption are examined first. Let us assume:

- All the conditions used to construct Fig. 8-39.
- The carbon dioxide intensities listed in Table 9-4 and the energy densities in Table 1-1.
- All the carbon dioxide emitted by the biofuels assumed in constructing Fig. 8-39 are recycled so that they produce no net increase in atmospheric carbon dioxide.
- GHGs other than carbon dioxide cause an increase in the greenhouse effect of seven percent.

The results are given in Fig. 9-27 and show a one gigatonne per year reduction in transportation carbon dioxide equivalent emissions between 2006 and 2040. This reduction gives a 55 percent decrease from the 2006 amount. It is difficult to estimate how much the GHGs other than carbon dioxide contribute to the atmospheric heat retention, but overall they are responsible for an increase of about six or seven percent. For transportation, much of the other GHG emissions are caused by leakages from pipeline and storage facilities, including vehicle fuel tanks. Biofuels and petroleum derived fuels other than gasoline, jet fuel and diesel are not considered in this example.

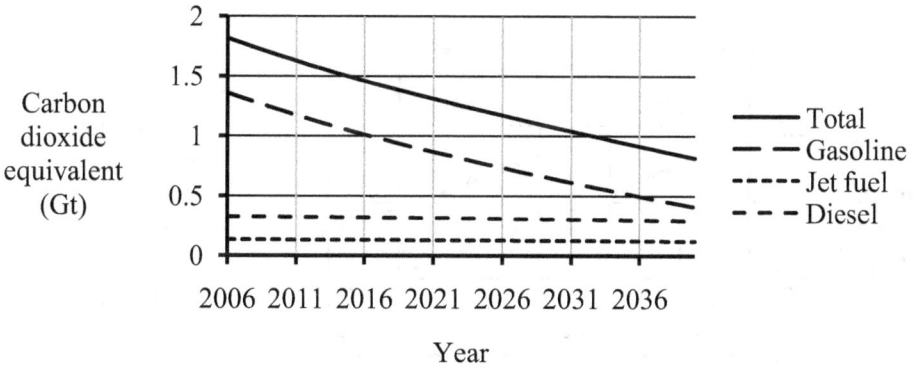

Fig. 9-27 Transportation GHG emissions assuming the reduced consumption in Fig. 8-39.

Another example examines the decrease in emissions obtained by replacing three-fourths of the gasoline powered vehicles by electric vehicles. The assumptions are:

- Three-fourths of all cars and light trucks are converted to electric vehicles. (The Electrification Coalition, which is made up of industries interested in promoting electric vehicles, considers converting three-fourths of all car and light truck vehicle-miles to vehicle-miles powered by electricity a reasonable goal.)
- The additional electricity needed for the electric vehicles is supplied by sources that emit no GHGs.
- All the conditions used to construct Fig. 8-39 except for those indicated by the first assumption. This includes the condition that the per capita total vehicle-miles for cars and light trucks are held constant at their 2006 value between 2006 and 2040.
- The conversion starts in January 1, 2011, and the number of vehicles converted each year is the number of years since the beginning times one-thirtieth times the total number of vehicles in use in that year.
- GHGs other than carbon dioxide cause an increase in the greenhouse effect of seven percent.

The corresponding reductions in carbon dioxide equivalent emissions are provided in Fig. 9-28. As with the previous example, only gasoline, jet fuel and diesel are taken into account. The total GHG emissions decrease 1.13 Gt/yr, 62 percent, between 2006 and 2040.

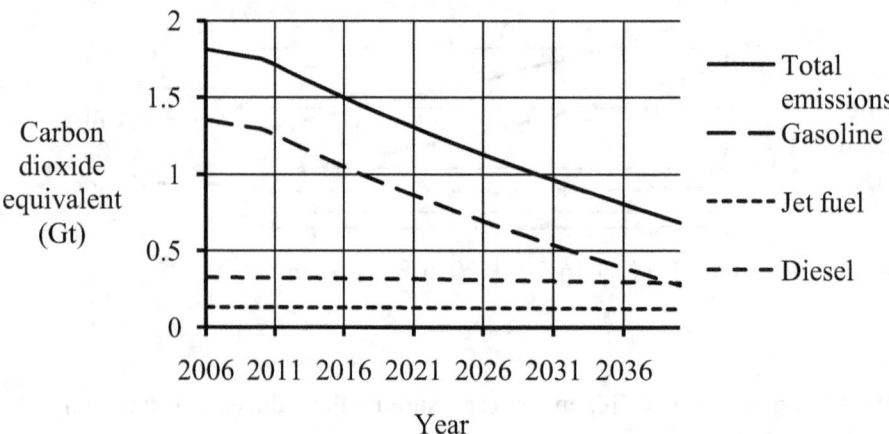

Fig. 9-28 Emissions assuming Fig. 8-39 and 75 % of gasoline vehicles are replaced by electric vehicles.

If all 2006 coal-fired electrical generating plants had been replaced by plants powered by natural gas with both types of plants being 33 percent efficient, then there would have been a reduction in GHG emissions from 2.46 Gt/yr to 1.6 Gt/yr, a decrease of 0.86 Gt/yr. If the natural gas plants had used combined cycle generation having 45 percent efficiency, then the reduction would have been 1.27 Gt/yr. But the past cannot be changed. However, suppose that:

- All coal-fired plants are replaced with nuclear, geothermal, solar and wind powered plants.
- The growth of electricity usage is 74.68 GkWh per year as indicated in Fig. 7-20.
- The current efficiency of the generation and distribution system is 31 percent.
- The replacement is to begin January 1, 2011, and the replacement increments are similar to those used in the previous example.
- Petroleum and natural gas generating facilities maintain the same percentages of total electrical energy production as they had in 2006 throughout the period 2010 to 2040.
- GHGs other than carbon dioxide cause an increase in the greenhouse effect of four percent.

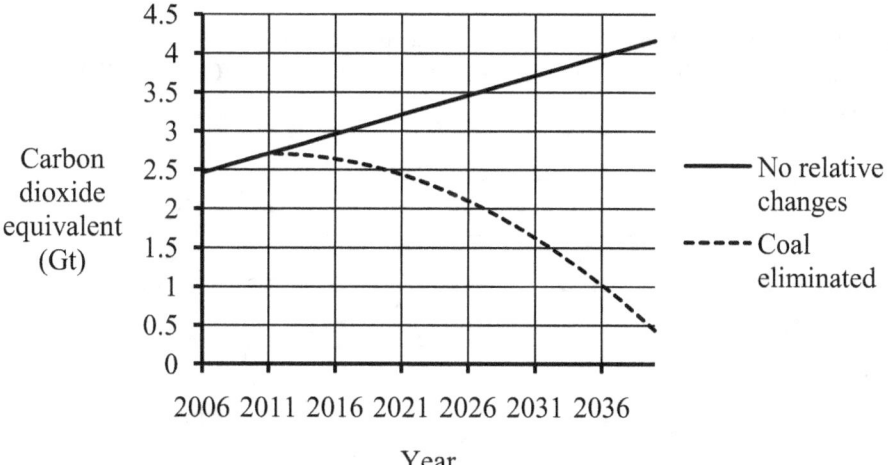

Fig. 9-29 Emissions assuming coal-fired electricity generation is eliminated.

The results of this scenario are given in Fig. 9-29 along with the emissions that would result if there were no changes in the way we currently generate electricity. By replacing coal with nuclear or renewable energy sources there would be a reduction of 2.02 Gt of carbon dioxide equivalent gases from the 2006 amount (a 82 percent decline) by 2040. The emissions by 2040 would be about ten percent of what they would have been if coal-fired plants had maintained their 2006 share of production.

The remaining scenario concerns the heat energy used directly by the residential, commercial and industrial sectors. It primarily projects what would happen if the direct means of providing space heating, water heating and industrial process heat were replaced with heat from electricity that is generated by facilities that are nonpolluting or carbon neutral renewable sources. The assumptions are:

- For the residential and commercial sectors 80 percent of the fossil fuel heat is replaced by heat produced by electricity or carbon neutral sources and for the industrial sector 50 percent of the fossil fuel heat is replaced by electrically produced heat or carbon neutral heat sources.
- The replacement is to begin January 1, 2011, and the replacement increments are similar to those used in the previous examples.
- The electricity needed to supply the heat is generated by nuclear or carbon neutral renewable means.
- GHGs other than carbon dioxide increase the greenhouse effect due to the residential and commercial sectors by four percent and the greenhouse effect due to the industrial sector by seven percent.

It is seen from Fig. 9-30 that there is a decline of only 0.78 Gt/yr carbon dioxide equivalents. The reduction is from 1.88 Gt/yr in 2006 to 1.10 Gt/yr in 2040, a 42 percent decrease.

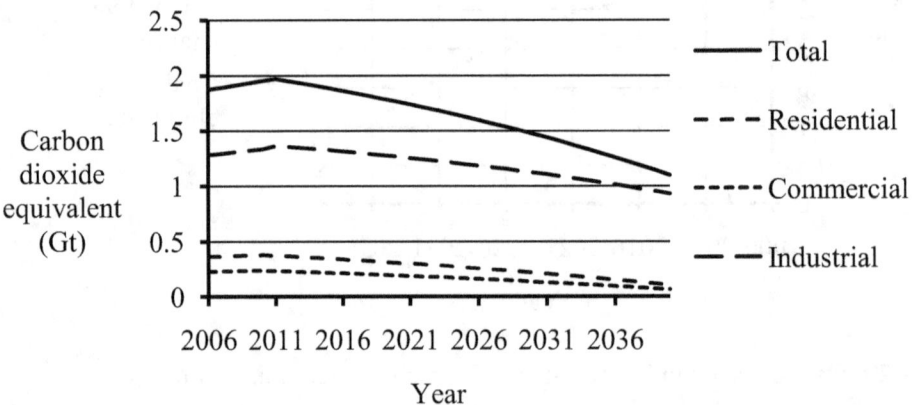

Fig. 9-30 Emissions assuming 80 percent of residential and commercial fossil fuel and 50 percent of industrial fossil fuel heat is replaced with means that do not emit GHGs.

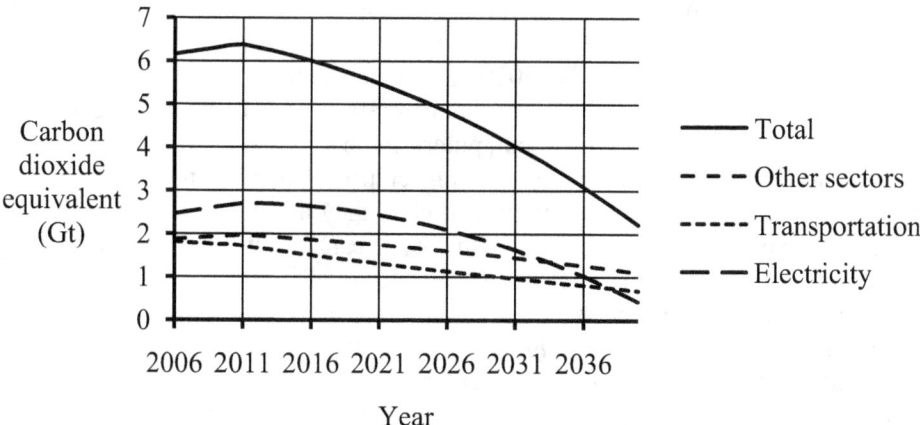

Fig. 9-31 Total emissions taken from Figs. 9-28, 9-29 and 9-30.

Figure 9-31 shows the sum of the carbon dioxide equivalents resulting from the amounts given in Figs. 9-28, 9-29 and 9-30. The total decrease is a 64 percent decline during the 2006 to 2040 period. For all the scenarios, the decreases are prevented from being greater by the projected increased demand for energy by the United States increasing population. The projected emissions per capita in the United States would be about 5.3 tonnes, of which 4.4 tonnes would be carbon dioxide. The average American would still be emitting as much as the world's 2008 per capita average. Although a 64 percent savings may seem impressive in light of an increasing population, such gains made in the United States and other developed nations could be easily overwhelmed by the world's poorer nations. The underdeveloped and developing nations have rapidly increasing populations that expect the better standards of living enjoyed by the developed nations.

As briefly discussed in Chap. 6, another area in which there may be appreciable savings in electricity consumption, and hence a reduction in pollution, is electronics. The growth in the use of electronics has been dramatic and this usage is likely to continue to grow exponentially for some years to come. Fortunately, the manufacturers have realized the importance of reducing the amount of electricity their products consume. The most affected sectors are the residential and commercial sectors and the primary products being targeted are those related to information and communication technology (ICT), particularly those related to the internet. An article in the January, 2010, issue of *Computer* entitled "Project Greenlight: Optimizing Cyber-infrastructure for a Carbon-constrained World" by Larry Smarr indicates that two to three percent of the 2007 global GHG emissions were caused by the use of ICT equipment. This article also indicates that these emissions are increasing at the compounded rate of six percent annually. In 2007, the telecommunications infrastructure was responsible for 37 percent of these ICT emissions, data centers were responsible for 14 percent and personal computers and their peripherals for 49 percent. By 2020 these percentages are expected to be 25, 18 and 57, respectively. Therefore, the concentration has been on reducing the electricity consumption by personal computers and peripherals.

One of the main electricity-draining culprits has been the practice of not completely turning off personal computers when they are not in use. In order for a computer to have an internet presence so that it can detect when it is receiving a message, it is maintained in a sleep mode and some of its electronics is left on. By assigning the task of monitoring the internet to specially designed low-power devices, a practice called *proxying*, a considerable amount of electricity can be saved. Advances in proxying and other measures for making electronics more efficient are expected to make a small but significant reduction in the residential and commercial consumption of electricity. It is hoped that these savings will offset the growth of ICT usage. There is some consideration being given to an ICT cap and trade system for large installations (see the March, 2010, issue of *Computer*).

Figure 9-32 shows the projected demand for generated electricity that would be needed if the present trend continues, the additional amount that would be required if the savings indicated in Fig. 9-31 are to be met and the corresponding total amount. These amounts assume:

- The linear growth rate shown in Fig. 7-20 and used in Fig. 9-29.
- The delivered electrical energy is used with the same efficiency as the direct heat energy.

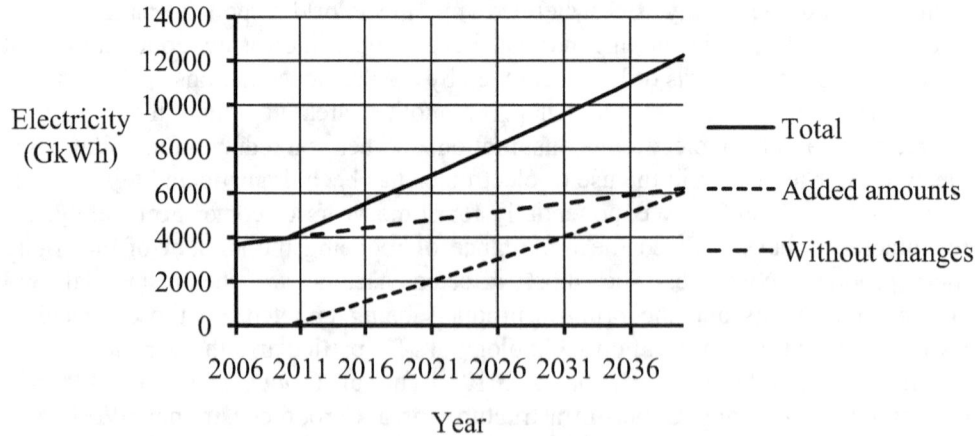

Fig. 9-32 Additional electricity needed to implement reductions assumed in Fig. 9-31.

- Half of the replacement of fossil fuels used in the residential, commercial and industrial sectors is electricity and the other half is carbon neutral renewables.
- Only 60 percent as much overall energy is needed to power electric cars as cars with internal combustion engines (this includes the energy required to generate the electricity using the 2008 energy sources).

The projected required electricity is 3.34 times the 2006 demand and twice the amount that would be needed without the changes corresponding to Fig. 9-31.

Now let us determine how the additional electricity demand indicated in Fig. 9-32 is to be generated provided that:

- Thirty percent of the total electricity is generated by wind turbines with a total energy output that is 30 percent of their maximum capability (i.e., the output attained if they operate full time at their maximum capacities).
- Twenty percent of the total electricity is generated by solar panels with a total energy output that is 30 percent of their maximum capability.
- Ten percent of the total electricity is to be generated by hydroelectric, geothermal and carbon neutral biomass sources.
- The 2010 natural gas, petroleum, nuclear and renewable generation facilities are retained with a total energy output that is 45 percent of their maximum capability.
- The remainder of the demand is met by additional nuclear facilities with a total energy output that is 90 percent of their maximum capability (nuclear facilities are the last to reduce their output when demand is low—they averaged operating at 90 percent of capacity in 2007).
- Except for 20 percent of the additional solar generation capacity, which is considered to be onsite, nine percent is added to the electrical energy that must be delivered to the user to account for distribution losses.
- The added capacities are to begin January 1, 2011 and be added in equal annual increments.

The capacities for the various types of generation facilities are graphed in Fig. 9-33. The "Added other" category includes hydroelectric, geothermal and other carbon neutral renewable facilities. "Other" indicates the existing capacity other than the coal-fired capacity. The added wind capacity could be met with 505,000 three megawatt wind turbines. This would require the addition of 16,800 turbines per year. If 33.3 acres per megawatt is required (the amount assumed in Chap. 6), then these turbines would be spread over 204,000 km^2 (78,800 mi^2), an area roughly 12 percent larger than North Dakota. If 95 percent of the land is still usable, then the actual footprint of the wind facilities would be 10,200 km^2 (3,940 mi^2). The solar panels would require 17,400 km^2 (6,700 mi^2), an area the size of six percent of

Arizona. Twenty percent or more of the solar panels may be onsite, presumably on rooftops. The nuclear facilities would require 481 nuclear reactors, each capable of generating 1,000 MW, or 128 nuclear generating plants the size of the Palo Verde facilities in Arizona, currently the largest nuclear facility in the United States. The nuclear generating capacity in the United States would be 5.3 times what it is today.

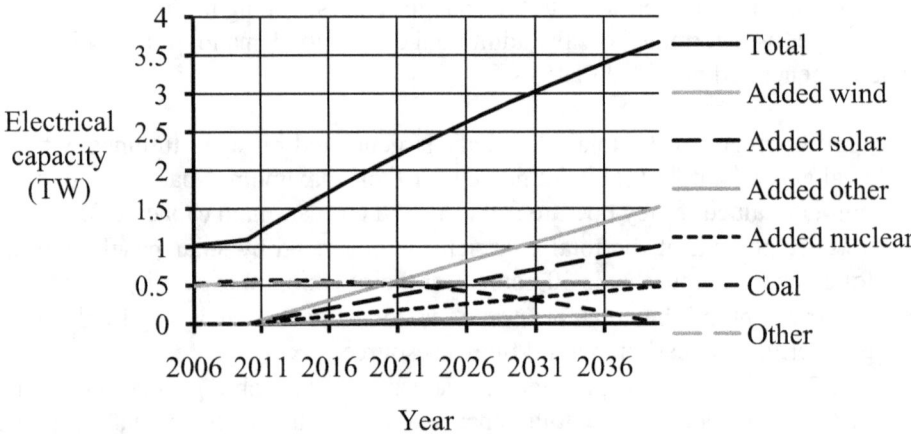

Fig. 9-33 Required electrical capacities corresponding to the demands in Fig. 9-32.

If one half of the coal-fired plants were retrofitted with sequestration equipment instead of being eliminated and are used to reduce the number of additional nuclear plants, only 197 1,000MW reactors (53 plants the size of Palo Verde) would be needed by 2040.

The problem of storing large amounts of electromagnetic energy in an electromagnetic form or another form that can be efficiently returned to electricity is partially solved by a unique marriage of wind energy and electric vehicles. Wind energy, unlike solar energy, is available at night when the batteries in most electric vehicles would be charged. The prospect of having a large percentage of vehicle batteries available for storing electrical energy at night would make a significant difference in our electrical storage capacity. Some studies have indicated that electric cars consume between 0.15 and 0.25 kWh/km (0.24 and 0.4 kWh/mi) depending on size, weight and body design. The April, 2011, issue of *Consumer Reports* states that in its test the Nissan Leaf consumed 0.22 kWh/km (0.33 kWh/mi). Using the EPA's formula, this is equivalent to 101 mpg. The Volt, a bigger car that can run as a hybrid, tested at 0.31 kWh/km (0.5 kWh/mi) with an EPA rating of 65 mpg. Suppose that:

- On average cars consume 0.2 kWh/km (0.32 kWh/mi) and travel 80 kilometers (50 mi) per day.
- One hundred million cars in the United States (roughly 70 percent) are electric or plug-in hybrids and charge their batteries at night.

Then each electric car would need to store 16 kWh each day and all of them would store 1.6 GkWh/day. For comparison, in 2008 the average amount of electricity consumed per day was 10.2 GkWh. Therefore, roughly one sixth of the amount of electricity consumed per day could be stored in car batteries at night. Because, in the past, electricity consumption at night has been much less than that during the daytime, using car batteries for storage would tend to equalize consumption and reduce the overall required electrical capacity as well as provide storage.

Figure 9-34 illustrates the 2040 energy-related emissions by their sources and the sectors that emit them assuming the scenarios discussed above. The renewable and nuclear energy sources are responsible for some emissions, but their contributions are very small, less than one half of one percent. Note that the industrial sector would emit almost as much as the other three sectors combined. In

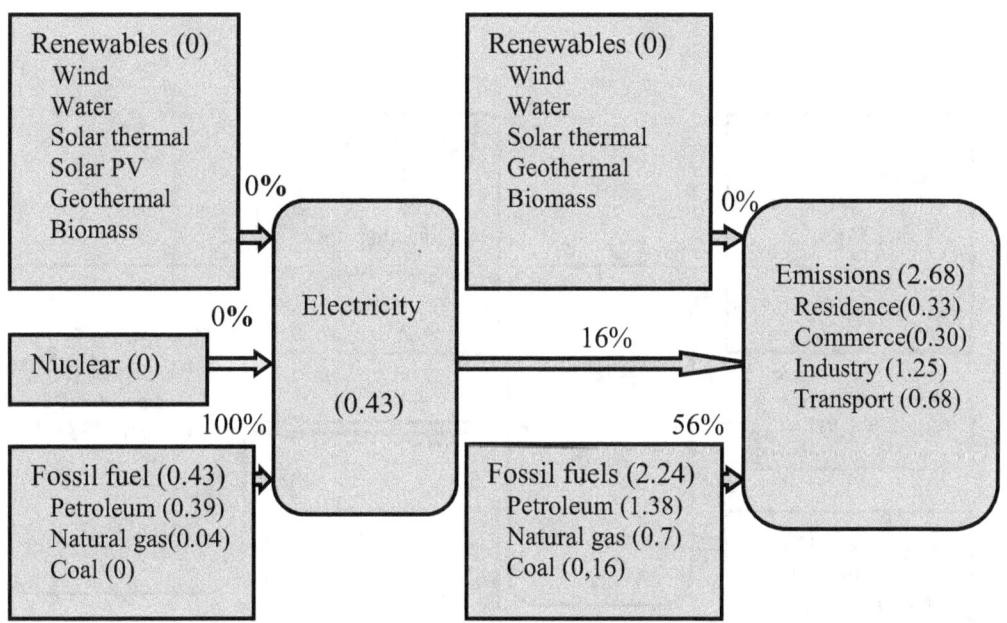

Fig. 9-34 2040 emissions contributions in Gt assuming the above scenarios.

contrast, the 2006 emissions were 1.26 Gt (20 percent) for the residential sector, 1.13 Gt (18 percent) for the commercial sector, 2.12 Gt (33 percent) for the industrial sector and 1.81 Gt (29 percent) for the transportation sector. Of the total 6.32 Gt energy-related emissions in 2006, 2.46 Gt (39 percent) originated with the generation of electricity and coal was a major cause of atmospheric GHGs. In 2040 electricity would be responsible for only 16 percent despite the considerable in-crease in electrical energy. The reduction would be primarily due to the elimination of coal-fired plants and the heavy reliance on renewable sources.

The energy consumption by sources and sectors corresponding to the emissions in Fig. 9-34 are depicted in Fig. 9-35. In contrast with the 2007 amounts given in Fig. 8-38, electricity usage would rise from 40 percent to 54 percent and usage of fossil fuels would drop from 85 percent to 32. The total energy consumption would remain about the same even though the estimated population increase is 41 percent. There are two reasons for the per capita decrease in energy consumption. One is that the energy inputs to wind, solar and hydro-facilities that were assumed to replace fossil fuel facilities were considered to be zero. The other is that the overall efficiency of electric cars, including the energy needed to generate the required electricity, is more than one and a half times that of internal combustion engines.

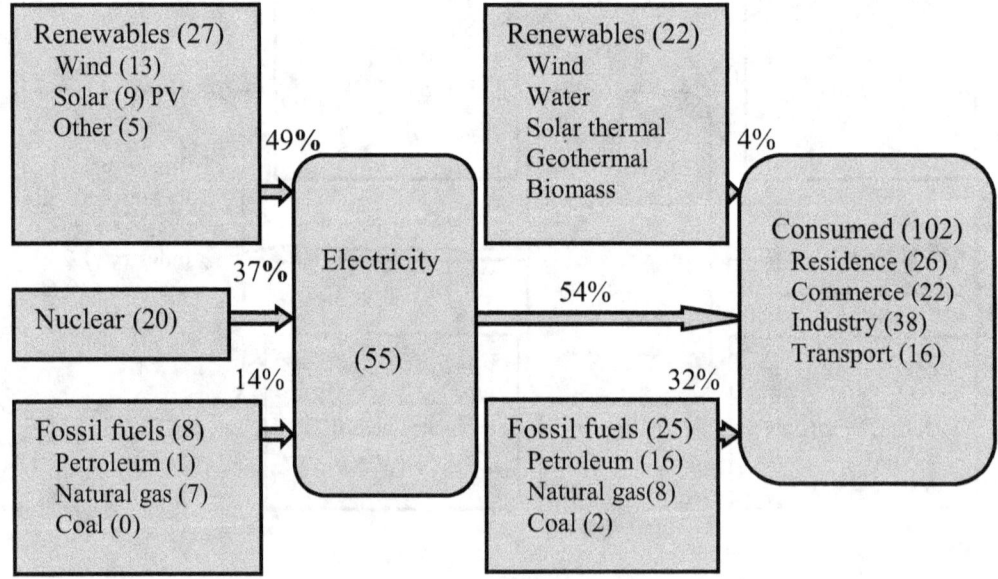

Fig. 9-35 2040 energy usage in PBtu assuming the above scenarios.

There is some question as to whether or not these scenarios are reasonable. Is it plausible to assume that by 2040:

- Seventy-five percent of the gasoline powered vehicles will be electrically powered?
- All coal-fired plants will be eliminated, or at least be replaced by plants that have sequestering facilities?
- Eighty percent of the heat needed by the residential and commercial sectors will be supplied by electricity and carbon neutral renewables?
- Fifty percent of the heat needed by the industrial sector will be supplied by electricity and carbon neutral renewables?
- The required wind, solar, other carbon neutral renewable and nuclear facilities will be constructed?

It is possible to technologically fulfill these proposals, but it would require enormous and expensive changes in our energy-related infrastructure and a drastic shift in jobs. Not only would there be a shift from petroleum powered to electric powered cars and light trucks, but a new means of refueling cars and light trucks would be necessary. Not only would new equipment and generating plants need to be constructed, but means of storing greater amounts of energy and distributing electricity would be required. It should be noted that these scenarios are just simplified examples that have not included other possibilities such as using biological sequestration, CHP, using more biofuels for powering cars and light trucks and using more efficient lighting and insulation.

In the context of the above examples, most of these alternatives may tend more toward alleviating the burden placed on electricity than on further reducing emissions. However, lessening the necessary infrastructure changes would certainly make these scenarios more appealing. Although the proposed changes may not be realized, other possibilities could still make a 50 percent reduction in GHG emissions by 2040 a realistic goal. Also not considered are conservation measures such as simply traveling less, turning thermostats down in winter and up in summer and turning off electrical devices when they are not in use.

Using biological sequestration may be a particularly important addition to the above scenarios. The Environmental Protection Agency (EPA) estimates the net average absorption of a pine plantation is roughly 250 tonnes of carbon (920 tonnes of carbon dioxide) per square kilometer per year (2,400 tonnes of carbon dioxide per square mile per year). Pine plantations equaling half the area of Montana could absorb 0.18 Gt per year. But trees absorb very little carbon dioxide right after they are planted and their absorption tapers off as they mature. It would take decades for trees planted now to reach their maximum growth. There is also the question of what to do with the trees when they are harvested. Presumably they would be used for lumber, which releases its carbon very slowly, or for fuel in a carbon neutral manner. If it is used for fuel it could reduce the need for electricity or fossil fuels or provide some of the energy for generating electricity. The EPA estimates another 25 to 75 tonnes of carbon, 92 to 275 tonnes of carbon dioxide, could be sequestered per square kilometer per year (238 to 712 tonnes per square mile) by using better tillage methods.

On the other hand, the IPCC claims that the deforestation of tropical forests could account for as much as 20 percent of the increase in atmospheric carbon dioxide. Worldwide the IPCC estimates that between 10 and 20 percent of the world's fossil fuel emissions could be offset by forest preservation, tree planting and improved agricultural methods.

The above examples are presented primarily to demonstrate the difficulties involved in reducing GHG emissions. They were chosen because they are reasonable, but there are other possibilities that are just as reasonable. Certainly, the percentages of energy sources, efficiencies and so on could be shifted around and still be reasonable. Although the proposed changes would not be easy, they could be made with minimum damage to the economy. Considering the current damage to our environment, such changes may have an overall positive effect on our economy even if the possibility of global warming is not taken in account.

President Obama in 2009 proposed a 17 percent reduction from the 2005 emissions by the year 2020. Under the scenario indicated in Fig. 9-31, the reduction would be 10.8 percent by 2020. Achieving a 17 percent reduction within the next ten years would be very difficult. Perhaps such a reduction could be reached by using sequestration in addition to the changes proposed above. The more time we leave ourselves to replace the current infrastructure, the more plausible the often proposed 80 percent reduction in emissions by 2050 becomes. The completion of a complex project typically follows a time curve such as the one given in Fig. 9-36. If this curve is followed, then an 80 percent reduction by 2050 could be achieved by a scenario similar to the one corresponding to Fig. 9-31. In the beginning, the change is slow because of planning and building the necessary manufacturing facilities, but becomes faster with experience and as the required manufacturing capability is completed. Change slows as completion is approached because there is often difficulty in finding the means to approach the final goal or there is a lack of urgency. The curve in Fig. 9-36 is a projection of emissions reduction assuming there is a 58.5 percent reduction by 2040, the percentage reduction reached by the above scenario. If the emissions reduction follows this curve, then there would be a 13.5 percent reduction by 2020, an 80 percent reduction by 2050 and a 92.5 percent reduction by 2100.

It should be noted that an alternative to reducing GHGs to combat global warming is to increase the reflectivity of the atmosphere. Particulates cause more of the sun's radiation to be reflected back into space before it reaches the Earth's surface, thereby cooling the Earth's surface. Some consideration has been given to intentionally adding particulate matter to the atmosphere to counteract the effects of global warming. One proposal is to shield the Earth from some of the sun's radiation by creating a cloud of reflective chaff between the Earth and sun. However, the possible outcomes of such actions are not understood and could be much worse than the current global warming. Attempts to negate the effects of global warming by human intervention are referred to as *geoengineering*. In any case, because geoengineering projects would be difficult or impossible to reverse if they did not work as predicted, they are not considered further here.

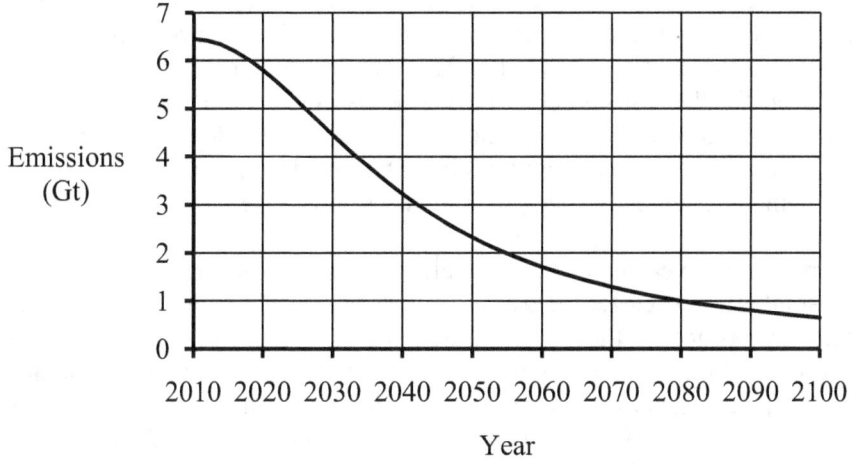

Fig. 9-36 Representative curve for completion of large projects such as reducing emissions.

The anticipated consequences of global warming are principally sea level rise and erratic weather behavior. Based on six different computer models, the IPCC predicts a sea level rise of between 0.2 and 0.6 m (8 in and 24 in) by 2100, but some scientists believe that the sea level rise could be much greater. Most of the water would come from the melting of the ice sheets covering Greenland and the Antarctic and the expansion of seawater as the temperature increases. Although the melting of glaciers and ice caps are an important indicator of global warming, they would contribute little to an increase in sea level because they contain relatively small amounts of water. Since 1850 the oceans have risen about 1.5 millimeters (0.06 inches) per year. In recent years this rate has increased to 3.1 millimeters (0.13 inches) per year. A 0.6 m (2 ft) rise would not severely affect the United States, but would have a profound effect on island nations such as Kiribati and low-lying nations such as Bangladesh. A one meter (3.3 ft) rise would have serious consequences for the outer banks of North Carolina, the Everglades of Florida and the Mississippi delta and would devastate island nations. Few expect an increase of more than one meter. However, if all of Greenland's ice sheet were to melt the sea level increase would be about six meters (20 feet). If all of the ice locked in the ice sheets of both Greenland and Antarctica were to melt the sea level would rise almost 68 meters (210 feet). Fortunately, no one is predicting such an occurrence, but prolonged global warming could cause the sea level to eventually rise by six meters. A six meter rise would flood much of the United States southern and eastern coastlines, submerge much of Florida, southern Louisiana and Manhattan and essentially eliminate some low-lying countries.

Large swings in sea level are not uncommon. In the distant past, there have been very great changes in the world's sea level. Since the beginning of the present interglacial period, the sea level has risen 120 m (390 ft). There have been five such extreme changes in the last 450,000 years. However, ice volume is near the lowest point it has been for the last 450,000 years. Because the present amount of ice is relatively small, an additional sea level increase of the magnitude of those that have occurred following the ice ages is not possible. But this does not mean that it is not possible for human activities to precipitate a catastrophic rise in sea level of six or more meters.

Unpredictable weather patterns would increase the severity of hurricanes, making the southern and eastern coastlines much more vulnerable to devastation, turn a significant portion of the United States into desert and cause serious water shortages in some areas. No one knows just how severe weather conditions might be or where the weather changes might be the most dramatic, but some fear that Great Plains farming areas could face the dust bowl conditions of the 1930s. Some are even more pessimistic and believe the southeastern United States could become a desert.

The loss of the world's vegetation would exacerbate the problem by resulting in less carbon dioxide being absorbed and the GHGs accumulating at an even faster rate. Because the solubility of water decreases as temperature increases, an increase in temperature causes carbon dioxide to be released from the oceans where most of the carbon dioxide is stored. Higher temperatures also would cause methane to be released from the permafrost of northern Canada, Alaska and Siberia. The addition of carbon dioxide and methane from their natural reservoirs would set off a spiraling positive feedback effect in which an increase in GHGs would cause an increase in temperature which would cause a further increase in GHGs and so on. Such a dire scenario may continue until the stored carbon dioxide and methane is depleted to the point at which there is once again a balance between GHG release and absorption.

Almost all scientists agree that the GHGs released by human activities have some effect on global warming, but there is considerable disagreement on the severity of the effect. There are just too many variables to accurately predict the consequences of anthropogenic-generated GHGs. Some of the complex relationships affecting our atmosphere are well understood, but others are not. For example, the frequencies absorbed and emitted by specific gas molecules are known accurately, but the global effect of the release of a specific quantity of a gas is not well known. As time goes by and more data is collected, predictions will become more reliable. But can we afford to wait before changing our habits? There is a point at which the greenhouse effect may become irreversible. An often quoted point of no return is an increase of 2 °C (3.8 °F) over that of pre-industrial revolution times. The current global average temperature indicates an increase of about 0.8 °C (1.44 °F) has already occurred. Some believe that the carbon dioxide already in our atmosphere will eventually cause an increase of 2.5 °C (4.5 °F). Even if the chance of irreversible damage happening were low, do we really want to take a chance on it not occurring?

It is becoming increasingly apparent that global warming is going to cause significant and expensive damage to our environment and we are at least partly responsible. Although there are some technological problems that need to be overcome, alleviating the human contribution to global warming is basically a matter of will, economics and politics. To make the necessary changes may require some sacrifice, but we would have the benefit of a cleaner, more healthful environment in addition to slowing global warming. In the end, the benefits may significantly outweigh the sacrifices and the world may become more prosperous than ever before.

In addition to the cited references given in this chapter, it should be noted that much of the general information on global warming was obtained from two series of lectures made available by The Teaching Company (*www.teach12.com*). One was entitled "Earth's Changing Climate" and was presented by Professor Richard Wolfson of Middlebury College. Professor Wolfson has also written the book *Energy, Environment and Climate* published by W. W. Norton & Company. The other was entitled "How the Earth Works" and was presented by Professor Michael E. Wysession of the University of St. Louis. For further information, Professor Wysession recommends the National Aeronautics and Space Administration (NASA) website, *www.nasa,gov*, the National Oceanic and Atmospheric Administration (NOAA) website, *www.noaa.gov*, and the United States Geological Survey (USGS) website, *www.usgs.gov*. Also, additional figures concerning global warming and climate change may be viewed at the website *www.globalwarmingart.com*.

10

POLICY

It is not the intent of this chapter to examine the complete body of the United States energy legislation in detail. Such an in-depth examination would require volumes, not a brief chapter. It is the intent to outline a sufficient amount of this legislation to reflect the changing energy policy of the United States over the last century and relate this policy to events both inside and outside the United States. In discussing the legislative acts that have been enacted, the amounts given are to indicate relative sizes only. The exact amounts, including cost amounts, are typically not given in the legislation, but are only maxima or ranges of amounts and the actual amounts are left to agencies that are to administer the programs named in the legislation.

Policy is an overall plan for meeting intended goals using acceptable procedures. Our interest is in the policies of one or more governments or organizations with regard to energy and its side effects. There are four international organizations that have important impacts on the production and consumption of energy. They are the United Nations (UN) and the three organizations introduced in Chap. 2, OPEC, OAPEC, and OECD. The UN is a large organization that includes all of the nations of the world and has policies related to many concerns of global interest. Currently, its primary concern related to energy is global warming. Its policy is to reduce GHGs in an equitable way so that the developed nations are not unduly harmed economically and the developing and underdeveloped nations can proceed to raise the standards of living of their citizens. As mentioned in Chap. 2, the policy of OPEC is to stabilize the price of oil and assure its members a reasonable price for their production. The policy of OAPEC is to promote the common cultural and Islamic traditions of its members that go beyond the exportation of oil, but oil is one of its primary concerns. The membership of OCED consists of the developed nations and its policy is to promote democracy and a global market economy. More specifically, it supports sustainable economic growth, increased employment, higher living standards, economic stability, economic development and the growth of world trade. Although energy is not mentioned, energy is a key element in achieving its goals.

In attempting to reach its energy-related environmental goals, the UN has convened a series of conferences. The first of these conferences was the *United Nations Conference on the Human Environment* held in Stockholm in 1972. It was the first UN conference that considered the global effects of pollution and it established the United Nations Environmental Program (UNEP). Many environmental conferences were to follow. They included the *1979 Geneva Convention on Long Range Transboundary Air Pollution*, the *1985 Helsinki Agreement* by 21 nations to reduce sulfur dioxide emissions, the *1988 Montreal Protocol on Substances that Deplete the Ozone Layer* and the *1989 Basel Convention on Transboundary Movements of Hazardous Wastes*. In 1983 the UN created the World Commission on Environment and Development, also known as the Brundtlund Commission, and gave it the task of studying the relationship between the effects of energy consumption and economic development. In 1987 this commission published its report entitled *Our Common Future*.

But it was the 1992 *United Nations Conference on Environment and Development (UNCED)* in Rio de Janeiro that (after a number of scientists began to associate the rapid rise in the amount of carbon dioxide in the world's atmosphere with the world's temperature rise) began to focus on the relationship between GHGs and global warming. This conference, which was on the 20th anniversary of the Stockholm conference, became known as the *1992 Earth Summit*. It set in motion the idea of binding agreements between the world's nations that would reduce the amounts of GHGs in the atmosphere. Representatives from 178 nations and other interested organizations attended the conference. It brought about the non-binding, non-ratified agreements by several developed as well as other nations on the amounts they were willing to reduce their carbon dioxide emissions. The non-binding environmental treaty produced by the UNCED is known as the *United Nations Framework Convention for Climate Change (UNFCCC)*. Beginning in 1995, the members have met annually in *Conferences of the Parties (COPs)* to assess the progress made toward realizing their proposed goals and consider other related global environmental problems.

To give the UNFCCC a legally binding standing, the nations met again in Kyoto, Japan, in 1997 and produced the *Kyoto Protocol*. Under this protocol, 37 nations agreed to reductions in the four major GHGs, carbon dioxide, methane, nitrous oxide and sulfur hexafluoride and the sets of gases called hydrofluorocarbons and perfluorocarbons. Because carbon dioxide is the dominate GHG being put into the atmosphere by human activity, the emphasis has been on carbon dioxide. There are now 194 members of the UNFCCC, 193 nations plus the European Union as a whole. The members were divided into two main categories, Annex I consisting of the industrialized nations and the group called Developing Nations, and one subcategory of Annex I known as Annex II. Some nations may declare themselves to have *economies in transition*, which means they are in the process of going from centrally controlled economies to free market economies. The nations with economies in transition are primarily the former communist nations. The Annex II nations are the Annex I nations that do not have economies in transition. The memberships in the categories are not fixed. A nation may join or be removed from the list of economies in transition. Initially there were 40 nations in Annex I and 23 in Annex II. Annex II nations are expected to take on the burden of helping the Developing Nations defray the costs associated with reducing GHG emissions.

Being a member of the UNFCCC and signing the protocol is not the same as ratifying the legally binding agreements. For a nation's agreements to become binding its appropriate political bodies must ratify them. Most notably, the United States has not ratified its commitments for reductions and the Bush administration stated in 2001 that it would never sign the Kyoto Protocol. China has not agreed to any goals with regard to its emissions. Also, for any of the agreements to be binding, it was decided that at least 55 nations must have ratified the treaty and at least 55 percent of the GHG emissions must be due to activities in the nations that have ratified the treaty. These requirements were satisfied in December, 2005. It was decided that 1990 would be the base year and all commitments would be made relative to the amounts of pollutants the various nations emitted during that year. The 1990 emissions, the nations' committed emission amounts by 2012 and the 2006 emissions for some of the signatories are given in Fig. 10-1. Neither China nor India has made any commitments and both had more than doubled their carbon dioxide emissions by 2006. The United States has not ratified its agreements and its

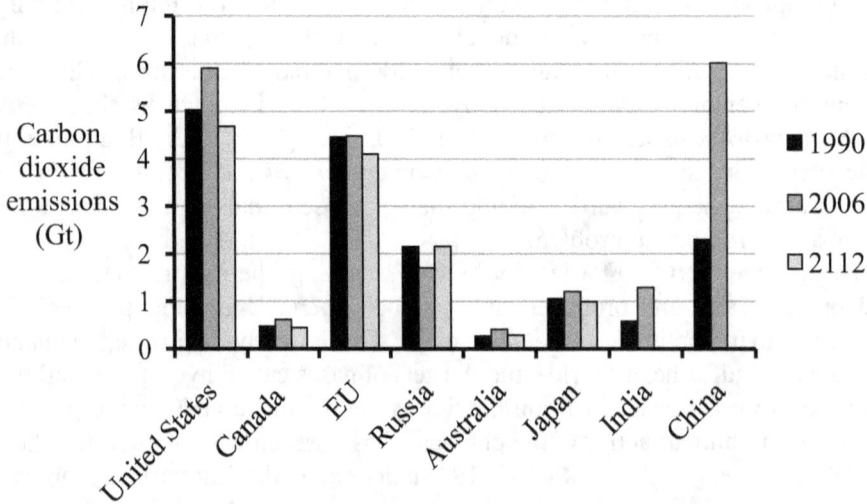

Fig. 10-1 Carbon dioxide emissions for some of the signatories to the Kyoto protocol [EIA].

emissions increased 17 percent between 1990 and 2006, although its 2009 emissions were only 12 percent above those of 1990. The severe recession that occurred during 2008 and 2009 contributed to keeping emission amounts from being higher. Between 1990 and 2006 the world's carbon dioxide emissions increased by 35 percent, from 21.9 Gt to 29.2 Gt. The EIA predicted in 2008 that the world's emissions in 2030 would be double those of 1990.

The principal means for limiting atmospheric pollution has been what is known as "cap and trade" systems. A *cap and trade system* is one consisting of a specified group of entities (e.g., countries, corporations or individuals), a group of pollutants, a credit value for each pollutant, a credit value limit for each pollutant set on each entity in the specified group and rules for trading, or perhaps buying, credit values. In addition, a cap and trade system may include provisions for exchanges outside the specified group. Rationing is a form of cap and trade if individuals are allowed to trade or buy rationing coupons. However, large scale cap and trade systems for countries or corporations were not considered until the late 1960s and the first law to include a restricted form of a cap and trade system was the United States Clean Air Act of 1990 that placed corporate limits on sulfur and nitrogen oxides. Although there has been considerable discussion of a cap and trade law for carbon emissions in the United States, as of 2010 no such law has been forthcoming.

The most contentious aspect of the Kyoto Protocol is the insertion of the "flexible mechanisms," that provide loopholes for the nations that do not meet their commitments. The flexible mechanisms are referred to as emissions trading, joint implementation and the clean development mechanism. They fall into one of two categories, one in which the credit exchanges are within the specified group and one in which the exchanges may be with entities outside the specified group. *Emissions trading* allows nations to buy GHG emissions permits (credits) from other nations that do not need them in order to cancel out their excess emissions (credits). *Joint implementation* permits an Annex I nation to reduce its obligation by investing in emissions reduction projects (thus gaining credits by reducing emissions) in another Annex I nation. The third is the *clean development mechanism* (*CDM*), which allows an Annex I nation to reduce its obligations by buying GHG credits from developing nations by helping to finance emission reduction projects. The argument against these mechanisms is that, although they do tend to reduce future increases, they do not guarantee the overall reduction of GHG emissions. They tend to let the developed nations continue with business as usual. However, the limits may be decreased in the future, thereby reducing the overall pollution. There is also a means by which organizations or individuals can buy credits with the intent of not using them, thus decreasing the number of credits available to polluters.

All nations that ratified their agreements are required to submit annual reports to be judged by the other nations. The IPCC is the main agency for setting the guidelines for these reports. The reports are designed to reflect the total emissions less the total sequestration of the reporting nations. They must include data on Land Use, Land Use Change and Forestation (LULUCF) as well as emissions data related to energy consumption. The LULUCF portions of the reports show overall emissions reductions through reforestation and better land use practices as well as net emissions increases due to deforestation and other decreases in biomass absorption. LULUCF considers six types of land: forest land, cropland, grassland, wetlands, settlements and other. The "other" category is to account for the difference between the total area of a nation and the sum of the areas of the other five categories. The changes from one type to another indicate the changes in sequestration and emissions. Although land use is important, except for large changes in the Earth's forest areas these changes contribute only a small amount to changes in GHG emissions. LULUCF projects may be used as CDM projects by the Annex I nations. The main problem with the LULUCF reports lies in the inaccuracies in estimating the absorption rates and amounts of the different types of vegetation.

The structuring of cap and trade laws is complicated and enforcement is very difficult. The limits placed on the various entities must first be decided and then the emissions must be closely monitored and penalties in terms of credits must be assessed. Setting limits is particularly difficult between countries that have diverse climates, economic conditions and backgrounds. An alternative to cap and trade would be to simply tax the use of fossil fuels and give credits for sequestration. Such taxes would serve to equalize the costs of using fossil fuels with those of using the more expensive renewables and could be used to subsidize the necessary new infrastructure.

In response to the Kyoto Protocol, the Clinton administration in 1998 requested the EIA to produce a report predicting the total amounts and sources of our energy if the United States were to emit seven percent less GHGs than in 1990. The principal results of this report are summarized in Fig. 10.2. This figure indicates a total energy consumption of 91.7 TBtu with petroleum, nuclear and renewables providing approximately the same amount of energy as they do today, natural gas providing much more and coal providing much less. As shown in Fig. 1-6, in 2008 the total energy consumed by the United States was 99,160 TBtu with 23 percent of the total originating from coal, 24 percent from natural gas and 37 percent from petroleum. Clearly, the report indicates that the likely emissions reduction will occur by shifting from the fossil fuel coal to the less polluting fossil fuels petroleum and natural gas. This, in fact, seems to be the current tendency because it is easiest to achieve. But, if we are to seriously attack the problem of air pollution, the shift must be toward increased usage of nuclear and renewable energy.

The *2009 United Nations Climate Change Conference*, also known as the *Copenhagen Summit*, was held in December, 2009. At the conference the United States stated its intention to agree to a 17 percent decrease from its 2005 emissions level (a four percent decrease from its 1990 level) by 2020 if a binding agreement could be reached. It also stated it would reduce its emissions to 42 percent of their 2005 level by 2030 and 83 percent by 2050. Again no binding agreement was in the offering because of the conflict between the Annex II nations and the developing

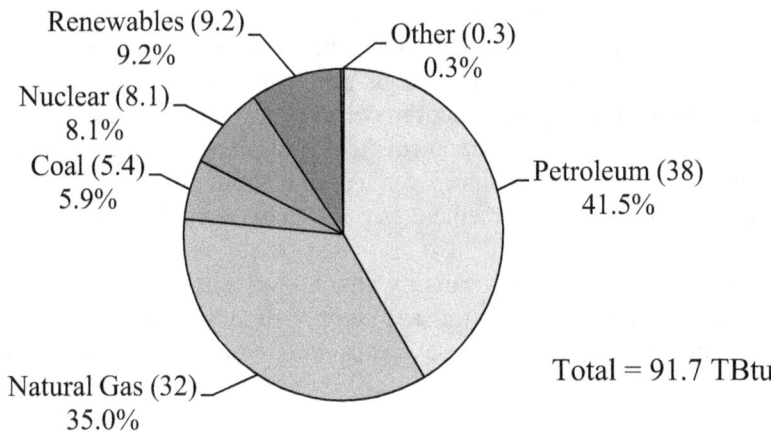

Fig. 10-2 Source percentages and amounts predicted for a seven percent reduction [EIA].

nations, primarily China and India. Many of the attendees considered the conference a failure because of its lack of concrete solutions to the problem of global warming despite the mounting evidence that it is in part anthropogenic. The annual COPs will continue, however, and there is a core of Annex I nations that have binding agreements. While the Obama administration supports GHG reductions, it has not formalized a plan on achieving them nor has it, as of 2011, asked the Senate to ratify the Kyoto Protocol. The body of work done under the UNFCCC has been enormous and those involved in the COPs have created a mountain of paperwork. It is hoped that definite progress will soon be forthcoming. Much of the future progress depends on the two greatest emitters, the United States and China. Progress hinges on allowing outside countries to inspect a country's books on its measuring, reporting and verifying (MRV) methodologies as well as its data.

In addition to governments, there have been several organizations participating in the global warming debates. Many of them are lobbying groups such as Greenpeace and the Sierra Club. But one organization worth special mention is the London based Carbon Disclosure Project (CDP). The CDP works with the world's largest corporations and their major investors to collect data on corporate GHG emissions. In 2008 it published emissions data for 1,550 of the largest corporations. Corporations, especially global corporations, have begun to take a serious interest in their emissions because of the binding commitments of many industrialized nations, possible future commitments by the United States and increased involvement of developing nations. The CDP is believed to have the largest collection of corporate emissions data in the world. Information on the CDP and their reports may be found at www.cdproject.net.

Although policy may provide guidance toward a goal, to have a meaningful effect it must be backed up by laws and enforcement. As is apparent from the above discussion, reaching a legal agreement among numerous political, economic and cultural entities is difficult and enforcement may be even more difficult. Nations are reluctant to follow through on their agreements if it is to put them at a disadvantage politically or cause their people hardship. Even within a nation, it is difficult to get a policy translated into law and enforcement may be uneven as the political winds change.

Table 10-1 summarizes the means by which the United States government can control the production, consumption and supply of our energy sources. The federal government has used all of these means in one form or another. Some of

Table 10-1   Government means of controlling the production,
consumption and supply of energy sources.

Subsidies
       Direct payments
       Research assistance
       Tax relief
              Tax credits
              Tax deductions, depletion allowance
Limiting corporate liability
Lease or sale of public lands and offshore areas
Licensing of production methods
Penalties
       Taxes
              Corporate taxes
              Retail taxes
       Fines
              Consumption
              Production
              Emissions
       Royalty payments
Restrictions
       Consumption limitations (rationing)
       Production limitations
       Emissions limitations
Emergency storage

---

these means, such as direct payments, research sponsorship and emergency storage, tend to be for short term adjustments while those that involve taxes, leasing, liability limits and licensing tend to extend over long periods of time. Restrictions may be short term, such as rationing to rectify a temporary situation, or long term such as placing caps on emissions. In the past the most effective laws have been those concerning taxes, fines, liability limits and leasing, but in the near future direct payments and research assistance may be needed in addition to tax manipulation and fines to redirect our consumption away from fossil fuels. Limiting corporate liability is primarily related to the nuclear power industry, and emergency storage buildup or release is essentially the buildup or release from the SPR discussed in Chap. 2.

The history of the United States energy legislation has extended over almost a century. Until the 1970s, the United States legislation has favored the expansion of energy usage. During the early part of the 20th century as the use of motorized vehicles grew, legislation encouraged the production of petroleum and the use of electricity. World War I, the Great Depression and World War II required the United States to consume more and more energy and that required an increase in production. The relationship between economic growth and energy became increasingly obvious and the growing demand for energy has extended through to the present. However, Hubbert warned of the diminishing production of petroleum in 1956 and by 1970 Hubbert's predictions began to become true. During the 1970s, Americans became painfully aware of their dependence on petroleum imports and their own shrinking production. The emphasis in the 1970s was on the conservation of petroleum. During this period laws were passed to restrict petroleum consumption and, as acid rain became an issue, laws were passed to limit the emission of sulfur and nitrogen oxides. By 1980 scientists began to be concerned about the global effects associated with the consumption of fossil fuels and the legislation encouraging the production of fossil fuels began to be moderated. Beginning in the early 1990s, global warming became the focal point of worldwide concern. Within the United States, global warming has been debated within several state governments as well as in the executive administrations and halls of Congress, but serious enforceable legislation was not enacted until 2007.

Table 10-2 lists the more prominent energy laws that have been passed by the United States federal government over the past century. Continuous corporate income taxation began with the Corporate Franchise Act in 1909, which allowed corporations to be taxed on one percent of their net income. This allowed taxes to be used to reward or penalize corporate activities. The law was clarified in 1913 and depletion allowance first became part of the law with the passage of the Tariff Act of 1913. *Depletion allowance* is a tax deduction for the mining of minerals, including petroleum and natural gas. It has been among the most contentious and debated features of the energy-related laws ever since. The Tariff Act of 1913 permitted up to five percent of the value of output to be deducted from a mining corporation's

Table 10-2 United States important energy laws.

Year	Act
1909	Corporate Franchise Act
1913	Tariff Act of 1913
1916	Revenue Act of 1916
1916	Stock Raising Homestead Act
1918	Revenue Act of 1918
1920	Federal Power Act
1920	Revenue Act of 1920
1924	Revenue Act of 1924
1926	Revenue Act of 1926
1932	Revenue Act of 1932
1935	Public Utility Holding Company Act of 1935
1936	Rural Electrification Act
1938	Natural Gas Act
1944	Flood Control Act of 1944
1944	Synthetic Liquid Fuels Program
1946	Atomic Energy Act of 1946
1954	Atomic Energy Act of 1954
1957	Price-Anderson Nuclear Industries Indemnity Act
1966	Uniform Time Act
1969	Revenue Act of 1969
1970	Clean Air Act of 1970
1973	Trans-Alaskan Pipeline Authorization Act
1974	Energy Reorganization Act of 1974
1975	Energy Policy and Conservation Act
1975	Tax Reduction Act of 1975
1977	Department of Energy Reorganization Act of 1977
1978	National Energy Act of 1978
1980	Energy Security Act
1982	Nuclear Waste Policy Act
1987	National Appliance Energy Conservation Act
1989	Natural Gas Wellhead Decontrol Act of 1989
1990	Clean Air Act of 1990
1992	Energy Policy Act of 1992
2000	Commodity Exchange Act
2005	Energy Policy Act of 2005
2007	Energy Independence and Security Act of 2007
2008	Food, Conservation, and Energy Act of 2008
2009	American Recovery and Reinvestment Act of 2009

taxes. But, the total deductions were limited to 33 percent of the original capital investment. In the Revenue Act of 1916 the law was changed to permit deductions not to exceed the larger of the capital invested in the property or the fair market value of the property on March 1, 1913. The depletion allowance was modified again in the Revenue Act of 1918 to be based on discovery value instead of discovery cost. It is this change that gave corporations the right to make large continuous deductions that has been the most controversial aspect of depletion allowances. The Revenue Act of 1920 limited the total reductions to the total net income of the property and the Revenue Act of 1924 set the limit to 50 percent of net income. The discovery value limitation was replaced by a percentage of gross income from the property by the Revenue Act of 1926, and for petroleum and natural gas producers this percentage was set at 27.5. The 50 percent of net income limitation remained.

In the Revenue Act of 1969, the 27.5 percentage was reduced to 22 percent, but the deduction also became subject to an alternative minimum tax (AMT). In addition to the percentage method of computing taxes, corporations have been allowed to use a cost basis computation. A cost computation consists of dividing the investment amount by total amount of mineral to be extracted multiplied by the amount of the mineral sold during the year for which the deduction applies. The Tax Reduction Act of 1975 no longer allowed large integrated corporations to use percentage deductions. They were required to use cost deductions, thereby limiting their total deductions to their actual costs. Until 1984, small companies and individuals that invested in marginal properties could still use either method, but the percentage was lowered to 15. The Energy Policy Act of 1992 repealed the AMT and for marginal properties other small changes were made in 1990 and 2001. The Government Accounting Office (GAO) has estimated that between 1968 and 2000 the total lost revenue due to the depletion allowance was approximately $82 billion in year 2000 dollars.

One of the oldest histories of legislation related to energy is the one associated with the leasing of federal property to private companies for the extraction of oil, natural gas and uranium. Because most of this property falls under the jurisdiction of the Department of the Interior (DOI), it is the DOI that is responsible for overseeing the leasing of almost all federal lands, Indian tribal lands and offshore areas. Offshore areas are the ocean areas extending from the United States shoreline to 320 km (200 mi) into the ocean. In the past essentially all offshore leases were on the continental shelf where the ocean is less the 200 m (660 ft) deep, but deeper areas are currently being explored. For petroleum and natural gas the leases are on the basis of bonuses, rents and royalties submitted during a bidding process. The bonuses, most often called bonus bids, are lump sum amounts given to the government upon a company being awarded a lease, rent is an annual per acre amount for land within the lease and royalties are the amounts collected as the mineral is extracted. The royalties are generally one-eighth of the wellhead value of the mineral extracted for onshore leases and one-sixth this value for offshore leases, although the DOI may demand a higher or lower rate if there are extenuating circumstances. The royalties may be collected in cash or as "royalties in kind" (RIKs), in which case the government simply takes its percentage of the minerals as payment. For a RIK payment the government and the company involved must agree on a place of delivery. RIK payments for petroleum are normally sent to the SPR.

The Minerals Management Service (MMS) is responsible for the collection and distribution of the royalties. Most of the money collected is given to the Department of Treasury, but some is given to Indian tribes and the various states in which the leases lie. The total cash and RIKs paid to the government by the oil and natural gas companies are valued at between six and nine billion dollars annually, about 80 percent of which is royalty payments. Because of the large amounts of money involved, the royalty percentage decisions have always been among the most controversial governmental energy decisions. The oil and gas companies normally press for smaller percentages by pointing out the high risks and costs they are incurring.

Leasing federal lands and setting royalty amounts has not been without corruption. The most famous case occurred in 1921 during the term of President Harding and is known as the Teapot Dome Scandal. By executive order, President Harding shifted control of the naval reserves at Teapot Dome in Wyoming and Elk Hills in California to the DOI. Subsequently, Albert Fall, the then Secretary of the Interior, leased the mineral rights to Teapot Dome to Harry Sinclair and the rights to Elk Hills to Edward Doheny without competitive bidding. At the same time, both Sinclair and Doheny gave Fall large interest free loans. Fall was convicted of accepting bribes and Sinclair was convicted of contempt for tampering with a jury. The mineral rights were later returned to the government.

The importance of the 1916 Stock Raising Homestead Act, which provided settlers with 640 acres of land, is that it was the first act to separate surface rights from subsurface rights. Thus, the owner of surface land does not necessarily own the subsurface mineral rights. The 1920 Federal Power Act created the Federal Power Commission (FPC) and charged it with regulating interstate electric power and natural gas activities and coordinating hydroelectric power projects. Following the Federal Power Act there have been several acts creating hydroelectric facilities in various regions of the United States. The most notable are the Tennessee Valley Authority Act of 1933 that created the Tennessee Valley Authority (TVA), the Bonneville Project Act of 1937 that created the Bonneville Power Administration (BPA) and the Flood Control Act of 1944 that provided for the later creation of the Southeastern Power Administration (SEPA) in 1950 and the Alaska Power Administration (APA) in 1960.

The Public Utility Holding Company Act of 1935 was designed as a trust-busting measure by placing limitations on electric power companies and large holding companies to prevent them from monopolizing the interstate natural gas industry. The 1935 Rural Electrification Act (REA) was to expand electrical service to rural areas by subsidizing the high costs of providing electricity to sparsely populated areas.

Although the Bureau of Mines had experimented with coal liquefaction since 1928, these activities were not given congressional authorization until 1944 when the Synthetic Liquid Fuels Act created the Synthetic Liquid Fuels Program. This program was administered by the Bureau of Mines and was for building laboratories to conduct research into the manufacture of synthetic fuels. It went through several changes and was abolished in 1985.

The United States legislative history of natural gas has been long and varied and dates back to the middle of the 19th century. In the beginning of its usage, methane was obtained locally, not from natural gas wells, but by processing coal. Cities found that as demand was established, monopolies began to form that controlled the production and distribution of methane and it became necessary to pass laws to regulate these monopolies. It was a classic case of a demand being created for a product that would then become a necessity and then be controlled by a monopoly. Regulation in such cases is required, but how much regulation is too little and how much is too much? Other examples of such situations are in the electrical and communications industries. As natural gas began to be transported between cities, state governments began to pass laws regulating the distribution and sale of natural gas. By 1900 most natural gas came from wells and was widely used throughout the United States. The problem of regulating natural gas became even more complex when interstate pipelines began transporting natural gas over long distances. From 1911 to 1928 individual states passed laws regulating these pipelines, but the Supreme Court of the United States declared these laws to be unconstitutional and it was up to the federal government to control the production, distribution and sale of natural gas. As indicated above, the Public Utility Holding Act of 1935 was passed in order to prevent unregulated monopolies in the interstate natural gas industry. The first law to restrict the prices charged by interstate pipeline companies was the 1938 Natural Gas Act. This act gave the Federal Power Commission (FPC), created in 1920, regulatory powers over interstate natural gas sales and limited certification powers over new pipeline construction. These powers were extended to all new pipeline construction in 1942.

In the early 1940s, a decision by the Supreme Court of the United States gave the FPC the right to control the wellhead price (the price the producer charges the pipeline company) if the producer and the pipeline companies were affiliated, and a 1954 decision by the Court gave the FPC the complete authority to control the wellhead price of natural gas. At first, the FPC tried to set ceiling prices with producers individually, but a large backlog of rate increase requests proved this approach untenable. In 1960, the FPC divided the United States into five regions and set a ceiling price for each region, but determining a fair price for entire regions was more difficult than anticipated. By 1974, a nationwide price ceiling was set that was much higher than the previous ceilings. All attempts at regulating prices for natural gas transported across state lines failed, however. Even the nationwide prices were too low to justify the risk of finding the new wells needed to satisfy the interstate market. This caused shortages and price imbalances between the producing states and many of the heavily consuming states.

To alleviate the problems associated with price ceilings the Natural Gas Policy Act of 1978, which was part of the omnibus National Energy Act of 1978, was passed. This Act replaced the FPC with the Federal Energy Regulatory Commission (FERC) and was to equalize intrastate and interstate natural gas markets. The price ceilings set by the FPC were to be phased out for most new wells. The removal of these price ceilings caused prices to rise and demand to shrink, resulting in a natural gas glut by 1985. This excess of production caused producers to require the pipeline companies to sign multiyear "*take or pay*" contracts that obligated the pipeline companies to pay for natural gas that they did not need. Because large quantities of natural gas are difficult to store, pipeline companies must pass natural gas on to users as quickly as possible and "take or pay" contracts placed a considerable burden on the pipeline companies. In 1987, the District of Columbia Circuit Court of Appeals forced FERC to allow the pipeline companies to pass some of their costs along to their customers and eliminated price restrictions on some new wells. The Natural Gas Wellhead Act of 1989 was to deregulate all wellhead prices and deregulation was fully accomplished by 1993. Also, in 1992, FERC required the pipeline companies to unbundle their distribution and sales services, thereby allowing their customers to choose these services from any provider in any quantity. This order gave all natural gas sellers equal access to end-users, primarily electric utilities and local distribution companies, and permitted end-users to receive natural gas on demand to meet peak requirements.

As became apparent at the beginning of the 21st century, too much deregulation has its pitfalls. In 2000, the Congress passed and President Clinton signed the Commodity Exchange Act that partially exempted energy trades from government regulation. This exemption became known as the *Enron loophole* and led to highly speculative trading by the Enron Corporation and to Enron's eventual downfall. The Enron loophole was closed by a clause in the Food, Conservation, and Energy Act of 2008 when congress passed the act over President George W. Bush's veto. All of the twists and turns related to natural gas regulations highlight the difficulties associated with legislating control over a commodity so embedded in our economy, even though some regulation is clearly needed.

The Atomic Energy Act of 1946 created the Atomic Energy Commission and gave it the responsibility of administering both nuclear weapons development and peaceful nuclear uses, thereby transferring the management of nuclear control to civilian authority. The advent of nuclear energy being used by private industry to generate electricity made it necessary to amend this act in 1954 to better regulate the development and handling of nuclear materials and facilities. This was followed in 1957 by the Price-Anderson Nuclear Industries Indemnity Act which partially indemnified private nuclear power companies against law suits. It set up a ten billion dollar fund paid for by the nuclear industry and provided a commitment by the government to pay all awards exceeding ten billion dollars.

The Energy Reorganization Act of 1974 split the responsibilities of developing and producing nuclear weapons and the civilian use of nuclear materials. An agency called the Energy Research and Development Administration was to supervise nuclear weapons production and other nuclear-related work and the Nuclear Regulatory Commission (NRC), which replaced the Atomic Energy Commission, was to oversee the regulation of nuclear facilities. An amendment to this act provided employees that filed complaints with the Occupational Safety and Health Administration (OSHA) with legal protection against retaliation. The Department of Energy Reorganization Act of 1977 created the Department of Energy (DOE), a cabinet level department which is an umbrella organization that is responsible for other energy-related matters as well as nuclear matters. As numerous onsite above-ground repositories were being created to temporarily store nuclear waste, it became obvious that permanent storage sites would be needed in the near future. The first legislation to deal with nuclear waste was the Nuclear Waste Act of 1982. This act created a schedule and procedure for building underground facilities capable of storing radioactive waste for thousands of years. It also provided for building monitored retrievable storage facilities where waste could be stored until it could be later reprocessed or moved to permanent storage. Unfortunately, as of 2010 little progress has been made in finding locations for storing nuclear waste and most of it is still resting in temporary facilities.

Although daylight saving time had been used in the past to reduce the consumption of electricity, the 1966 Uniform Time Act required each state to decide whether or not it wanted to participate in daylight savings and set the beginning and end times for daylight savings. It is not required for a state to participate in daylight savings, but if it does it must participate as a whole. Arizona is an exception because the Navajo Nation (Indian tribes were allowed to decide separately) chose to go on daylight savings time and the State of Arizona did not. The idea behind daylight savings is that electricity is saved if nighttime is shifted so that there is more daylight in the evening and less daylight in the morning.

After the discovery of large deposits of petroleum on the North Slope of Alaska in the 1960s, ways of getting the oil to the lower 48 states were studied intensively. Despite considerable resistance by environmental groups, in 1973 the Trans-Alaskan Pipeline Authorization Act was passed by Congress and signed by President Nixon. The production in this area is now in decline and the oil companies want to lease areas within the Arctic National Wildlife Reserve (ANWR). There is forceful environmental opposition to granting leases in the ANWR and such leases have thus far not been forthcoming. However, there are also large amounts of natural gas throughout the North Slope area and plans are being drawn up to build a natural gas pipeline through Canada to the lower states. As of 2010 no plans have been approved and there will undoubtedly be strong opposition to any pipeline plans even though there are fewer environmental risks associated with natural gas pipelines than oil pipelines.

In October, 1948, the industry town of Donora, Pennsylvania, was covered by a thick cloud of air pollution that remained over the town for five days. The cloud killed 20 of the town's residents and sickened many others. In 1952, a similar incident in London, England, was responsible for over 3,000 deaths. Such incidents caused alarm in both Europe and the United States. In 1955, the United States enacted its first federal legislation dealing with air pollution, the Air Pollution Control Act. This Act recognized air pollution as a problem and provided some money to study the health effects of air pollution, but left most of the responsibilities to state and local governments. It was amended in 1960 and 1962, and in 1963 the first legislation bearing the name Clean Air Act became law. Although the Clean Air Act of 1963 provided federal money for research into air pollution by the states and pressed for emission standards for vehicles as well as stationary pollution sources, it still left most of the responsibilities with state and local governments. It was amended in 1965, 1966, 1967 and 1969. Each time it increased the involvement of the federal government in the control of air pollution.

In 1970, the federal government accepted a central role in the control of air pollution by passing the Clean Air Act of 1970. This Act established the Environmental Protection Agency (EPA) and charged it with creating National Air Quality Standards for the chemicals most responsible for polluting the air. The EPA was also to establish New Source Performance Standards that were to set allowable amounts of pollutants for various industries. Recognizing that much of our pollution came from vehicles, the Act also set strict standards for vehicular emissions. Under the Act, states were to develop EPA approved State Implementation Plans (SIPs) for monitoring their pollution and enforcing the EPA standards. The Act was strengthened several times as more was learned about pollution and its effects. One more important provision of the Act was the requirement to phase out leaded gasoline and by 1996 lead was no longer used in fuels.

By 1990, the constituents and sources of air pollution were well understood and a much more definitive act was needed. In addition to the states, the Clean Air Act of 1990 gave tribal nations the right to develop separate air quality management programs. Although carbon dioxide was beginning to be recognized as a global problem, the 1990 Act concentrated on the six pollutants, particulate matter, ground-level ozone, carbon monoxide, sulfur oxides, nitrogen oxides and lead. Despite allowable standards being set for these pollutants by the Clean Air Act of 1970 and its amendments, millions of people still lived in areas with unhealthy levels of pollution. The 1990 Act gave particular emphasis to particulate matter, sulfur and nitrogen oxides and the interstate and international spread of the pollution from single sources.

Borne on the wind, pollutants may travel for thousands of miles. For example, the pollution from India, China and the southeastern Asian nations sometimes covers much of the Indian Ocean. Part of the 1990 Act specifically targeted the power plants and industrial complexes that emitted sulfur and nitrogen oxides. These oxides are what cause acid rain and other forms of spreading acidic material. Also, there is the problem of widely distributed particulate matter. Particulate matter not only severely reduces visibility in urban areas, but can cause severe breathing problems for the elderly, children and asthmatics. When it was discovered that particles less the 2.5 micrometers were the ones most likely to cause health problems, the EPA set limits on fine particulate matter. The 1990 Act provides for interstate commissions to develop regional plans for limiting widespread pollution. Meaningful fines are levied on industries and localities for not obeying the limitations set by these commissions.

Since 1970, special attention has been given to vehicle emissions, fuels produced by refineries and service stations that dispense the fuels. The emission standards for vehicles have been tightened several times and light trucks, because their mpg ratings are less, have more restrictive limits than cars. Lead and mercury have almost been eliminated from vehicle exhausts and the nitrogen oxides that produce ground-level ozone have been substantially reduced by the use of catalytic converters. According to the EPA, a new car produced today will emit only ten percent as much pollution as a new car produced in 1970. About 50 percent of the nitrogen oxides in our atmosphere are emitted by vehicles and 40 percent is emitted by power plants. There are now regulations strictly limiting the particulate emissions from diesel vehicles, and service stations are required to limit their volatile organic compounds (VOCs) that are due to the evaporation of fuel, especially gasoline. The VOCs are highly toxic and are known to cause cancer. Beginning in 2006, refiners had to reduce the sulfur in gasoline by 90 percent because sulfur inhibits catalytic converters from operating properly. Individuals also have responsibilities in maintaining clean air. Familiar to most of us are the annual inspections that verify that the emissions from our vehicles are in compliance with EPA standards. There are also regulations that restrict the emissions from new equipment such as lawn mowers. Small internal combustion engines are much worse polluters per gallon of fuel than their larger counterparts. Air quality will continue to improve as old equipment and vehicles are replaced by new equipment that must meet continually improving standards.

As a direct result of the petroleum shortages in the 1970s there were two important acts regarding our dependence on foreign petroleum enacted into law. They were the 1975 Energy Policy and Conservation Act (EPCA) and the National Energy Act of 1978. The EPCA is the act that created the SPR, but is best known for establishing the Corporate Average Fuel Economy (CAFE) standards that required the car and light truck manufacturers to meet fleet average mpg ratings. After excluding vehicles with gross vehicle weight ratings (GVWRs) over 6,000 pounds, for each year each manufacturer had to meet a fleet average mpg rating for its fleet of cars and a separate fleet average mpg rating for its light trucks or pay a penalty. The GVWR limits were later raised to 8,500 in 1980 and then, in 2007, to 10,000 pounds for light trucks, but not SUVs. The averages were harmonic averages and computed as follows:

$$\frac{\text{Total number of vehicles}}{\dfrac{\text{No. of A}}{\text{mpg of A}} \; + \ldots + \; \dfrac{\text{No. of N}}{\text{mpg of N}}} \; ,$$

where A,…,N represent different models of cars or light trucks. For example, suppose that a manufacturer produced four models of light trucks, A, B, C, and D with mpg ratings of 18, 22, 24 and 12, respectively. Also, the numbers produced were 360,000, 220,000, 180,000 and 100,000, respectively. All of the vehicles had GVWRs less than 10,000 pounds except model D. Then model D would be excluded from the calculation and the average used would be

$$\frac{760,000}{20,000 + 10,000 + 7,500}$$

or 20.3 mpg. Separate, more generous computations were made for vehicles that used or partially used alternative fuels such as alcohols.

The fleet average requirements for the period 1978 through 2011 are given in Fig. 10-3. The penalty in dollars for not meeting these requirements is currently computed by taking 55 times the amount the manufacturer is under the officially set average times the total number of vehicles, cars or light trucks, produced by the manufacturer during the year. For the above example, if the CAFE standard mpg were 19, then the penalty would be:

$$(20.3\text{-}19) \times 55 \times 760,000 = 54,340,000 \text{ dollars.}$$

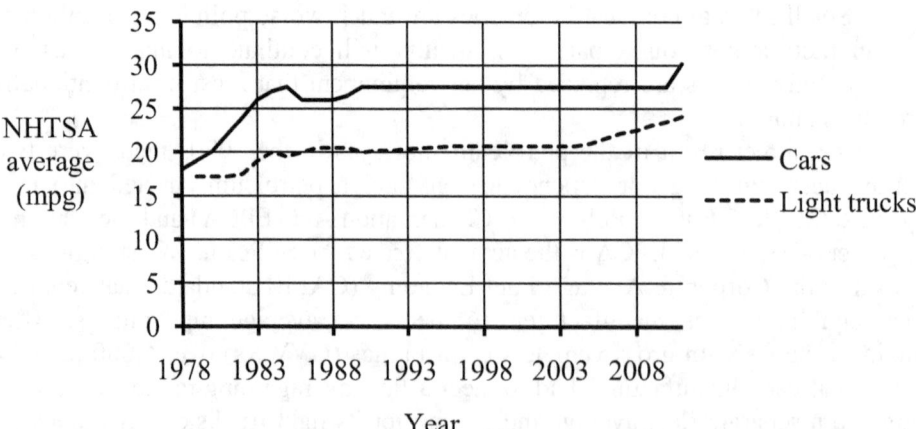

Fig. 10-3 NHTSA fleet average mpg requirements [NHTSA].

However, if a manufacturer is above the set average it will be given credits that can be used in lieu of any deficiencies accrued during the previous or following three years of the year in which the credits were earned. The calculation for credits is the same as that for penalties except that the set average is subtracted from the manufacturer's average.

Congress gave the National Highway Traffic Safety Administration (NHTSA) the task of establishing the limiting mpg averages for cars and light trucks and the EPA the responsibility of determining the mpg ratings for the various car and light truck models. The EPA measured both city and highway mpg values and gave the city values a 55 percent weighting and the highway values a 45 percent weighting in determining their final overall ratings. The averages set by the NHTSA could, however, be adjusted by Congress. In establishing the averages the NHTSA was to set the averages at their maximum feasible levels, but was to take into account:

- Technical feasibility.
- Economic practicability.
- Effect of other standards on fuel economy.
- Need for the nation to conserve energy.

The main purpose of the National Energy Act of 1978 was to avoid future shortages of petroleum and natural gas by using less of these fuels. The Natural Gas Policy Act mentioned above was only one of five acts in the National Energy Act of 1978. The other four are the Public Utility Regulatory Policies Act (PURPA), Energy Tax Act, National Energy Conservation Policy Act and Power Plant and Industrial Fuel Use Act. The intent of the PURPA was to encourage the development of renewable energy and more efficient power plants by forcing utility companies that controlled the distribution of electricity to buy electricity from independent "qualifying facilities" at the costs they would incur in generating the power themselves. These costs were essentially interpreted as the costs of the fuel needed by the utilities' electric power plants. In addition, ten percent tax credits, known as Investment Tax Credits (ITCs), were to be given to defray the costs of wind, solar and other generation facilities that consume less fossil fuel. The main result has been the increased construction of wind farms and cogeneration and combined-cycle power plants.

The Energy Tax Act included two laws of particular importance. One provided income tax credits to individuals who reduced fossil fuel consumption by installing renewable energy systems for heating and generating electricity. It offered tax credits equal to 30 percent of the cost of the systems up to $2,000 and 20 percent of the cost between $2,000 and $10,000. It also offered businesses tax credits up to a maximum of 25 percent of the cost of such systems. The other important law in the Energy Tax Act instituted what has come to be known as the "gas guzzler tax." This tax was on the purchase of new vehicles with GVWRs of less than 6,000 pounds. As was used to determine the CAFE ratings, the EPA mpg ratings were used. Since 1991 these taxes have been as shown in Fig. 10-4. There are no taxes on vehicles with mpg ratings greater than 22.5. Because vehicles over 6,000 pounds were exempted from these taxes, they have proved to be counterproductive. They encouraged people to buy SUVs, minivans, pick up trucks and other vehicles that weighed more than 6,000 pounds and consumed considerably greater amounts of fuel than the vehicles the taxes were designed to discourage.

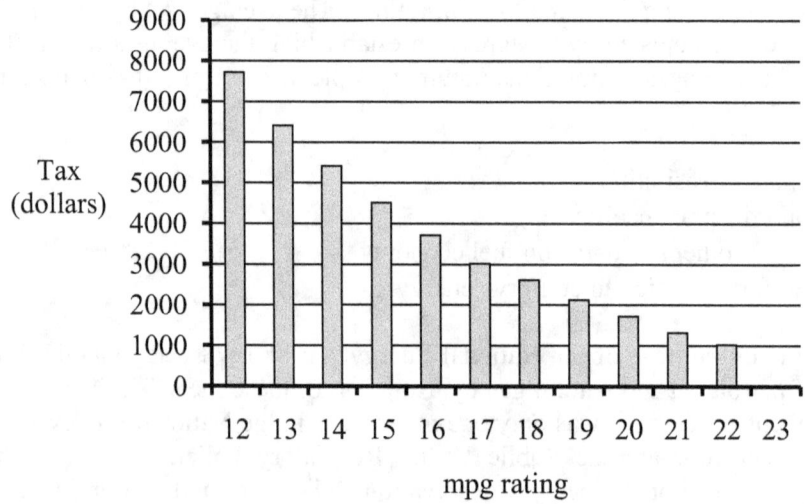

Fig. 10-4 Taxes on vehicles weighing less than 6,000 pounds [EPA].

The National Energy Conservation Policy Act required electric utility companies to offer energy conservation audits in order to slow the growth of electricity demand. Not yet realizing that global warming is associated with coal, the Power plant and Industrial Fuel Use Act encouraged the use of coal over that of petroleum and natural gas. The overriding concern at the time was our need to import increasing amounts of petroleum from countries with unstable governments and the intent was to reduce these imports by using more alternate fuels, particularly the United States most abundant and inexpensive energy source, coal. The Act severely curtailed the construction of large boilers fueled by petroleum or natural gas by electric companies and other industries. This, of course, contributed substantially to the early 1980s glut of natural gas. Many of the provisions of the Act were repealed by the Natural Gas Utilization Act of 1987, which initiated the rapid increase in the use of natural gas. Between 1988 and 2002 the consumption of natural gas for industrial processing and electricity generation went up by about 47 percent. This Act is another example of legislation with unintended results.

Following the National Energy Act of 1978 was the Energy Security Act signed into law by President Carter in 1980. It promoted the use of alternative energy sources through guaranteed loans. Most of the provisions of the Energy Security Act have been superseded by later laws.

The inefficiencies of household appliances had become a concern of law-makers as early as 1975, but the first serious legislation confronting these inefficiencies directly was the 1987 National Appliance Energy Conservation Act. This Act required that newly manufactured appliances meet standards set by the DOE. It authorized the DOE to update appliance standards according to schedules established by Congress and there have been several updates. As a result of this Act, the Federal Trade Commission (FTC) has required manufacturers of most major appliances to attach labels that estimate the appliances' energy efficiencies. Some of the appliances included in these regulations are refrigerators, freezers, room air conditioners, clothes washers and dryers, dishwashers, kitchen stoves, pool heaters, water heaters and light bulbs.

The Energy Policy Act of 1992 (EPACT1992) is divided into 30 titles. It covers all aspects of energy and was a response to the Persian Gulf War in 1991. Table 10-3 lists the topics included in the Act and the titles in which they are treated. Title I was entitled "Energy Efficiency" and is primarily concerned with the new construction of buildings, but contains two provisions on the efficiency of electrical and natural gas utilities. Title I requires each state, in consultation with the Council of American Building Officials (CABO) and the American Society of Heating, Refrigeration and Air Conditioning Engineers (ASHRAE), to create new building code standards. It also provides requirements for new residential and commercial equipment and industrial processes. Nine titles deal with fossil fuels, three of which are concerned with security. These titles also have provisions covering imports, exports, royalties and leasing. In order to encourage domestic production, they also allowed independent oil and gas producers greater deductions against the alternative minimum tax related to percentage depletion and intangible drilling costs. Six titles are in the area of nuclear power and relate to waste, enrichment, licensing and reprocessing. Only Title XII dealt with renewable energy sources. It mainly provided support for studies and research into the development of renewable sources. Three of the titles were related to electricity. The most notable was Title VII that reformed the Public Utilities Holding Company Act of 1935 and opened up the electrical grid to independent producers even more than previous legislation.

Four titles had to do with transportation and other fuel burning equipment, three with alternative and replacement fuels and one with electric vehicles. The alternative and replacement fuels considered in Titles III, IV and V included natural gas, liquid petroleum gas and electricity as well as alcohols and biodiesel. The principal provisions were for reducing the gasoline and diesel fuels consumed by

Table 10-3 List of topics and their corresponding titles that appeared in EPACT1992.

Titles	Topic
I	Efficiency: residential, commercial, construction and utilities
XXV, XXVI, XXVII	Fossil fuels
II	Natural gas
XV, XVIII	Petroleum
XIV, XX	Security
XIII	Coal
XXVIII, XXIX	Nuclear
VIII	Waste
IX	Enrichment
XI	Health, safety and environment
X	Revitalization
XII	Renewables
VII, XVII, XXIV	Electricity
	Transportation and other fuel burning equipment
III, IV, V	Alternative and replacement fuels
VI	Electric vehicles
XXI	Environment
XVI	Climate change
XXII	Economic growth
XIX	Revenue
XXIII	Administration
XXX	Other

fleets of vehicles owned by private companies and federal, state and local governments. They mandated that these fleets reach levels of alternative fuel use by schedules determined by the DOE after consulting with the states and other major fleet owners. Credits could be gained by those exceeding the limits and a fleet owner could use the credits later or trade them with other fleet owners. Title VI was designed to promote the development of electric vehicles and for supporting transportation-related development and research projects.

There were two titles to reduce the undesirable environmental impact of energy usage. Title XVI was the first recognition by Congress that GHGs were a potential global problem. Much of this title was to provide for retroactive as well as future collection of GHG data. However, there were provisions for promoting the export of renewable energy, energy efficiency and clean coal technologies and the creation of a global climate change response fund to mitigate the effects of climate change both inside and outside the United States. Unfortunately, money was not to be put into the fund until the United States ratified the UNFCCC, something that the United States has never done. Title XXI was concerned with several aspects of energy efficiency and advanced electricity generation and distribution, particularly advanced nuclear reactors.

The effects on the federal government's revenue due to its energy policy are considered in Title XIX. This title is concerned mainly with the loss of revenue due to tax deductions and credits given to promote conservation and clean energy sources. Title XXII supports the development of energy-related technologies and education programs that will improve economic growth. The support is to encompass materials as well as efficient equipment and the production of alternative fuels. The remaining titles consider the administration of the proposed programs and a variety of miscellaneous topics.

A list of the titles in the massive Energy Policy Act of 2005 (EPACT2005) and the topics contained in them is given in Table 10-4. This act includes numerous tax breaks and some royalty suspension for exploration. It is more specific with regard to its incentives than EPACT1992. Title I sets more restrictive mandatory energy standards for a number of residential and commercial appliances and other equipment. Title XIII provides tax credits to businesses and individuals for investment in energy efficient properties, properties that use renewable energy sources, and a variety of energy efficient appliances. It also gives a tax credit of $1,000 per home to builders of manufactured homes that are 50 percent more efficient than those built in accordance with the 2003 code and $2,000 per home for other homes meeting the same conditions. Residential taxpayers may get up to 30 percent of the total cost as a credit for the installation of solar cells (limited to $2,000), solar heating equipment (limited to $2,000) and fuel cells (limited to $500). Title XIII also provides business tax credits of 30 percent toward the purchase of fuel cells and 10 percent toward the purchase of micro-turbines, and tax credits up to $1.80 per square foot for installing more efficient lighting and air conditioning. Subsidies for industrial processes were fairly limited. However, there was provision for federally funded projects to increase the use of fly ash and blast furnace slag in the production of cement and tax credits given to certain producers of coke or coke gas. The cement production projects were intended to increase the use of these normally wasted minerals and reduce the amount of carbon dioxide released in the production of cement. The tax subsidy for coke and coke gas producers was a modification of an earlier similar subsidy.

There were numerous sections of the EPACT2005 related to transportation. Some of the more important are:

Table 10-4 List of topics and their corresponding titles in EPACT2005.

Titles	Topic
I	Efficiency: appliances, equipment and construction
V	Fossil fuels
III	Natural gas
III	Petroleum
IV	Coal
VI	Nuclear
II	Renewables
XII	Electricity
VII	Transportation and other fuel burning equipment
XV	Alternative and replacement fuels
VIII	Hydrogen
XVI	Climate change
IX	Research and development
XVII	Innovative technologies
XI	Personnel and training
XVIII	Studies
XIII	Tax incentives
X	Administration
XIV	Other

- Tax credits to the purchasers of certain lean burn or hybrid vehicles. The amount is determined by the vehicle's model and is phased out by lowering the amount each quarter after the initial production of the model.
- It is mandated that the production of renewable fuels increase to 7.5 billion gallons per year by 2012 and that 0.25 billion gallons per year be from cellulosic biomass.
- Because of the hazards associated with MTBE, the oxygen content requirement for renewable fuels was eliminated.
- A tax credit of $0.51 per gallon for the production of ethanol was extended through 2008. Tax credits of $1 per gallon for the production of renewable biodiesel and $0.50 per gallon for biodiesel made from recycled oil or yellow grease were extended through 2008.
- Reimbursements to the states for enforcing the laws regarding underground storage tanks.

To allow the United States to become more energy independent there are several provisions to promote the production and refining of fossil fuels. Some of them are:

- Petroleum refiners could immediately depreciate 50 percent of the cost of refinery expansions if the expansions increased the capacity by at least five percent.
- The final decision on LPG port facilities was given to FERC, but FERC was required to work with the states with respect to safety.
- Royalty relief was granted for marginal wells and productive wells discovered in difficult locations. Royalties were suspended for productive Gulf of Mexico wells discovered within five years of the enactment of EPACT2005 and located in waters exceeding 400 m (1300 ft) in depth.
- Certain requirements for the leasing of federal lands for the production of coal were eliminated and Production Tax Credits (PTCs) were made available for the production of coal on Indian lands.
- A total of $1.3 billion in ITCs were made available for new or refurbished coal-fired generation projects, coal gasification projects and projects that remove 99 percent of sulfur dioxide and 90 percent of mercury from the emissions of electrical generating plants. To be eligible for the ITCs, refurbished coal-fired plants must improve their efficiencies between four and seven percent.
- ITCs of 20 percent were made available for qualifying Integrated Gasification Combined-Cycle (IGCC) projects. ITCs of 15 percent could be given for other qualifying advanced technology projects.

Several congresspersons, including Senators Clinton and McCain, voted against the Act because they felt these provisions encouraged the continued reliance on fossil fuels and it subsidized coal and oil companies that did not need the incentives. In fact, in 2007 and 2008 the oil companies reported record profits. But, the Act also offered enough incentives for conservation and improved efficiencies that the then Senator Obama voted for the bill.

Several provisions targeted electricity generation and distribution and nuclear power facilities. Some of them are:

- Public electric utilities were required to offer net metering to anyone who requested it. *Net metering* records the difference between the energy supplied by the grid and the energy supplied to the grid. This allows residences and businesses that have solar panels or wind turbines to get credit for the energy they generate but do not need.
- In light of the grid availability to independent power companies, the Public Utility Holding Company Act of 1935 was repealed.

- In order to avoid blackouts such as the one in 2003, improved reliability and operational standards for the grid were mandated.
- To encourage the improvement of the grid, the capital recovery period for new distribution installations was reduced from 20 years to 15 years.
- The 1957 Price-Anderson Nuclear Industries Indemnity Act that protects nuclear power plants from excessive liability was updated and extended through 2025.
- Authorizes up to a total of two billion dollars to cover cost overruns for up to six new nuclear reactors.
- Provides a PTC estimated to be worth 1.8 cents per kilowatt-hour for new nuclear reactors. It may be claimed for the first eight years of a reactor's operation, but is limited to a total of $125 million per gigawatt of capacity per year.
- Requires the federal government to buy at least 7.5 percent of its electricity from plants that use renewable sources by 2013 if it is deemed economically feasible.
- Eligible new renewable generation facilities may claim a 1.5 cent per kilowatt-hour PTC for the first ten years of operation.

Finally, EPACT2005 made available a variety of grants, tax incentives and guaranteed loans for the development of innovative technologies related to:

- Renewable energy systems.
- Advanced fossil fuel processes.
- Hydrogen fuel cells.
- Carbon capture and sequestration.
- Efficient generation and distribution of power.
- Fuel efficient vehicles.
- Pollution control.
- IGCC plants that meet certain pollution standards.

As indicated in its preamble, the Energy Independence and Security Act of 2007 was to provide "… energy independence and security, to increase the production of clean renewable fuels, to protect consumers, to increase the efficiency of products, buildings, and vehicles, to promote research on and deploy greenhouse gas capture and storage options, and to improve the energy performance of the Federal Government …." It revisited many of the provisions contained in the earlier Energy Policy Acts and made them more compatible with the conditions at the time of its enactment. It tended to be more definitive in the limits it set and concentrated on reducing the consumption of fossil fuels. Some of its more important provisions are:

- Cars and light trucks are to have average CAFE ratings of 35 mpg by 2021, but manufacturers will be permitted to trade credits.
- Boilers that are powered by fossil fuels are to have efficiencies of at least 80 percent by 2012.
- For the first time serious consideration is given to improving the fuel consumption per ton-mile of large trucks, semis and trains, although no goals were set.
- Limits on lighting were set that will essentially eliminate the use of incandescent bulbs in most places by 2013.
- Additional and extended funding for research and development of renewable energy sources, carbon sequestration and biofuels.
- Education for those who are to work in energy efficiency and renewable energy fields.
- Modernization of the electricity grid.
- Support for weatherization of residences. The total funding is to vary from $0.75 billion in 2008 to $1.4 billion in 2012.
- Definite goals for reducing the amount of fossil fuel used in public institutions, particularly federal and state institutions. New federal buildings built in 2010 are to consume 55 percent less fossil fuel as compared to similar buildings constructed in 2003. Those built in 2020, are to consume 80 percent less fossil fuel.
- Infrastructure grants of up to 33 percent of their cost (limited to $180,000) for service stations to modernize their facilities to handle renewable fuel gasoline blends that are between 11 and 85 percent and renewable fuel diesel blends of at least 10 percent.
- Grants, loans and other incentives are to be given to small businesses to reduce their energy consumption, use renewable energy sources, produce renewable energy products or do renewable energy research. The object is to get more small business involvement in reducing energy demand and creating renewable or energy efficient products.

EPACT2005 specifically refers to carbon emissions with regard to the development of carbon capture and sequestration technologies and both EPACT2005 and EPACT2007 do contain a number of provisions that indirectly affect the reduction of carbon dioxide through improved efficiencies and the use of renewable fuels. In order to gauge the effects of these provisions on future emissions, note that in 2008 the United States emitted 5.81 Gt of carbon dioxide, of which 1.93 Gt was emitted directly by vehicles, 1.53 Gt by the other three sectors and 2.36 Gt by electricity generation.

The principal relationship the Food, Conservation and Energy Act of 2008 has to energy is through the relationship between agriculture and biofuels. This Act created the Biomass Crop Assistance Program (BCAP) for the purpose of subsidizing the raising and processing of energy crops. BCAP pays up to 75 percent of the cost of establishing an eligible biomass crop and up to $45 per ton for harvesting, storing and transporting the biomass to a biomass processing facility. A companion provision pays up to 30 percent of the cost of development and construction of demonstration-scale biorefineries and provides guaranteed loans up to 90 percent of the cost. The Act also provides grants for biomass research, updating biorefineries to use less fossil fuel and improving the efficiencies of farms, ranches and other rural entities.

The American Recovery and Reinvestment Act of 2009 (ARRA2009) was an economic stimulus package designed to bring the United States out of the recession occurring at that time by pumping money into the economy and creating jobs. Energy expenditures were to account for only about $84.7 billion, or eleven percent, of the total funded amount of $787 billion. Of the $84.7 billion, $36.7 billion was directed to the DOE. The distribution of the funds to be received by the DOE is shown in Fig.10-5. Of these funds, 46 percent, were to be spent providing incentives for energy efficiency and renewables. The remaining $48 billion was for rebuilding the United States transportation infrastructure and most of it, 57 percent, was for the construction and repair of highways and bridges. The distribution of these funds is given in Fig. 10-6.

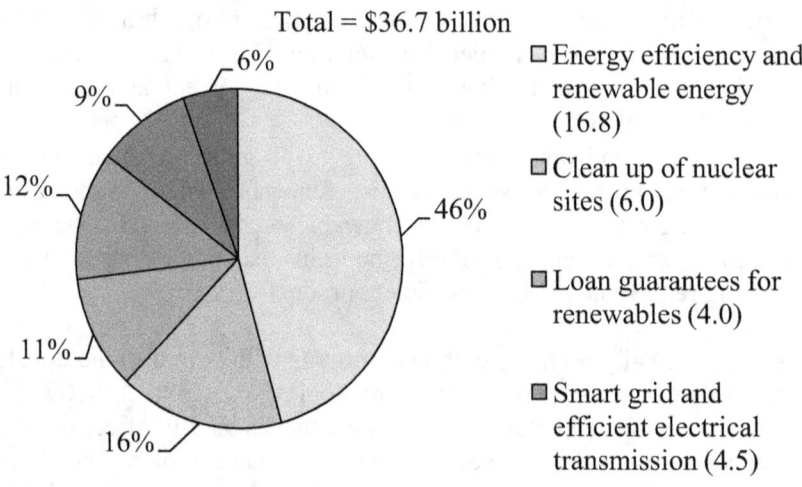

Fig. 10-5 Distribution of funds given to the DOE from ARRA2009 [DOE].

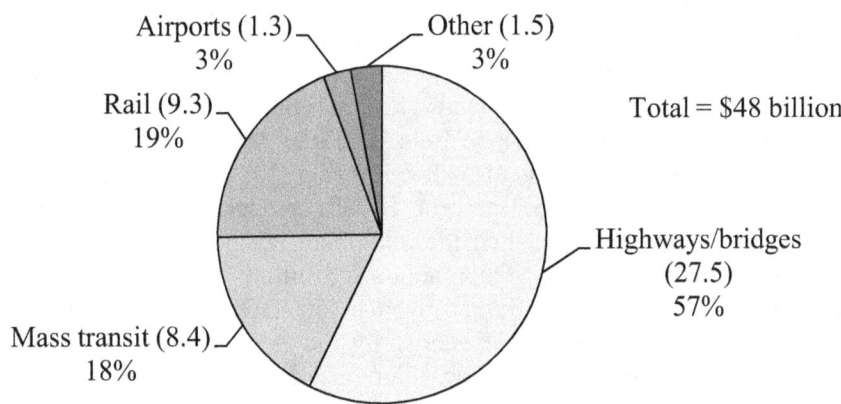

Total = $48 billion

Fig. 10-6 Distribution of funds from ARRA2009 for transportation infrastructure.

The tax that is most sensitive to the average consumer is the excise tax on gasoline. The federal excise tax on gasoline was first introduced in the Revenue Act of 1932 and was initially one cent per gallon. As of 2010, the last time it was changed was in 1994 when it became 18.4 cents per gallon. The history of the federal excise tax is shown in Fig.10-7. Perhaps the more interesting history of this tax is shown in Fig. 10-8, which gives the tax as percentages of the average annual

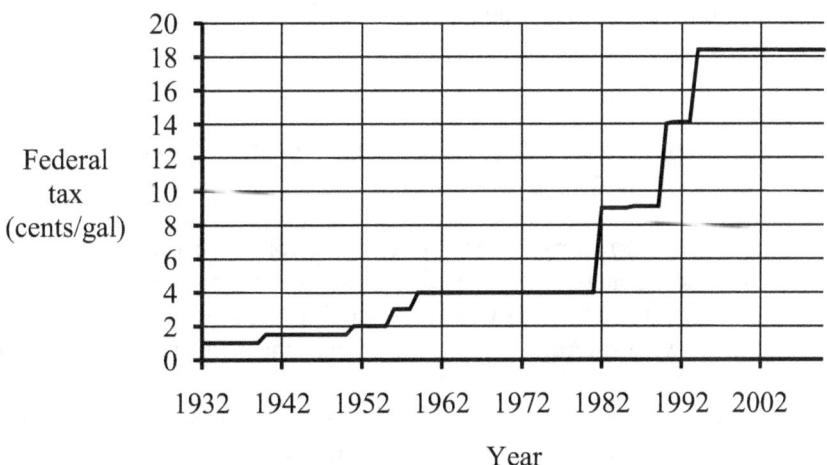

Fig. 10-7 Federal excise tax on gasoline.

prices. Note that as a percentage of price the federal tax is essentially the same in 2008 as it was in 1949 and was near its all time low.

In addition to the federal tax, each state has placed a tax on gasoline. The amount varies from state to state and with time, but the average tax over all states has remained almost constant since 1994. The total tax, including the average state tax, as a percentage of price is also shown in Fig. 10-8. The average percentage of the total tax since 1994 is 25.3 percent. If the percentage of total tax per gallon in 2008 had been this average, then the additional tax collected on the 137 billion gallons consumed would have been about $18 billion. This amount is about one billion dollars more than the funds given to the DOE to subsidize programs for energy efficiency and renewable energy by the ARRA2009. The percentages of taxes on diesel fuel are slightly higher than those for gasoline. In Europe and Japan, the taxes on petroleum fuels are several dollars per gallon.

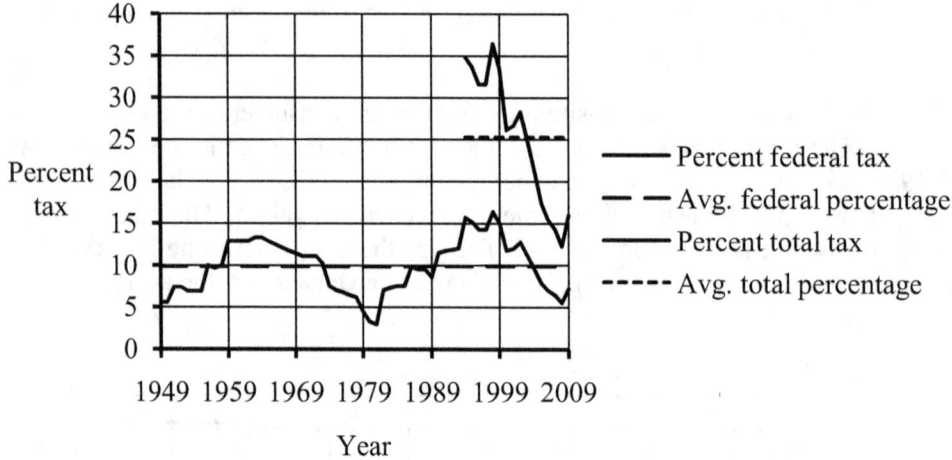

Fig. 10-8 Federal and total taxes on gasoline as a percentage of price.

As one reviews the history of the above legislation, it becomes apparent that it has been reactive, not proactive. Most of the legislation has been in response to an existing problem or crisis. Rarely has Congress acted to prevent a potential problem from occurring, even when there has been ample warning well in advance. So far, Congress' practice of reacting to past events has not caused an economic or environmental catastrophe, although it may have sometimes been counterproductive or made the solution more difficult. Presently, we are faced with two problems that could be catastrophic, the finiteness of petroleum and global warming. The lack of a continuing abundant supply of petroleum is primarily an economic problem that, for the United States, could be solved by Congress. Global warming, on the other hand, is an international problem that can only be solved by nations working together. While the United States alone cannot provide a solution to this problem, as the nation with the largest GDP and second greatest carbon dioxide emissions, it should take a leading role in global warming debates and set an example for other nations to follow. Although the recent legislation has been in the right direction, especially that beginning with the Energy Policy Act of 2007, more remains to be done and more definite plans need to be made. Perhaps it is time to rely more on what we already know and do more manufacturing and constructing while continuing our studying and researching.

Underlying the world's energy difficulties is population growth and the desire for improved standards of living by the developing and underdeveloped nations. Population growth and the aspirations of the poorer nations will have an enormous impact on how our energy future plays out. Hopefully, through diligent planning and deliberate execution, they will not be the undoing of our energy future.

A

HISTORY AND SCIENCE OF ENERGY

So what is energy? And when did we first realize the concept of energy? The concept of energy began to evolve when the ancient Greeks, particularly Aristotle, began to think about the nature of motion, but their explanations were based on crude observations and not verified by scientific experimentation. It was not until Galileo Galilei used experimentation to justify his ideas in the 16th century that the science of motion, and hence energy, began to take shape. Because motion, which could be observed directly, was the primary interest during the 16th and 17th centuries, it was mechanical energy that was studied first. It is believed that one of Galileo's earliest results concluded that, in the absence of air, all bodies fall equal distances during equal time periods regardless of their weight. This was contrary to the belief held by Aristotle that had been accepted for centuries.

During the 17th and early 18th centuries, Galileo's work was followed by that of Isaac Newton, Rene Descartes, Christiaan Huygens and Gottfried Leibniz. In parallel with their scientific discoveries they developed the necessary mathematics, particularly calculus and graphics, needed to describe their findings. It was these four who first expressed their conclusions precisely using calculus and mathematical formulas. It was from their studies that force and motion were related mathematically.

Newton is credited with the three basic laws that relate force and motion. The first states that, if the forces on an object cancel one another, there is no change of motion. His second law states that, if the forces do not cancel, then the net, or *resultant*, force is equal to the mass of the object times its acceleration (i.e., the rate of change of its velocity). In either case, note that it is the change in motion that is important, not the motion itself. The second law is mathematically written $F=ma$, where F is the resultant force, m is the mass and a is the acceleration. From this equation, the *inertial mass*, or simply *mass*, of an object is defined as the constant of proportionality between the force on the object and the object's acceleration. If the mass is in kilograms (kg) and acceleration is in meters per second per second (m/s^2), then the force is in *newtons* (*N*). That is, a newton is the force needed to accelerate one kilogram one meter per second per second (kg/s^2). Newton's third law states that whenever one object exerts a force on a second object, then the second object exerts an equal but opposite force on the first object.

In addition, Newton proposed the *Law of Gravitation*, which states that all particles have a force of attraction that is directly proportional to the product of their masses divided by the square of the distance between them. The constant of proportionality is called the *gravitational constant* (*G*) and, if the force is in newtons, the masses are in kilograms and the distance is in meters, then $G \approx 6.673 \times 10^{11}$. For objects of nonzero dimensions, the distance used is the distance between their centers of gravity. The *weight* of an object is the force exerted on the object due to gravity. Although mass is a basic property of an object, weight varies with location. The weight of an object on Earth is the force (in newtons) equal to G times the mass of the object in kilograms times the mass of the Earth in kilograms divided by the square of the distance between the gravitational centers of the Earth and the object in meters. Because the mass of the moon divided by the moon's radius squared is about one-sixth the Earth's mass divided by the radius of the Earth squared, the weight of an object on the moon is about one-sixth its weight on Earth.

These four laws are still used today and the physics that is based on them is referred to as *Newtonian physics*. They are considered accurate as long as the relative velocities involved are small compared to the velocity of light (approximately 300,000 m/s) and the gravitational forces are not extreme. Such extreme conditions cause the relativistic effects predicted by Einstein to affect the fabric of our universe in non-intuitive ways and are not considered applicable to the presentation given in this book. However, they must be taken into account when performing astronomical measurements or planning space flights.

Huygens and his student Leibniz meticulously studied the motion of objects and Huygens found that, when non-elastic balls collide, the sum of the products of their masses and respective velocities squared is the same before and after they collide. The fact that this sum remained unchanged before and after a collision was the earliest form of the *conservation of energy principle*. But it was Leibniz, who coined the term *vis viva* to describe the mass times the velocity squared, whose experiments led to the modern equation $E=0.5mv^2$, where E is the *kinetic energy* (the modern name for one half the vis viva) of an object, m is the object's mass and v is the object's velocity. If m is in kilograms and v is in meters per second, then E is in joules. A *joule* (*J*) is the kinetic energy of a one-kilogram (kg) object moving at the square root of two meters per second (m/s). Despite the fact that considerable research into the nature of force and motion was achieved during the 17th and 18th centuries, it was not until 1807 that the word "energy", the Greek word for "work", was introduced into our vocabulary by Thomas Young.

In 1803, L. N. M. Carnot had noted that a raised object, if dropped, had the capacity to attain kinetic energy. He called this capacity *latent vis viva*, which is now called *potential energy*. Young realized the need to bring the various notions of energy under a common umbrella and, recognizing the relationship between latent vis viva and the vis viva of motion, defined *energy* to be the ability to do work. In a mechanical system, *work* is defined to be the resultant force on an object times the distance traversed by the object due to the force. This is expressed mathematically by the equation $W=Fx$, where W is the work, F is the force, and x is the distance. If F is in newtons and x is in meters, then W is in newton-meters (N-m). If a force is used to raise an object to a height x, then the object will have the potential energy Fx. When the object is dropped and returns to its starting point, by assuming a force ma and using elementary calculus it can be shown that the object will attain a velocity v such that $v^2=2ax$. Thus the kinetic energy is $0.5mv^2=max=Fx$, and the kinetic energy returned to the object is the same as the work required to produce it. These equations also show that a joule is the same as a newton-meter.

Clearly, potential energy is a form of stored energy. By raising an object and keeping it at its elevated height its potential energy may be used later to perform useful work. Elevators have offsetting weights that go up when the elevator goes down and vice versa. When an elevator is resting at the lowest floor, the weights are resting at the highest floor. Then, when the elevator needs to go up the weights use their potential energy to help pull it up. Similarly, the water raised by the sun and stored behind a dam has potential energy that can be used to drive a water wheel to grind corn or a hydro-turbine to generate electricity.

On the other hand, it is impractical to use, at a future time, the kinetic energy of an object moving in a straight line, the movement implied above. However, $0.5mv^2$ also applies to rotational movement. The velocity of an object moving in a circle is 2π times the radius of the circle times u, the number of revolutions per second. Therefore, for rotational movement $0.5mv^2$ becomes $0.5m(2\pi ru)^2$. A spinning object, such as a flywheel, may be viewed as being made up of infinitesimal particles. Elementary calculus may be used to sum up the kinetic energies of these particles and arrive at the overall kinetic energy of the rotating object. This overall energy is $0.5I(2\pi u)^2$, where I is the *moment of inertia* and is the integral sum of the mr^2 values of the particles. Although it is impractical to save the kinetic energy of an object moving in a straight line, the kinetic energy of a rotating object (e.g., a flywheel) may be put to useful work at a later time.

While some were developing the concept of mechanical energy, others, including Francis Bacon, Joseph Black, John Locke, Robert Boyle, Jacques Charles, Benjamin Thompson (Count Rumford) and James Joule were investigating heat. Although the notion that heat is due to motion was proposed by Bacon as early as 1620, the belief that heat is an invisible, elastic fluid, called *caloric*, persisted as late as 1800. In the 18th century, Black showed that heat and temperature are not the same by applying the same amount of heat to equal weights of water and iron and found that the iron became much hotter than the water. However, this alone did not dispel the notion that heat is caloric and, in fact, Black supported the caloric theory. In the 18th century, such notables as Locke and Boyle expressed doubts about the caloric theory and believed Bacon's concept that heat is caused by motion. By the early 19th century Rumford had shown that the weight of a liquid does not change as it is heated and that enough heat is created during the boring of a cannon to boil water. With the emerging strong indications that all matter consists of different combinations of a few fundamental particles, around the turn of the 19th century evidence was mounting that heat is due to the agitated motion of these particles.

The increased interest in heat in the 18th century gave rise to the laws of Boyle and Charles. *Boyle's law* states that, if temperature is held constant, then the pressure of a gas in a closed container is inversely proportional to its volume, and *Charles' law* states that, if the pressure of a gas is held constant in a closed container, then the volume of the gas is proportional to its temperature. By combining these laws, Joseph Gay-Lussac concluded that when the volume of a gas is held constant the pressure is proportional to its temperature. This conclusion strongly supported the concept that, when pressure is increased, the resulting temperature rise is caused by the increased accumulated kinetic energy of the particles in the gas. Furthermore, in 1811, Amedeo Avogadro published his principle stating that equal volumes of different gases at the same pressure and temperature contain the same number of particles (now known to be molecules). This principle allowed scientists to use the laws of motion to provide a mathematical basis for the ideas surrounding heat and temperature and relate the laws of motion to the heat in a gas.

It is now known that, even in a solid object with a fixed lattice of molecules, heat is the vibration of the molecules within the lattice and an increase or decrease in the heat of the object causes an increase or decrease in the object's temperature. Be it a gas, liquid or solid, heat is the motion of the particles in the substance. Temperature is a measure of a portion of the motion in the object and depends on the substance that makes up the object. As Black had discovered, when water and iron contain the same amount of heat energy, the iron is much hotter than the water. The exact relationship between heat and temperature is complicated by the fact that the motion of the molecules is not just back and forth motion, but three-dimensional rotational motion as well. Heat is a measure of all motion, while temperature is primarily an indicator of oscillating and free motion. How much of the motion in a substance is revealed by its temperature depends on the shape and structure of its molecules. The change in the energy per degree of temperature per amount of mass in a substance is called the *specific heat capacity* of the substance and may vary with temperature. The specific heat capacities of many substances tend to be constant over the range 0 °C (32 °F) to 100 °C (212 °F), the temperature range over which most measurements were made at the time of the early experiments.

During the first half of the 19th century, the *first law of thermodynamics* was established. It states that the change in internal energy of a closed system is equal to the heat energy added to the system minus the work expended by the system and is equivalent to the conservation of energy principle for a system involving both heat and mechanical motion. By the middle of that century, the equivalence of heat and mechanical energy was demonstrated by J. Robert Mayer and James Joule. In one of Joule's many experiments, he used falling weights to stir an insulated container of water. By knowing the specific heat capacity of water, the energy change in the water could be determined from its temperature change. The mechanical energy expended to churn the water was calculated using the formula $E=Fx$ and the energies were found to be the same. These experiments provided further evidence that the conservation of energy principle includes general systems involving both mechanical and heat energy.

Experiments by Sadi Carnot showed that mechanical work could be done by heat only if the temperature decreases. This conclusion led to the formulation of the *second law of thermodynamics*, which, in effect, states that a system that does mechanical work and, at some point, involves heat cannot be 100 percent efficient. Because all systems that do useful work involve heat in one way or another, this law implies that there cannot be a perpetual motion machine. It also implies that, as the temperature differences in the universe continually decrease, the universe will eventually lose its ability to do work. This is sometimes colloquially described by "you can't win, you can't break even and you can't get out of the game."

There are three mechanisms for transferring heat: conduction, convection and radiation. With *conduction*, heat moves through matter by the free or oscillating particles colliding with one another. *Convection* is the physical movement of heated matter. *Radiation* transfer occurs when the electromagnetic waves (discussed later) created by the motion of particles in one heated object cause particle motion in another, perhaps distant, object. With radiation, the energy transmission actually is achieved by the heat being transformed into electromagnetic energy and then back into heat, i.e. particle motion. Conduction is what permits the heat of a flame to move through a metal pan or pipe to heat water on the other side. The convection of steam is often used to move heat from a boiler to where the heat is actually used. When a system of steam radiators is used to heat a building, the convection of the steam carries the heat to the radiators and conduction heats the iron in the radiators from which the electromagnetic radiation warms the surrounding air.

With time, the many ways we developed for creating and transforming heat and the versatility of heat made it more than a scientific curiosity. By the late 18th century, engineers as well as scientists were examining the possibilities of heat. During this period, James Watt invented the steam engine and the industrial revolution began.

In the latter half of the 19th century, Josef Stefan, John Tyndall and Ludwig Boltzmann determined that the energy radiating from an object is proportional to its *absolute temperature* in degrees Kelvin, °K, (°K=°C+273), raised to the fourth power. The constant of proportionality depends on many of the object's properties, including its nature and surface area. If only heat energy is considered, the first law of thermodynamics dictates that the energy entering an object must equal the energy absorbed by the object plus the energy output from the object. As heat accumulates in the object its temperature increases according to its specific heat capacity. But at the same time energy radiates from the object as its temperature rises. Sooner or later a temperature is reached at which the object is in a balanced state with the radiating energy equaling the input energy. Of current concern is the change in the balance point temperature of the Earth and its atmosphere as the makeup of the Earth's atmosphere changes. Basically, a change in the constitution of the atmosphere changes the constant of proportionality that is multiplied by the fourth power of the absolute temperature. If the balance temperature rises, so does the temperature of the Earth's lower atmosphere and the result is global warming.

Although the study of chemical energy progressed in parallel with that of motion and heat, the real breakthroughs in chemistry did not take place until the 19th century when the theory of particles began to be substantiated by more accurate experimentation. The idea that everything in the universe is composed of a few *elements* is quite ancient. In the sixth century B.C., the Greek scholar Thales hypothesized that everything is composed of different forms of water. In the fifth century B.C., Empedocles thought everything was made from the four elements fire, air, earth, and water, and Democritus introduced the concept that all matter is made of small, indivisible bundles, which he called *atomos*, or *atoms*. These ideas resurfaced during the renaissance, and by 1803 John Dalton used the idea of atoms to explain why certain chemical reactions occurred in simple proportions. From the writings of Dalton and others, Dmitri Mendeleev composed the first periodic table in 1869, and in 1913, Niels Bohr proposed his famous model in which an atom consists of a nucleus in the center with electrons orbiting around it. Then by 1932, the work of Ernest Rutherford, Niels Bohr, and James Chadwick had established that the nucleus consists of protons and neutrons. Today, atoms are known to contain other small particles, but in this book, only protons, neutrons and electrons need to be considered.

Although the atoms of Democritus were indivisible, modern day atoms are not indivisible and even protons, neutrons and electrons are not the fundamental entities we seek. Ironically, Thales' idea that everything may be composed of different forms of a single entity is seriously being researched today. Today, instead of water, the entity is a tiny packet of energy called a *string*.

The modern periodic table contains over one hundred different kinds of atoms. For our purposes, each atom consists of a nucleus of protons and neutrons surrounded by electrons. The two most important properties of these three major constituents are mass and charge. Protons have a positive charge, neutrons have no charge and electrons have a negative charge that is equal in magnitude to that of the proton. In its stable state, each atom has an equal number of electrons and protons and is, therefore, electrically neutral. The atom is held together by the mutual attraction between the positive protons and oppositely charged electrons and a force, known as the *strong force*, between the neutrons and protons. A substance that consists of only one kind of atom is referred to as an *element*. Each type of atom, or element, is given an *atomic number* that indicates the number of protons in the atom. The elements of interest in this book and the symbols used to identify them are summarized in Table A-1. From this table, it is seen that hydrogen has the atomic number 1, indicating that it has one proton, and is the simplest of the atoms.

Table A-1: Elements important to energy use.

Element	Symbol	Atomic number	Mass number
Hydrogen	H	1	1
Hydrogen			2
Hydrogen			3
Helium	He	2	4
Helium			3
Lithium	Li	3	7
Lithium			6
Carbon	C	6	12
Nitrogen	N	7	14
Oxygen	O	8	16
Sulfur	S	16	32
Krypton	Kr	36	84
Krypton			92
Barium	Ba	56	137
Barium			141
Platinum	Pt	78	195
Uranium	U	92	238
Uranium			235
Uranium			236
Neptunium	Np	93	237
Neptunium			239
Plutonium	Pu	94	244
Plutonium			239

Although an atom is identified by its number of protons (its atomic number), it may contain different numbers of neutrons. These different forms of the same type of atoms are called *isotopes*. The common isotope of hydrogen has no neutrons, the deuterium isotope of hydrogen has one neutron and the tritium isotope has two neutrons. Because different isotopes of the same element have different masses, they may have different properties. Each isotope is given a *mass number*, which is the total number of protons and neutrons in its nucleus. Mass numbers are also given in Table A-1. Frequently, it is not necessary to specify the isotope, but when an isotope is specified, its mass number is written as a superscript preceding the element's symbol. The isotope deuterium is written ^{2}H.

A set of atoms may bond together to form a *molecule*, and a substance consisting entirely of one type of molecule is called a *compound*. Water is a compound that is made up of an oxygen atom bonded with two hydrogen atoms. The abbreviation for a compound, which is referred to as the *compound's formula* or *chemical formula*, is constructed by writing the symbols for its atoms side by side and using subscripts to indicate the number of atoms of each type. The formula for water is H_2O and a molecule of ethane, which has two carbon atoms and six hydrogen atoms, is indicated by C_2H_6.

Sometimes it is useful to additionally show the structure of a compound by showing the atoms individually and using lines to represent the bonds between them. The resulting diagram is called a *Lewis formula*. The Lewis formula for water is

$$H \!-\! O \!-\! H$$

and that of ethane is

$$\begin{array}{c} \text{H} \quad\ \text{H} \\ | \quad\ | \\ \text{H} \!-\! \text{C} \!-\! \text{C} \!-\! \text{H} \\ | \quad\ | \\ \text{H} \quad\ \text{H} \end{array}$$

In contrast to nuclear reactions, in chemical reactions the nuclei of the atoms are indivisible. In a *chemical reaction*, one set of compounds is converted into a different set of compounds in such a way that the number of each type of atom remains the same before and after the reaction. A chemical reaction may be written as a *chemical equation* by writing the formulas for the *reactants* to be converted on the left separated by plus signs, and then writing a right-pointing arrow followed by the formulas for the *products* on the right separated by plus signs. In addition, the numbers of molecules or relative amounts of the compounds involved are given to the immediate left of each compound's formula. Hydrogen and oxygen normally exist as *diatomic* molecules (i.e., H_2 and O_2) so that the chemical equation for burning hydrogen to form water is

$$2H_2 + O_2 \rightarrow 2H_2O$$

Note that the number of hydrogen atoms is four and the number of oxygen atoms is two on both sides of the equation. Because the numbers of each type of atom are the same on both sides of a chemical equation, chemical equations are said to be *balanced*. The numbers beside the formulas for the compounds are not necessarily integers, they may indicate arbitrary amounts, e.g.,

$$10.4H_2 + 5.2O_2 \rightarrow 10.4H_2O$$

In any case, a chemical equation is balanced.

Sometimes a molecular formula includes symbols giving the state of the compound. The letters "g", "l" and "s" respectively indicate "gas". "liquid" and "solid", so that $H_2O(l)$ indicates liquid water as opposed to ice or steam, and

$$2H_2(g) + O_2(g) \rightarrow 2H_2O(l)$$

shows that two gases are being combined to form liquid water.

Although chemical equations are balanced with respect to the numbers of atoms, the total mass is not necessarily the same on both sides of the equation. Thus, according to *Einstein's equation, $E=mc^2$*, energy may be absorbed or released during a chemical reaction due to the breaking and reforming of the bonds between the atoms. If energy is released by a reaction, mass is lost and the reaction is *exothermic*, and if energy is absorbed, mass is gained and the reaction is *endothermic*. For chemical reactions, the mass loss or gain is extremely small. Many substances, commonly referred to as *fuels*, may be burned in the presence of oxygen to produce heat. Such reactions are referred to as *combustions* and are exothermic. An example of an endothermic reaction is the one that absorbs heat to produce carbon monoxide and hydrogen from carbon and water, i.e.,

$$C + H_2O \rightarrow CO + H_2$$

Energy is needed to cause such a reaction and this energy is stored in the increased mass of the reaction's products.

Hydrogen, carbon, nitrogen and oxygen are among the most abundant elements and, while hydrogen fuels the sun, it is carbon compounds that provide most of our earthbound fuels. Carbon is unique in its ability to readily bond with itself and other elements. Early on, it was thought all compounds containing carbon were from living things and so such compounds were given the name *organic compounds*. A *covalent bond* is one in which the bond is due to the sharing of an electron and, because carbon can easily share four electrons, it can easily form four such bonds. Note the four bonds made by each carbon atom in the Lewis formula for ethane given earlier. The Lewis formulas for carbon monoxide, carbon dioxide and methanol are:

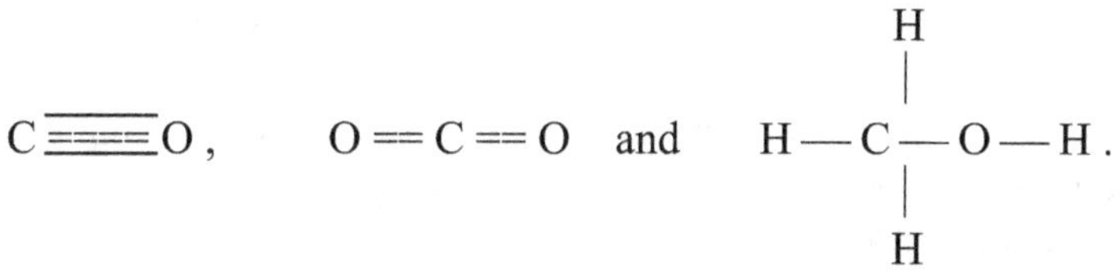

$$
\begin{array}{ccccc}
& & & & H \\
& & & & | \\
C\!\!\equiv\!\!\equiv\!\!O\,, & O\!=\!C\!=\!O & \text{and} & H\!-\!C\!-\!O\!-\!H\,. \\
& & & & | \\
& & & & H
\end{array}
$$

Because carbon can so readily bond with other atoms, it appears as the backbone of all complex, biologically important molecules, such as carbohydrates, amino acids and proteins. But of special interest to us are *hydrocarbons*, which are the compounds that consist of carbon and hydrogen only, and alcohols, which are the organic compounds that contain an oxygen-hydrogen, OH, group, e.g., methanol (shown above) and ethanol,

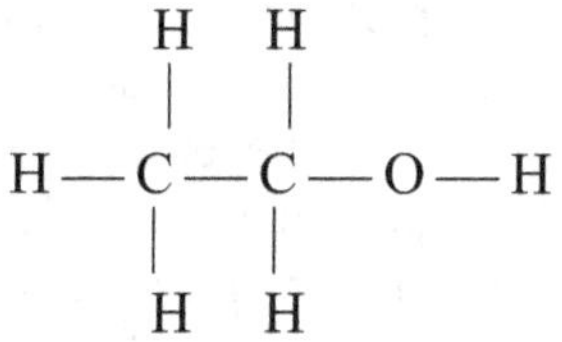

$$
\begin{array}{ccc}
H & H & \\
| & | & \\
H\!-\!C\!-\!C\!-\!O\!-\!H & \\
| & | & \\
H & H &
\end{array}
$$

Note that ethanol is ethane with an oxygen atom inserted in one of the carbon-hydrogen bonds. Hydrocarbons, alcohols and *carbohydrates*, which have formulas of the form $C_m(H_2O)_n$, account for almost all, but not all, of our chemical fuels.

When the Earth's atmosphere was first formed, one of its main constituents was carbon dioxide. As plant life began to evolve, the carbon dioxide provided plants with the energy they needed to grow through a process called photosynthesis. *Photosynthesis* uses the light energy from the sun to convert carbon dioxide and water to carbohydrates and oxygen,

$$
\text{Light} + nCO_2 + nH_2O \rightarrow \overset{\text{Photosynthesis}}{\underset{\text{Carbohydrate}}{C_nn(H_2O)}} + O_2
$$

(n is an integer). It is a high-efficiency, endothermic reaction that stores more than 90 percent of the light energy into the carbohydrates. As plant life spread, more and more carbon dioxide was taken from the atmosphere and more and more oxygen was added to the atmosphere. It is believed that this oxygen is what enabled the development of animals, including humans, who breathe in oxygen and breathe out carbon dioxide. Today, animals and plants have a symbiotic relationship in which plants provide the oxygen and animals provide the carbon dioxide. But there are other sources of both carbon dioxide and oxygen that affect the balance between the two gases. The latest concern is that the production of excess carbon dioxide by burning fossil fuels, which are primarily hydrocarbons, combined with the reduction of the Earth's forests is likely to cause serious global warming.

As discussed above, in an atom's normal state it has the same number of electrons as protons and is electrically neutral. But under certain conditions in which energy, such as heat, is applied to a substance, the externally applied energy may overcome the attraction between the electrons and protons, causing the electrons to escape and become *free electrons*. What remains behind is a positively charged atom called an *ion*. While electrons are free to move about even in some solids, ions in a solid are held relatively fixed.

In 1864, James Maxwell set down the electromagnetic equations that bear his name and relate electricity and magnetism to each other and to their changes with time. A space in which a magnetic force (i.e., a magnetic attraction or repulsion) exists is called a *magnetic field*. An *electric field*, which is related to the attraction and repulsion of charged particles, is similarly defined. Maxwell's equations indicate that a changing magnetic field causes an electric field and vice versa. Two aspects of these equations can be easily demonstrated by creating a current by passing a wire loop through a magnetic field and by constructing a magnet by passing a current through a coil of wire. Electric and magnetic fields have both intensity and direction.

There is a quantity at each point within an electric field known as the point's *potential*. Potential in an electric field is analogous to pressure in the atmosphere or a body of water. The difference between the potentials at two points in the field is called the *potential difference*, or *voltage difference* or simply *voltage*, between the points and is measured in *volts* (V). An electric field may cause electrons to become free from their atoms and flow in the direction opposite to that of the field, thereby creating an *electric current*. The magnitude of this current is measured in *amperes* (A). This flow may be impeded by collisions with the molecules in the substance through which the electrons are traveling. This impediment is called the *resistance* (R) of the substance and is measured in *ohms* (Ω). *Conductors* are substances with little resistance, *insulators* have high resistance and allow very little, if any, flow, and *semiconductors* fall somewhere in between. Copper is a good conductor and silicon is a good insulator, even though silicon doped with antimony is a semiconductor. *Superconductors* have no resistance but can be maintained at only extremely low temperatures.

Although the resistance of any substance may vary with temperature and other physical quantities, when these quantities are held constant a simple equation relates the voltage applied to a substance to the amount of current produced by the voltage. Early in the nineteenth century, G. S. Ohm established that the voltage between two points in volts is equal to the resistance in ohms times the current in amperes. The equation $V=IR$ is known as *Ohm's Law*. Although this equation predates Maxwell's equations, it is a special case that can be derived from them. Ohm's Law means that, for a given substance, as the voltage increases the current increases proportionately according to the equation $I=V/R$. The relationship between voltage (i.e., voltage difference) and current is analogous to that between pressure difference and water flow. As one increases the pressure difference between the two ends of a hose, the water flow increases. Or, if the resistance of the hose is decreased by using a larger hose, then the water flow increases even when the pressure difference is kept the same.

Both electric and magnetic fields contain energy. The collisions within a conductor cause some of the electric energy to be converted into heat energy. The rate at which energy is converted from one form to another is called *power*. Power is measured in *watts* (W), with one watt being the conversion of one joule of energy per second (J/s). Another important equation states that the power (P), the rate at which this energy is converted to heat in watts, is equal to the current (I) in amperes, times the voltage (V) in volts, i.e., $P=IV$. Together with Ohm's Law, this equation can be used to produce the equation $P=IV=I^2R$, which states that the rate of heat generated by the flow of electrons is equal to the square of the current in amperes times the resistance in ohms. This relationship is named after James Joule, who measured the heat that was dissipated when a voltage causes a current to flow through a resistance. Through such experiments, the conservation of energy principle was expanded to include electrical energy as well as mechanical and heat energy.

Any part of an electrical system that outputs electrical energy is called an *electrical source*, or simply *source*, and any part of an electrical system that consumes energy is considered to be a *load* to the remainder of the system. A single piece of equipment, such as an electric motor or coffee maker, may be a load or everything that consumes energy connected to an electrical energy source may be referred to as the load of that source.

From the equations $P=IV=I^2R$ it is apparent why electric power companies use high voltages to transmit energy through power lines. Suppose that a specific amount of power must be provided at the end of a transmission line having a resistance R. A voltage V is applied across the transmission line to produce the current I needed to supply the power. Then the power lost by the line (i.e., the rate at which heat is lost by the transmission) is I^2R. If a thousand times as much voltage is used, then only one thousandth as much current is required to provide the same power to the load because $VI=1,000V(I/1,000)$. Therefore, the power lost by the line is $(I/1,000)^2R=I^2R/1,000,000$ and the heat lost by the transmission line is one millionth that lost if the voltage V is used. Unfortunately, most electrical equipment cannot operate at extremely high voltages and *transformers* must be included with the load equipment to lower the voltage. The equipment in your home normally requires 110 volts (220 volts in most other countries), but your power may be transmitted a long distance using a transmission voltage of 110,000 volts or more. The transmission voltage must be lowered to your home voltage by a series of transformers located in the vicinity of your home. Also, the voltage of an electrical source, such as a generator, is normally much lower than the transmission voltage and there must be transformers to increase the voltage output by the source. The savings in line losses is partially offset by the losses in the transformers, but the total transformer losses are still much less than the line losses avoided by using the higher transmission voltage.

An *electrical generator* is a machine that creates a changing magnetic field, which, in turn, causes a voltage between its output terminals, i.e., electrical connection points. When a load is connected between the terminals a current will flow through the load and electrical energy is provided to the load. Since energy is output by the generator and there are energy losses within the generator, the energy needed to drive the generator must exceed the energy output to its load. This energy normally is provided by a steam, wind or water turbine. In a large electrical system in which several generators use an array of transformers and transmission lines to deliver electrical energy to a large number of distributed loads, the energy lost within the system is typically five to ten percent of energy used to drive the generators. This means that the efficiency of the electrical generators and distribution system is 90 to 95 percent. However, if a steam turbine is used to drive the generator, the mechanical energy supplied to the generator by the turbine is only about one third of the heat energy of the fuel needed to produce the steam. Therefore, the overall efficiency of a steam turbine, generator and distribution system combination may be only 30 to 32 percent.

When the intensity and direction of an electric field is held steady, the current in the field does not change and is called a *direct current* (*dc*). On the other hand, if the direction of the field alternates, then the resulting current is an *alternating current* (*ac*). The advantage of an ac electrical system is that *transformers* can be used to increase or decrease voltage and thereby reduce transmission losses. Transformers require a changing field in order to produce an output and, therefore, do not work in a dc system. After the invention of the light bulb by Thomas Edison in 1879, the first commercial electrical systems began to be constructed. These systems were quite small and the transmission lines were short. As a result, the line losses were small and there was no need to increase and then decrease the voltage. As the systems grew larger, longer lines were needed and using ac power became more and more important. Today, all commercial electrical systems use ac. Although ac power is delivered by the power company, the end user may need some dc to drive local electronics or machinery. If so, a *converter* is used to change the ac to dc. Conversion may be done electronically or by a motor-generator set that uses an ac motor to drive a dc generator. No matter how the conversion is attained, there is some energy loss. Motor-generator sets tend to be over 90% efficient, but an electronic converter is typically much less efficient.

Sometimes the generated electricity is dc, such as the case in which the electricity is generated by solar cells. In such cases, it may be necessary to convert dc to ac. A device for converting from dc to ac is called an *inverter*. (Some people refer to both converters and inverters as inverters, but both terms are used in this book to clearly distinguish between the two types of conversion.) As with a converter, an inverter may be motor-generator set or an electronic device. In either case, there will be a loss of energy.

In an ac system, both the voltage and current are *sine waves* that may or may not be aligned so that they reach their peaks at the same time. The typical sine waves $\sin(t)$ and $0.8\sin(t - 0.1\pi)$ are shown in Fig. A-1. The magnitude is along the vertical axis and the time t is along the horizontal axis. Sine waves are repetitious and two cycles are shown in the figure. Because the voltage and current are not aligned and the *effective* (with respect to power) values of the voltage and current are less, the equation $P=IV$ must be modified to indicate the average power $P=IV\times\cos(\alpha)$, where V and I are the effective values of their respective sine waves and α is the angular difference between the two waves. The effective value of a sinusoidal voltage or current is the maximum value divided by the square root of two. The cosine of α is called the *power factor* and is always less than one. In Fig. A-1 the angular difference is 0.1π (18 degrees) and the associated power factor would be 0.96. Some loads may cause the supply voltage and current to be misaligned (i.e., the power factor to be less than one) and, therefore, extra current must be supplied to the load in order to satisfy its power requirement. Power companies charge large corporations that use a lot of power according to the power factor of the electricity

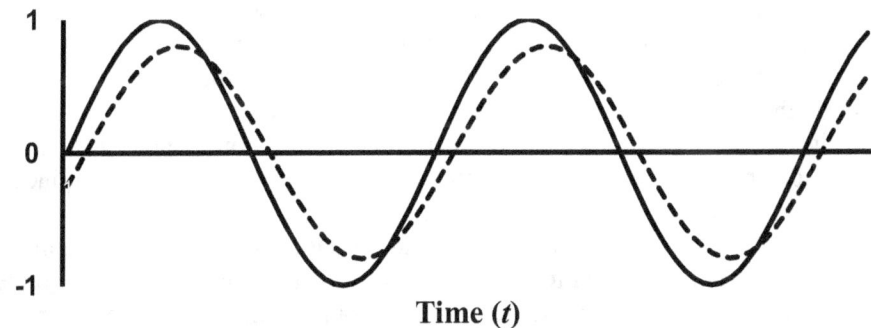

Fig. A-1 Typical sine waves.

being supplied as well as the amount of electrical energy they consume. However, there is equipment that may be attached to the load presented by the corporation that allows the supply voltage and current in the transmission line to be aligned. Thus, $P=IV$ for the transmission line providing the power. Although power factor correction equipment consumes energy, as with transformers the energy lost is less than the extra energy that would be lost in the transmission lines if the voltage and current were misaligned.

There is more to electric and magnetic fields than moving electrons through conductors. In accordance with Maxwell's equations, moving electrically charged particles create electromagnetic fields of energy. If a charged particle is moving or oscillating, then this energy radiates outwards from the particle until it strikes another particle and is absorbed. Also, subatomic changes may cause electromagnetic energy to be emitted. One of science's greatest dilemmas has been whether this radiated energy should be viewed as small bundles of energy, called *photons*, or as waves. When viewed as waves, the waves are sine waves and have the form shown in Fig. A-1. As noted above, sine waves are cyclic and how often they repeat themselves is called the wave's *frequency*. Figure A-1 shows two complete cycles of each sine wave so that both have the same frequency even though they have different magnitudes and are displaced from one another. If the frequency were doubled, twice as many cycles would occur in the same amount of time. The frequencies emitted by a particle are determined by its oscillations and structural changes and, if several particles are emitting energy at different frequencies, the resulting electromagnetic waves are superimposed to form the radiation of a whole band of frequencies simultaneously.

Whether electromagnetic energy is viewed as consisting of photons or waves, it can travel through matter or space. When traveling through matter, some of the energy is absorbed by the matter and may heighten the energy level within the matter's atoms and cause some of the electrons to become free. By combining certain semiconductor materials, an electric field is created that will result in the free electrons becoming a current when an external conducting path is provided. This is the basis of a *solar cell*. Also, the absorbed energy may cause the molecules in the matter to vibrate, thereby converting the electromagnetic energy into heat. While some of the kinetic energy of the particles in the matter is transferred to other particles through collisions, some is returned to electromagnetic energy by the motion of the charged particles in the matter. In other words, not only does electromagnetic energy produce heat, but the heated matter also produces electromagnetic energy.

The nuclear reactions occurring in the sun produce a tremendous amount of electromagnetic energy of a wide range of frequencies that radiate into space. Some of this energy strikes the Earth and some of the lower frequencies, known as *radio* and *infrared frequencies*, are primarily the ones that are responsible for heating the Earth and its atmosphere. Some of the higher frequencies provide us with light. Some of the very high frequencies, the *ultraviolet frequencies*, are useful in some respects but, if they are too strong, they tend to be unhealthful and can cause skin cancer. The very high frequencies, *X-rays* and *gamma rays*, can be deadly. Fortunately, our atmosphere reflects most of this high frequency radiation. Although ozone is unhealthful to breathe, the ozone layer high in the atmosphere protects us from excessive ultraviolet radiation. The unit of frequency is *hertz* (*Hz*), which is the number of sine wave cycles per second. The important ranges of electromagnetic frequency are summarized in Fig. A-2.

Max Planck discovered that the energy in a photon is proportional to its frequency and the constant of proportionality, known as Planck's constant, is 6.626×10^{-34} joule-seconds. The frequencies that are absorbed or emitted by a molecule depend on the structure of the molecule, e.g. the frequencies absorbed by water are predominately in the radio spectrum. Although the higher frequencies contain more energy, it is the lower frequencies that are normally associated with heating because it is these frequencies that are absorbed by the molecules in plants and animals.

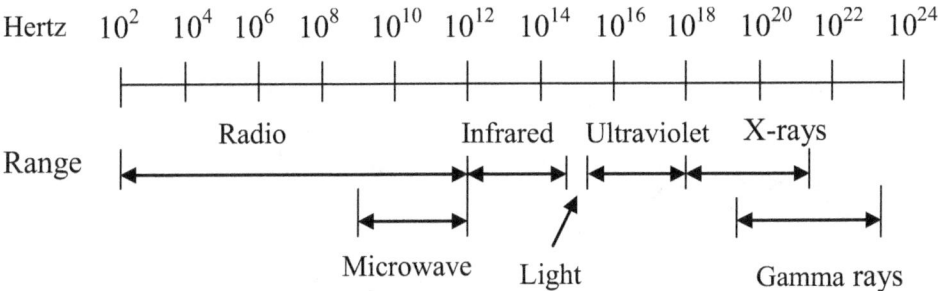

Fig. A-2 Important electromagnetic frequency ranges.

While solar cells have given us the ability to turn the energy from the sun directly into electricity, our learning to master the creation, control and retrieval of electromagnetic waves has given us radio, television, satellite communications, cell phones, microwave ovens, global positioning, radar, night vision equipment, medical imaging lasers, and many other modern conveniences. None of these modern conveniences existed before we learned how to control electromagnetism through the application of Maxwell's equations.

Nuclear reactions differ from chemical reactions in that the atoms actually change into different atoms. *Nuclear fission* occurs when an atom's nucleus is broken into two or more pieces, called *fission fragments*, and *nuclear fusion* occurs when two or more nuclei combine to become a single nucleus. Nuclear fission takes place naturally in heavy elements such as uranium, but normally occurs very slowly and causes the elements to undergo radioactive decay. Experiments performed by Lise Meitner, Fritz Strassman and Otto Hahn in the late 1930s confirmed that when the fission of a uranium atom occurs the energy of its fragments and the difference in mass between the uranium and the fragments satisfies Einstein's equation, $E=mc^2$. A typical fission is shown in Fig. A-3. This fission is that of a ^{235}U atom, which when struck by a neutron (n) of sufficient energy, temporarily becomes an unstable ^{236}U atom. The ^{236}U atom then splits into a barium, ^{141}Ba, nucleus, a Krypton, ^{92}Kr, nucleus and three neutrons. The total mass of the fragments is less than the mass of the ^{235}U atom and the original neutron, and the difference is accounted for by the mechanical motion of the fragments and radiated electromagnetic energy. If the neutron products have sufficient energy, they may cause additional fissions in nearby ^{235}U atoms. Under certain conditions in which a sufficient number of ^{235}U atoms are present a *chain reaction* may occur, and under carefully engineered conditions the chain reaction may become a violent explosion emitting an enormous amount of energy. Otherwise, the chain reaction may be controlled and the resulting energy used to generate the steam needed to drive a steam turbine.

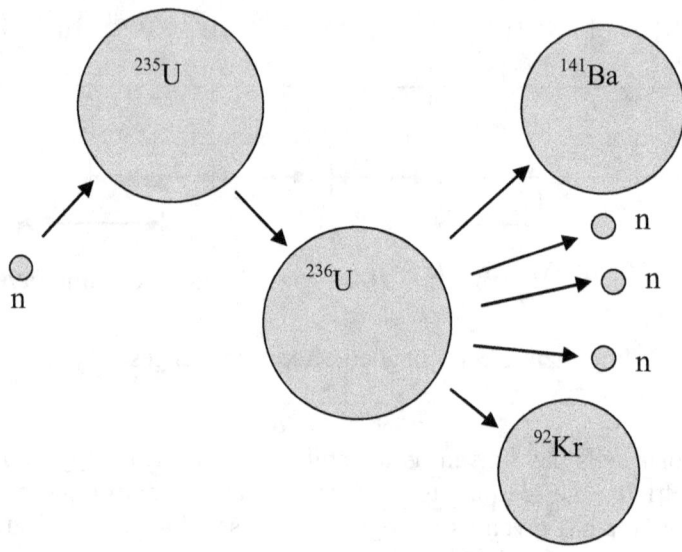

Fig. A-3 Fission of ^{235}U.

The most abundant fissionable material is ^{238}U, but ^{238}U cannot sustain a chain reaction. However, uranium ore is about 0.7 percent ^{235}U, which is capable of sustaining a chain reaction. It takes a complicated and expensive process that often involves a series of centrifuges to separate ^{235}U from ^{238}U. A second fissionable isotope capable of a chain reaction is plutonium 239, ^{239}Pu. This isotope may be produced by bombarding ^{238}U with neutrons using a *breeder reactor*. All of the commercial nuclear power plants in use today are fueled by either ^{235}U or ^{239}Pu and, therefore, rely on uranium ore to supply their energy.

There are two forms of controllable nuclear fusion being studied, *cold fusion* and *hot fusion*. The processes for achieving cold fusion have yet to produce a net energy gain. Hot fusion requires extremely high temperatures in the neighborhood of 120 million °C (216 million °F). Consequently, achieving a sustainable hot fusion reaction is a formidable task. The form of nuclear fusion that seems most promising as a usable energy source is shown in Fig. A-4. It is the fusion of a deuterium nucleus, ^{2}H, and a tritium nucleus, ^{3}H, to produce a helium nucleus, ^{4}He, and a neutron, n. Even if a practical method for creating the reaction is found, tritium does not occur naturally and must be produced. While 0.02 percent of the hydrogen in seawater is deuterium and seawater is plentiful, tritium is most easily produced from lithium ^{6}Li, which is of limited supply.

Be it fission or fusion, mass is lost to energy in the motion of its products and radiated electromagnetic energy. In addition to the primary fragments resulting from fission, *alpha particles* (helium nuclei), *beta particles* (high-energy electrons), *gamma rays* (high-energy photons) and *neutrinos* (very small, electrically neutral particles) are produced.

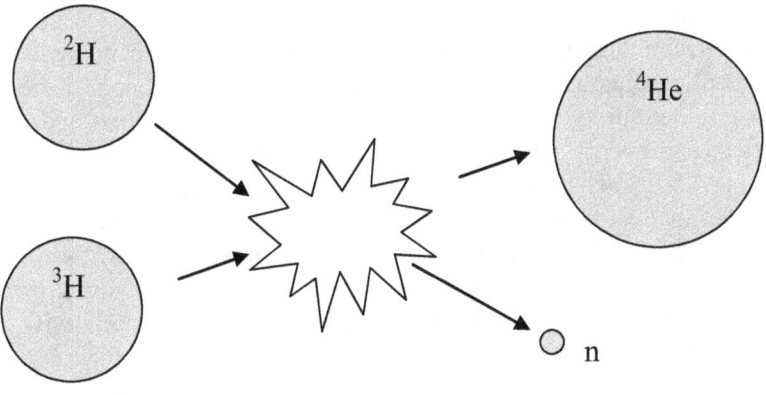

Fig. A-4 Fusion of hydrogen into helium.

It is interesting to note how most scientists believe the universe was formed. Following an extremely violent explosion, the big bang, some of the energy produced began to collect into subatomic particles and hydrogen, which through gravitational attraction began to form stars. As the stars became more and more dense, the pressures in the stars became so great that the hydrogen began to fuse into helium. Although some of the mass turned into energy, the pressures at the centers of the stars became progressively greater and further fusion caused atoms with higher atomic numbers to be created. For stars the size of the sun, fusion could continue until oxygen is formed, but even iron could be formed in the very massive stars. At this point the fusion process due to extreme pressure is no longer possible because the atoms with atomic numbers greater than iron do not fuse by pressure alone. Before a massive star exhausts itself completely by turning into a ball of iron, it explodes into a *supernova*. Some of the intense energy in the supernova is turned into the masses of the elements heavier than iron. The very heavy elements are unstable and some of their masses are returned to energy through nuclear fission. The natural fission of uranium occurs very slowly, but the atoms with greater atomic numbers fission very quickly. Although most of the heavy elements are thought to be created by supernovae, there are other processes that can produce these elements.

The discussion so far has concentrated on identifying the major categories of energy and the discussion of their relationship has been limited. In order for energy to be a fundamental universal quantity, these energy types must be shown to have a common mathematical basis that extends to the conversions between the various types. As indicated earlier, in the first part of the 19th century heat and electric current were found to be the kinetic energy of molecules and electrons, and heat, mechanical and electrical energy were found to be equivalent. As better experimental techniques were found and better equipment became available, more and more relationships among the different types of energy began to reveal themselves. Slowly but surely mathematical equations were developed and experimentally justified, and energy emerged as a central thread to many of these equations. Finally, in 1905, Albert Einstein produced his paper on special relativity and a separate paper that contained the landmark equation, $E=mc^2$. In particular, by taking into account the interchange between mass and energy indicated by this equation, the conservation of energy principle has been expanded to all forms of energy. In 1916, Einstein extended his work to the general theory of relativity. Although Einstein was a theorist as opposed to an experimentalist, his theories concerning relativity have since been verified by numerous experiments using techniques and equipment that were not available to him at the time.

When Galileo first began experimenting to determine the nature of his surroundings, he had no means of measuring temperature and only crude means of measuring time. As better, more accurate clocks, thermometers and other measuring apparatus were invented and improved upon, experimentation began to reveal our universe in greater detail. Today, high-speed computers, lasers, electron microscopes, space-based telescopes, atomic clocks, high-speed photography, particle accelerators and electromagnetic detection and imaging devices permit us to measure and view phenomena to a degree that is far beyond the imaginations of the early scientists. Science and experimental equipment, out of necessity, have grown up together. Better and better equipment has allowed us to verify the workings of our universe with greater and greater assurance. There is a limit, however. In 1927, Werner Heisenberg proposed his uncertainty principle, which places a limit on how accurately position and momentum (mass times velocity) can be measured. It is a result of the observer effect that no measurement can be exact because observation always affects the outcome of the experiment. Only the future will tell what discoveries will be possible as we continue to improve our instruments of measurement and observation.

B

UNITS OF MEASURE AND CONVERSIONS

Over the past three centuries, scientists have developed a standard set of units called the *International System of Units (SI)*, which are now used throughout the world. Scientists consider the most fundamental measurements to be those of length (space), mass and time. Basic to this system is the *meter* (m)-*kilogram* (kg)-*second* (s) (mks) system—meters for length, kilograms for mass and seconds for time. The units for volume, force, energy, power and other physical quantities are derived from these units. For example, the SI unit for volume is the meter-meter-meter, or *cubic meter* (m^3). Similarly, the SI unit for force is the *newton* (N), which is a kilogram-meter per second per second (kg-m/s^2). The SI unit for energy is the *joule* (J) and is a newton-meter (N-m). *Power* is the rate at which energy is expended and its SI unit is the *watt (W)*, which is a joule per second (J/s). A watt is also a newton-meter/second (N-m/s).

It has been only recently that scientists have settled on the SI system. While science and technology were developing, different countries used different units (e.g., while England used feet and gallons, France used meters and litres). Also, scientists, engineers and other professionals tended to create their own systems of measurements. Even today, engineers frequently use *British thermal units* (Btus) to measure energy, American construction workers use feet for distance and the oil industry uses *barrels* (bbls) instead of liters or cubic meters. The SI units and non-SI units encountered in this book are summarized in Table B-1 along with the symbols used to represent them. The conversion factors between the various units found in this book (some of which are close approximations) are given in Table B-2. The numbers in parentheses are the inverses of the conversion factors shown on the left. For example, one centimeter is 0.39370 inches and one inch is 2.540 centimeters.

Table B-1: Units of measure and their symbols.

Quantity	Unit	Symbol
Length	Meter	m
	Kilometer	km
	Foot	ft
	Yard	yd
	Mile	mi
Mass	Kilogram	kg
	Gram	g
Time	Second	s
	Hour	hr
	Day	da
	Year	yr
Area	Square kilometer	km^2
	Square mile	mi^2
	Hectare	ha
	Acre	ac
Volume	Cubic meter	m^3
	Liter	L
	Cubic foot	ft^3
	Cubic yard	yd^3
	Barrel	bbl
	Gallon	gal
Force	Newton	N
	Pound	lb
	Tonne	t
Energy	Joule	J
	British thermal unit	Btu
	Kilowatt hour	kWh
Voltage	Volt	V

(Cont'd on page 339)

(Cont'd from page 338)

Current	Ampere	A
Resistance	Ohm	Ω
Power	Watt	W
Temperature	Celsius	°C (liquid water at sea level, 0-100)
	Fahrenheit	°F (liquid water at sea level, 32-212)
	Kelvin	°K (0°K is absolute zero)
Pressure	Pascal	Pa
	Atmosphere	atm

Table B-2: Unit conversions and inverse conversions.

Length	m	= 1000 km	(0.001)
	m	= 1.0936 yd	(0.91441)
	m	= 3.2808 ft	(0.30480)
	m	= 100 cm	(0.01)
	cm	= 0.39370 in	(2.54)
	km	= 0.62137 mi	(1.6093)
Mass	kg	= 1000 g	(0.001)
Time	hr	= 3600 s	(0.0027778)
	da	= 86400 s	(0.000011574)
	yr	= 365.24 da	(0.0027379)
Area	km^2	= 0.38610 mi^2	(2.59)
	km^2	= 10000 ha	(0.0001)
	mi^2	= 640 ac	(0.0015625)
	ha	= 2.471 ac	(0.40469)

(Cont'd on page 340)

(Cont'd from page 339)

Volume	m^3	= 1000 L	(0.001)
	m^3	= 1.3080 yd^3	(0.76456)
	m^3	= 35.315 ft^3	(0.028317)
	m^3	= 6.2898 bbl	(0.15899)
	L	= 0.26417 gal	(3.7854)
	L	= 0.0062898 bbl	(158.99)
	bbl	= 42 gal	(0.023810)

Force	N	= 0.22480 lb	(4.4844)
	t	= 2204.6 lb	(0.00045360)
	ton	=2000 lb	(0.0005)
	ton	=1.0 short ton	(1.0)

Energy	J	= 0.00094781 Btu	(1055.1)
	J	= 0.27778×10^{-6} kWh	(3.6×10^6)
	Btu	= 0.00029308 kWh	(3412)

Temperature	°C = 5(°F − 32)/9
	°F = (9/5)°C + 32
	°K = °C +273

Pressure	atm = 101,325 Pa	(9.892×10^{-6})

It should be pointed out that weight is a force and mass is not the same as weight. On Earth, weight is a measure of the gravitational force the Earth exerts on an object. To a scientist, a kilogram is a measure of the fundamental property of matter called mass. But in everyday usage, a kilogram has come to be used as a measure of weight as well. It is the force that the Earth exerts on matter having one kilogram of mass and this force is equivalent to about 2.2046 pounds.

Because it is inconvenient to use the SI mks system to measure very small or very large amounts, we have units such as an erg, one ten-milllionth of a joule, and a tonne, one thousand kilograms. To avoid using powers of ten when discussing very small or very large quantities, the series of prefixes and abbreviations summarized in Table B-3 has been developed—e.g., *nano (n)* means one-billionth (10^{-9}) and one-billionth of a second is a nanosecond (ns). A billion (10^9) watts is a gigawatt (GW).

Table B-3: Unit prefixes and their symbols.

Unit multiple	Prefix	Symbol
10^{-9}	nano	n
10^{-6}	micro	μ
10^{-3}	milli	m
10^{-2}	centi	c
10^{-1}	deci	d
10^{1}	deca	da
10^{2}	hecto	h or C
10^{3}	kilo	k
10^{6}	mega	M or MM
10^{9}	giga	G
10^{12}	tera	T
10^{15}	peta	P

C

ACRONYMS

ac	alternating current
ALMR	Advanced Liquid Metal Reactor
AMT	Alternative Minimum Tax
APA	Alaska Power Administration
ARRA2009	American Recovery and Reinvestment Act of 2009
ASHRAE	American Society of Heating, Refrigeration and Air Conditioning Engineers
BCAP	Biomass Crop Assistance Program
BHE	Borehole Heat Exchanger
BIGCC	Biogas Integrated Gasification Combined Cycle
BLS	Bureau of Labor Statistics
boe	barrel of oil equivalent
BPA	Bonneville Power Administration
BTS	Bureau of Transportation Statistics
BWR	Boiling Water Reactor
CABO	Council of American Building Officials
CAFE	Corporate Average Fuel Economy
CDD	Cooling Degree Day
CFC	Chlorofluorocarbon
CDM	Clean Development Mechanism
CDP	Carbon Disclosure Project
CPH	Combined Power and Heating
COPs	Conferences of the Parties
dc	direct current
DOE	United States Department of Energy
DOI	United States Department of Interior

DOT	Department of Transportation
DMFC	Direct Methanol Fuel Cell
EGS	Enhanced Geothermal System
EIA	Energy Information Administration
EPA	Environmental Protection Agency
EPCA	Energy Policy and Conservation Act
EPACT1992	Energy Policy Act of 1992
EPACT2005	Energy Policy Act of 2005
EPICA	European Project for Ice Coring in Antarctica
EU	European Union
EV	Electric Vehicle
FBR	Fast Breeder Reactor
FERC	Federal Energy Regulatory Commission
FPC	Federal Power Commission
FTC	Federal Trade Commission
GAO	Government Accounting Office
GDP	Gross Domestic Product
GHG	Greenhouse Gas
GSHP	Ground Source Heat Pump
GVWR	Gross Vehicle Weight Rating
HDD	Heating Degree Day
HDI	Human Development Index
HHV	High Heat Value
HLW	High-Level Waste
HTF	Heat Transfer Fluid
IAEA	International Atomic Energy Agency
ICE	Internal Combustion Engine
ICP	In-situ Conversion Process
ICT	Information and Communication Technology
IGCC	Integrated Gasification Combined-Cycle
ILW	Intermediate-Level Waste
IMF	International Monetary Fund
IPCC	Intergovernmental Panel on Climate Change
IS	International System of units
ITCs	Investment Tax Credits
LED	Light-Emitting Diode
LHV	Low Heat Value

LIFE	Laser Inertial Fusion Engine
LLW	Low-Level waste
LPG	Liquid Petroleum Gas
LULUCF	Land Use, Land Use Change and Forestation
MCFC	Molten Carbonate Fuel Cell
mks	meter-kilogram-second
MMS	Minerals Management Service
mpg	miles per gallon
MRV	Measuring, Reporting and Verifying
MSW	Municipal Solid Waste
MTBE	Methyl Tertiary Butyl Ether
NASA	National Aeronautics and Space Administration
NEI	National Energy Institute
NERC	North American Electric Reliability Council
NGPL	Natural Gas Plant Liquids
NHTSA	National Highway Traffic Safety Administration
NIF	National Ignition Facility
NOAA	National Oceanic and Atmospheric Administration
NRC	Nuclear Regulatory Commission
NREL	National Renewable Energy Laboratory
NYMEX	New York MErcantile Exchange
NYSERDA	New York State Energy Research and Development Authority
OAPEC	Organization of Arabic Petroleum Exporting Countries
OECD	Organization for Economic Cooperation and Development
OPEC	Organization of Petroleum Exporting Countries
OSHA	Occupational Safety and Health Administration
PAFC	Phosphoric Acid Fuel Cell
PEM	Proton Exchange Membrane
PHWR	Pressurized Heavy Water Reactor
ppmv	parts per million by volume
PTC	Production Tax Credit
PUREX	Plutonium URanium EXtraction
PURPA	Public Utility Regulatory Policies Act
PWR	Pressurized Water Reactor
RIKs	Royalties in Kind
SEPA	Southeastern Power Administration
SIP	State Implementation Plan

SMES	Super Magnetic Energy Storage
SOFC	Solid Oxide Fuel Cell
SPR	Strategic Petroleum Reserve
SUV	Sport Utility Vehicle
TRI	Toxic Release Inventory
TTW	Tank To Wheel
TVA	Tennessee Valley Authority
UN	United Nations
UNCED	United Nations Conference on Environment and Development
UNEP	United Nations Environmental Program
UNFCCC	United Nations Framework Convention for Climate Change
UNEP	United Nations Environmental Program
UNDP	United Nations Development Program
USCB	United States Census Bureau
USGS	United States Geological Survey
VLLW	Very Low-Level waste
VOC	Volatile Organic Compound
WNA	World Nuclear Association
WTT	Well To Tank
WTW	Well To Wheel

INDEX

www.ingramcontent.com/pod-product-compliance
Lightning Source LLC
Chambersburg PA
CBHW081104170526

45165CB00008B/2325